Asia Conference on Mechanical Engineering and Aerospace Engineering (MEAE 2018)

MATEC Web of Conferences Volume 198 (2018)

Wuhan, China
23 - 25 June 2018

Editors:

Huafeng Ding
Hongliang Yuan

ISBN: 978-1-5108-7085-7

Printed from e-media with permission by:

Curran Associates, Inc.
57 Morehouse Lane
Red Hook, NY 12571

Some format issues inherent in the e-media version may also appear in this print version.

Printed by Curran Associates, Inc. (2018)

For additional information, please contact EDP Sciences – Web of Conferences
at the address below.

EDP Sciences – Web of Conferences
17, Avenue du Hoggar
Parc d'Activité de Courtabœuf
BP 112
F-91944 Les Ulis Cedex A
France

Phone: +33 (0) 1 69 18 75 75
Fax: +33 (0) 1 69 28 84 91

contact-edps@webofconferences.org

Additional copies of this publication are available from:

Curran Associates, Inc.
57 Morehouse Lane
Red Hook, NY 12571 USA
Phone: 845-758-0400
Fax: 845-758-2633
Email: curran@proceedings.com
Web: www.proceedings.com

TABLE OF CONTENTS

PREFACE

Welcome to 2018 Asia Conference on Mechanical Engineering and Aerospace Engineering (MEAE2018), which is held during June 23-25, 2018 at China University of Geosciences (Wuhan) in Wuhan, China.

The aim objective of MEAE 2018 is to provide a platform for researchers, engineers, academicians as well as industrial professionals from all over the world to present their research results and development activities on Mechanical Engineering and Aerospace Engineering. This conference provides opportunities for the delegates to exchange new ideas and application experiences face to face, to establish business or research relations and to find global partners for future collaboration.

The present volume contains the selected papers submitted to this conference and provides up-to-date, comprehensive and worldwide state-of-the art knowledge in the field of Mechanical Engineering and Aerospace Engineering. All papers included in this collection had undergone the careful peer-review by the conference technical committee members and international reviewers before they're selected for publication. Studies presented in this volume cover the following topics: Aerospace Mechatronics and Avionics Systems, Aerospace Communications, Mechanical Engineering in Aerospace, Aerospace Engineering and Management, Electronic Systems,

We'd like to express our gratitude to all the committee members for their great professionalism and efforts. And thank all the sponsors and participants for their valuable contributions and support to MEAE 2018.

MEAE2018
Conference Committee

Statement of Peer review

In submitting conference proceedings to MATEC Web of Conferences, I certify to the Publisher that I adhere to the **Policy on Publishing Integrity** of the journal in order to safeguard good scientific practice in publishing.

1. All articles have been subjected to peer review administered by the proceedings editors.
2. Reviews have been conducted by expert referees, who have been requested to provide unbiased and constructive comments aimed, whenever possible, at improving the work.
3. Proceedings editors have taken all reasonable steps to ensure the quality of the materials they publish and their decision to accept or reject a paper for publication has been based only on the merits of the work and the relevance to the journal.

Title, date and place of the conference

2018 Asia Conference on Mechanical Engineering and Aerospace Engineering (MEAE 2018)
June 23-25, 2018
Wuhan, China

Proceedings editor(s):

Prof. Huafeng Ding, China University of Geosciences, China

Prof. H.L.Yuan, Tongji University, China

Date and editor's signature

SINS/OD fault-tolerant navigation algorithm based on tracked vehicle kinematic constraints

Tang Xiaqing[1], Sun Zepeng[1], Zhang Huan[1] and Wu Meng[1]

[1] *Army Academy of Armored Forces, Beijing 100072, China*

Abstract. In this paper, for the problem that the accuracy declining of the SINS/OD navigation caused by the sliding and idling during the process of turning and straight running, and stopping output of the odometer, a kinematic constraint model of the tracked vehicle is established. The sliding and idling during the process of turning are compensated. In the filtering process, the difference of the displacement increment is used as the observed quantity, and a two-layer fault detection algorithm is used to detect the output fault of the odometer. The simulation experiments show that the SINS/OD integrated navigation algorithm assisted by the kinematic constraints of the tracked vehicle can detect the output fault of the odometer accurately and perform fault-tolerant navigation, which greatly improves the positioning accuracy after the odometer failure.

1 Introduction

With the change of the form of war, the modern war requires that the combat vehicles can be flexible in the wide battlefield, provide their own position information in time and accurately, and realize the real-time inspection and beating. The tracked vehicle navigation system composed of the SINS and odometer (OD) has the characteristics of complete autonomy, high accuracy and good real-time performance. SINS has the advantages of high accuracy and low failure rate. It is usually used as a reference system. The number of pulses output by OD unit time indicates the increment of total mileage, and it can't directly express the position of the carrier1. When the vehicle body's motion state is complex, its output can't describe the motion of the carrier. At this time, the rationality of the OD output should be detected2.

At present, the commonly used fault detection method in integrated navigation is χ^2 test, which can also be divided into state χ^2 test and residual χ^2 test. The state χ^2 test requires a longer inspection time for measuring faults, and it may diverge with time3. Therefore, residual χ^2 test is often used to detect and isolate faults4. In the process of the research, Zhao Hongsong 1detected three velocity components in the carrier coordinate system. Gong Jun4 designed a two-layer fault detection algorithm to detect the velocity components. All of them have achieved good results.

In the SINS/OD combination navigation, it is necessary to make a difference in the output of OD when measuring the speed, and it is easy to amplify the observation noise and then affect the filtering accuracy. For the position measurement, the dead reckoning (DR) error should be increased, the system dimension increases, and the error of the OD will accumulate gradually, causing the error to diverge. At the same time, the observation of the system is reduced, and it needs a long time to estimate the state. In this paper, the displacement increment in the unit time is selected as the observation measurement. At the same time, according to the characteristics of the tracked vehicle, the motion constraint model of the tracked vehicle is established, and the SINS/OD fault-tolerant navigation scheme based on the kinematic constraints of the tracked vehicle is proposed.

2 Characteristics of tracked vehicle movement

Tracked vehicles are different from wheeled vehicles. Tracked vehicles has no steering wheels for direct steering. Instead, they change the track ratio of the corresponding side through planet steering mechanism on the left and right sides, and use the speed difference between the two tracks to achieve steering. Two two-stage planet steering mechanisms are used on each of the left and right sides of a certain type tracked vehicle, and its joystick has three operating positions: the most former position, the position I and the position II. The corresponding motion states of the above three positions are as follows:

(1) The most former position: When one side joystick is at the most former position, the transmission ratio is approximately 1:1 and it is in a direct transmission state.

https://doi.org/10.1051/matecconf/201819801001

(2) Position I: When one side of the joystick is at the position I, the gearbox rotation speed is a constant, and the transmission ratio of this test is approximately 1.42:1, which is at a low speed.

(3) Position II: When one side of the joystick is at the position II, the driving wheel of this side is in a locked state and the transmission ratio is 0.

3 Kinematic constraints and OD correction of tracked vehicles

Define coordinate system m as the odometer coordinate system. The coordinate origin O is located in the installation position of the odometer. The x_m axis points to the right of the body transverse axis, the y_m axis points to the front of the body longitudinal axis and the z_m axis is perpendicular to the ox_my_m plane and vertical upward.

During turning, the movement of the high speed side track and the low speed side track is inconsistent, the high speed side track will idle, the low speed side track will slide5 and the lateral velocity of the vehicle is not zero, as shown in Figure 1. The kinematic constraint of ordinary wheeled vehicle can't meet the needs of tracked vehicle, so it is necessary to re-establish kinematic constraints and correct the OD output at the right time.

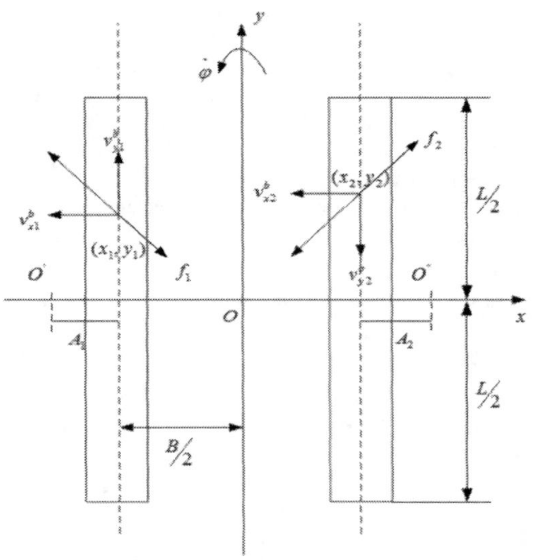

Figure 1. Kinematic relation of the two side tracks when turning.

According to the position of the two joysticks, the motion state of the tracked vehicle can be divided into five states: running straight line at high speed, running straight line at low speed, turning in situ, turning in curve and parking.

3.1 running straight line at high speed

When the two joysticks of the tracked vehicle are at the most former position, the planet steering mechanisms have a transmission ratio of 1:1, and the output of the OD is equal to the true speed. At this time:

$$\begin{cases} v_x^m = 0 \\ v_y^m = v_D \\ v_z^m = 0 \end{cases} \tag{1}$$

3.2 running straight line at low speed

When both sides of the joystick are at the position Ⅰ, the vehicle runs at low speed, and the transmission ratio on both sides is 1.42:1.
At this time:

$$\begin{cases} v_x^m = 0 \\ v_y^m = \dfrac{v_D}{1.42} \\ v_z^m = 0 \end{cases} \tag{2}$$

3.3 Turning in situ

When one joystick is in the position II and the other is in the most former position or position I, the vehicle performs turning in situ. At low speed side, the track driving wheel is in a locked state, the transmission ratio is 0, and the track on the other side is running in a high speed or a low speed. The turning radius R_{I} is half of two track spans B.

$$R_{\text{I}} = \frac{B}{2} \tag{3}$$

When sliding and idling occured in the process of turning in situ, the displacement of the vehicle body is small and the actual output of the odometer is larger, so the OD information should be abandoned and the navigation parameters are calculated only by using the inertial navigation data, and no measurement updates should be carried out until turning is completed.

3.4 Turning in curve

When one joystick is in the position I and the other joystick is in the most former position, the vehicle turns in curve. At this point the turning radius R_{II} satisfies:

$$\frac{R_{\text{II}} + B/2}{R_{\text{II}} - B/2} = \frac{v_{\text{II}}}{v_{\text{I}}} \tag{4}$$

Where v_{I} is the track speed on the low speed side and v_{II} is the track speed on the high speed side. Bring $v_{\mathrm{II}} = 1.42 v_{\mathrm{I}}$ into formula (4) :

$$R_{\mathrm{II}} = 2.88B \quad (5)$$

In the process of turning in curve , the displacement of vehicle is large. It is necessary to modify and compensate the sliding and idling during the turning process based on the kinematic constraint of the tracked vehicle. Ignoring the influence of the offset of the vehicle body centroid, the lateral speed of vehicle body during turning in curve is as follow:

$$v_x^m = \frac{v_{x1}^b + v_{x2}^b}{2} = -\dot{\varphi} L \quad (6)$$

Where $\dot{\varphi}$ is the steering angle rate and L is the track ground length. The forward speed is:

$$v_y^m = \frac{v_{y1}^b + v_{y2}^b}{2} = \frac{v_{II} + v_I}{2} + \frac{\dot{\varphi}}{2}\left(A_2 - A_1\right) \quad (7)$$

On the cement floor6, $A_1 \approx 0.4520$, $A_2 \approx 0.8615$, brings $v_{\mathrm{II}} = 1.42 v_{\mathrm{I}}$ into the formula (7):

$$v_y^m = \frac{v_{y1}^b + v_{y2}^b}{2} \approx 0.8521 v_D + 0.2048\dot{\varphi} \quad (8)$$

After correction, the speed output of tracked vehicle is $\mathbf{v}_D^m = \begin{bmatrix} v_x^m & v_y^m & 0 \end{bmatrix}^T$, and the increment of the position from t_{i-1} to t_i is $\mathbf{\Delta S_i^m} = [\Delta S_{xi}^m \quad \Delta S_{yi}^m \quad 0]$.

4 The model of integrated navigation system

The navigation coordinate system n is defined as the east-north-up (ENU) coordinate system, the carrier coordinate system b and the OD coordinate system m is the right front up (RFU) coordinate system.

Note: The vehicle's yaw is ψ , pitch is θ , roll is γ , attitude error is $\phi = [\phi_E \quad \phi_N \quad \phi_U]$, speed error is $\mathbf{\delta v^n} = [\delta v_E \quad \delta v_N \quad \delta v_U]$, position error is $\mathbf{\delta p} = [\delta L \quad \delta\lambda \quad \delta h]$. The installation deviation between OD and SINS are yaw deviation α_ψ , pitch deviation α_θ and roll deviation α_γ . There is a transformation matrix C_m^b between the coordinate system b and the coordinate system m . When the vehicle has no lateral displacement, the following relationship exists7:

$$\mathbf{\Delta S_i^b} = C_m^b \mathbf{\Delta S_i^m} = \begin{bmatrix} \sin\alpha_\psi \cos\alpha_\theta \\ \cos\alpha_\psi \cos\alpha_\theta \\ \sin\alpha_\theta \end{bmatrix} \Delta S_{yi}^m \quad (9)$$

The OD error is $\boldsymbol{\xi}_\mathbf{D} = [\delta\alpha_\psi \quad \delta K_{od} \quad \delta\alpha_\theta]^T$, which are the calibration error of yaw installation deviation, scale factor error and the calibration error of pitch installation deviation of the OD. Neglecting the second order small quantity, the actual displacement increment is as follows:

$$\mathbf{\Delta S_j^b} = \begin{bmatrix} \sin(\alpha_\psi + \delta\alpha_\psi)\sin(\alpha_\theta + \delta\alpha_\theta) \\ \cos(\alpha_\psi + \delta\alpha_\psi)\cos(\alpha_\theta + \delta\alpha_\theta) \\ \sin(\alpha_\theta + \delta\alpha_\theta) \end{bmatrix} (1 + \delta K_{od}) \Delta S_{yj}^m \quad (10)$$
$$\approx \mathbf{\Delta S_j^b} + \Delta S_{yj}^m M_\alpha \boldsymbol{\xi}_\mathbf{D} + \boldsymbol{\omega}_\mathbf{D}$$

Among them, $M_\alpha = \begin{bmatrix} \cos\alpha_\theta\cos\alpha_\psi & \cos\alpha_\theta\sin\alpha_\psi & -\sin\alpha_\theta\sin\alpha_\psi \\ -\cos\alpha_\theta\sin\alpha_\psi & \cos\alpha_\theta\cos\alpha_\psi & -\cos\alpha_\psi\sin\alpha_\theta \\ 0 & \sin\alpha_\theta & \cos\alpha_\theta \end{bmatrix}$, $\boldsymbol{\omega}_\mathbf{D}$ is the measurements noise of OD.

4.1 Measurement of displacement increment under the carrier coordinate system

The filtering period of integrated navigation is T , and the sampling period of SINS and OD is T_m .The relationship between them is as follow: $T = kT_m$. In the filtering period, the position increment of SINS and OD in coordinate system b are:

$$\mathbf{\Delta p} = \sum_{i=1}^{k} \mathbf{\Delta p_i} = \sum_{i=1}^{k} \left(C_{ni}^b \mathbf{v_i^n} + C_{ni-1}^b \mathbf{v_{i-1}^n} \right) \frac{T_m}{2} \quad (11)$$

$$\mathbf{\Delta p_D} = \sum_{i=1}^{k} \mathbf{\Delta p_{Di}} = \sum_{i=1}^{k} \mathbf{\Delta S_i^b} \quad (12)$$

Where C_{ni-1}^b and C_{ni}^b are the attitude matrix of time t_{i-1} and time t_i respectively. In the ideal case there is $\mathbf{\Delta p} = \mathbf{\Delta p_D}$, that is:

$$\sum_{i=1}^{k} \left(C_{ni}^b \mathbf{v_i^n} + C_{ni-1}^b \mathbf{v_{i-1}^n} \right) \frac{T_m}{2} = \sum_{i=1}^{k} \mathbf{\Delta S_i^b} \quad (13)$$

Considering the influence of δv^n , ϕ and ξ_D under actual conditions, the actual values of Δp and Δp_D are:

$$\Delta\mathbf{p} = \sum_{i=1}^{k}\left(C_m^b\mathbf{v}_i^n + C_{m-1}^b\mathbf{v}_{i-1}^n\right)\frac{T_m}{2} - \sum_{i=1}^{k}C_m^b(\mathbf{v}_i^n\times)\phi_i\frac{T_m}{2} - \sum_{i=1}^{k}C_{m-1}^b(\mathbf{v}_{i-1}^n\times)\phi_{i-1}\frac{T_m}{2} + \sum_{i=1}^{k}\left(C_m^b\delta\mathbf{v}_i^n + C_{m-1}^b\delta\mathbf{v}_{i-1}^n\right)\frac{T_m}{2} \quad (14)$$

$$\Delta\mathbf{p_D} = \sum_{i=1}^{k}\Delta\mathbf{S_i^b} \approx \sum_{i=1}^{k}(\Delta\mathbf{S_i^b} + \Delta S_{yi}^m M_\alpha\xi_{Di} + \omega_{Di}) \quad (15)$$

4.2 Combination scheme

The attitude error ϕ, velocity error $\delta\mathbf{v^n}$, position error $\delta\mathbf{p}$, gyro drift ε^b, and accelerometer bias ∇^b are selected as the state of SINS. The state of integrated navigation system consists of the state of the SINS and OD, that is, $\mathbf{x} = [\mathbf{x_I} \quad \xi_D]^T$. The equation of state is8

$$\begin{bmatrix}\dot{\mathbf{x}}_I\\ \dot{\xi}_D\end{bmatrix} = \begin{bmatrix}F_I & 0\\ 0 & 0\end{bmatrix}\begin{bmatrix}\mathbf{x_I}\\ \xi_D\end{bmatrix} + \begin{bmatrix}\omega_I\\ 0\end{bmatrix} \quad (16)$$

The difference between displacement increment of SINS and OD is taken as the observation. In a short period of time, consider that $\delta\mathbf{v}_i^n \approx \delta\mathbf{v}^n$, $\phi_i \approx \phi$, and $\xi_{Di} \approx \xi_D$. The observation equation as follow:

$$\mathbf{z} = \Delta\mathbf{p} - \Delta\mathbf{p_D} \approx \sum_{i=1}^{k}\{\left(C_m^b\mathbf{v}_i^n + C_{m-1}^b\mathbf{v}_{i-1}^n\right)\frac{T_m\delta\mathbf{v}^n}{2} - [C_m^b(\mathbf{v}_i^n\times) + C_{m-1}^b(\mathbf{v}_{i-1}^n\times)]\frac{T_m\phi}{2} - \Delta S_{yi}^m M_\alpha\xi_{Di}\} - \omega \quad (17)$$

Where $\omega = \sum_{i=1}^{k}\omega_{Di}$ is the observation noise.

5 Fault detection and tolerance

5.1 Conventional residual χ^2 test

The idea of residual χ^2 test is to get the residual error of step k according to the filtering equation.

$$\gamma_k = \mathbf{z_k} - \hat{\mathbf{z}}_{k,k-1} = \mathbf{z_k} - H_k\hat{\mathbf{x}}_{k,k-1} \quad (18)$$

Its covariance as follow:

$$P_{zz} = H_kP_{k,k-1}H_k^T + R_k \quad (19)$$

If there is no fault in the system, γ_k should be Gauss white noise with zero mean, that is, $\gamma_k \sim N(0, P_{zz})$; if γ_k does not satisfy this property, then the system will fail. Therefore, the fault detection function can be constructed as

$$\lambda_k = \gamma_k^T P_{zz}^{-1}\gamma_k \quad (20)$$

The threshold T_D should be set in advance. If $\lambda_k \le T_D$, the system is judged to be normal. Otherwise, the system is faulty.

5.2 Fault classification

The reasons for the abnormal OD output in the course of tracked vehicle motion are as follows: sliding, idling, no output of OD, turning in situ, turning in curve, bumping and jumping, etc.

The first type of fault: When the vehicle is sliding, idling and there is no output from the OD, the actual distance traveled by the vehicle between ΔS_{yi}^m are obviously inconsistent and is called the first type of fault.

The second type of fault: When the vehicle sidesliding, turning in situ or turning in curve, the crawler will slide or idle, causing $\Delta S_{xi}^m \neq 0$, which is called the second type of fault.

The third type of fault: When the road condition is poor, the vehicle will bump and jump, causing the vehicle to move from high to low, which is called the third type of fault.

5.3 Two-layer fault detection scheme

According to the tracked vehicle kinematics constraint model, when the vehicle is running normally, there is no lateral and vertical movement, that is, $\Delta p_x^b = \Delta p_z^b = 0$. When $\mathbf{z_k}$ is invalid according to the conventional residual χ^2 test, $z_k(1)$ and $z_k(3)$ will continue to be judged. If $z_k(1)$ and $z_k(3)$ are effective, the tracked vehicle kinematic constraints are used to assist SINS to navigate. Otherwise, the pure inertial navigation is carried out. The concrete steps are as follows:

Step 1. According to the Kalman filtering equation, the residual error $\gamma_k = \mathbf{z_k} - H_k\hat{x}_{k,k-1}$ and the corresponding detection function $\lambda_k = \gamma_k^T P_{zz}^{-1}\gamma_k$ are calculated. If the measurement is valid, the SINS/OD navigation is performed normally. Otherwise, the process jumps to step 2.

Step 2. Construct a new observation equation $\mathbf{z_{D,k}} = J_k \mathbf{x_k} + \boldsymbol{\omega}$, where J_k is the first and third line of H_k .That's means $\mathbf{z_{D,k}} = [z_k(1) \quad z_k(3)]^T$. Construct a new residual error $\gamma_{D,k} = \mathbf{z_{D,k}} - J_k \hat{\mathbf{x}}_{\mathbf{k,k-1}}$ and the new detection function $\lambda_{D,k} = \gamma_{Dk}^T P_{D,k}^{-1} \gamma_{D,k}$, where $P_{D,k}$ is the covariance matrix of $\gamma_{D,k}$ and the detection method is the same as in step 1. When the observation is valid, the measurement is updated based on $\mathbf{z_{D,k}}$ and J_k . Otherwise, go to step 3.

Step 3. Only the SINS information is used for purely inertial navigation and no measurement updates are performed.

6 Simulation

The simulation conditions are set as follows: the constant drift of the gyro is 0.02 /h, the noise is 0.02 /h, the accelerometer bias is 1×10^{-4}g, the noise is 0.5×10^{-4}g, the calibration error of OD is $\boldsymbol{\xi}_{\mathbf{D}} = [0.5^\circ \quad 0.02 \quad 0.5^\circ]^T$ and the initial misalignment angle is $\phi = \left[0.1^\circ, 0.1^\circ, 0.2^\circ\right]$ .The starting position of the vehicle is 39.84991°N, 116.17715°E and the initial height is 60 meters. The simulation time is 1685 seconds and the maximum speed during this period is 10m/s.

In order to verify the effectiveness of the algorithm, the faults in Table 1 is inserted during the simulation

Table 1 The type of faults and the time of its occurrence

Time/s	Driving State	Type of faults
240-245	Running straight with uniform velocity	idling
540-550	Turning in situ	Sidesliding and idling
840-845	Running straight with uniform velocity	No output
1110-1120	Turning in curve	Sidesliding and idling
1360-1365	Running downhill with uniform velocity	sliding

The attitude error, velocity error, and position error of the SINS/OD integrated navigation are given in Figure 2, Figure 3 and Figure 4 respectively. The red curve represents the result of the algorithm without fault detection and fault tolerance. The black curve represents the result of the two-layer fault detection and fault tolerance algorithms under the vehicle kinematic constraints in the SINS/OD integrated navigation. And the moments of failure are indicated by the vertical dotted lines.

It can be seen from the yaw error in Fig. 1, the northward speed error in Fig. 2 and the latitude and longitude error in Fig. 3, the OD output information can't match the actual motion of the vehicle when a fault occurs. Conventional SINS/OD integrated navigation can't find the faults and isolate them in time, leading to sudden jumps in various errors and the quality of navigation sharply decreasing. In the conventional SINS/OD integrated navigation mode, the errors can't converge to the ideal accuracy in a short time even after the fault recovery. Fault-tolerant navigation is able to detect and isolate the OD faults effectively by using two-layer fault detection algorithm under the tracked vehicle kinematic constraints. It can be seen from the errors, there is no obvious changes of the accuracy of fault-tolerant SINS/OD navigation when the fault occurs, and the accuracy always maintained at a high level.

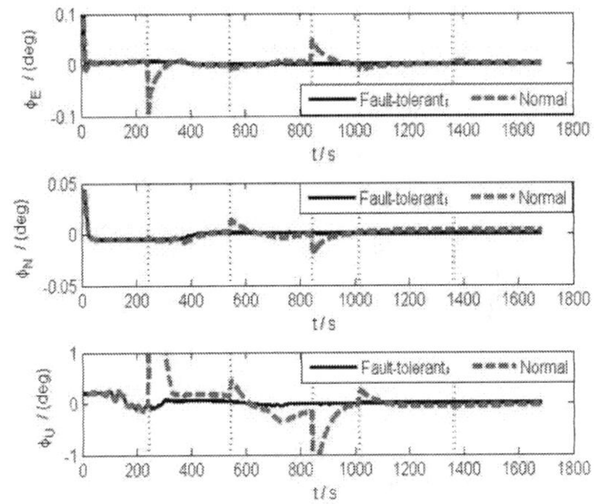

Figure. 2 The estimation of attitude error

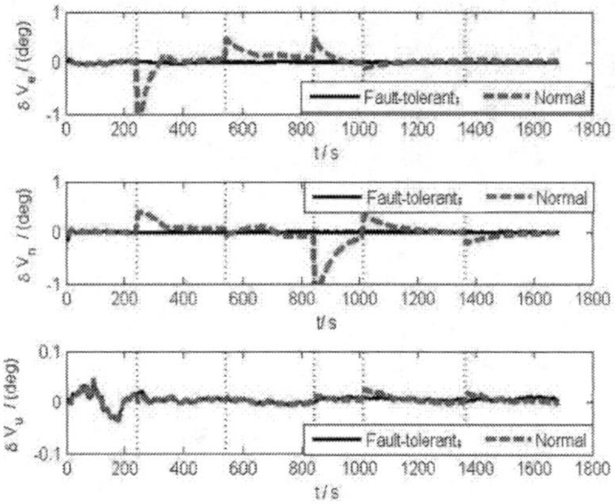

Figure. 3 The estimation of speed error

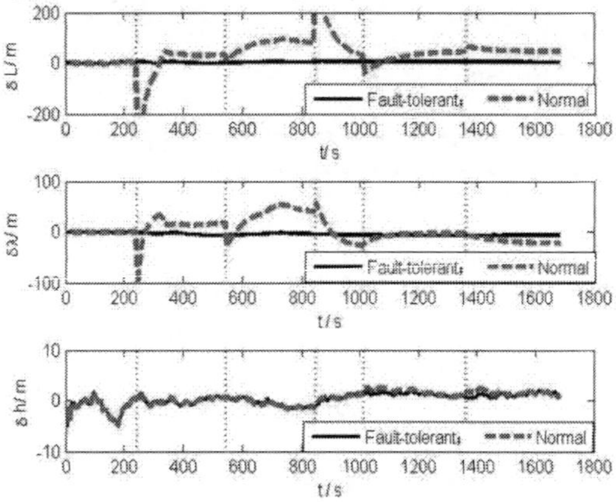

Figure. 4 The estimation of position error

7 Conclusion

In this paper, a kinematic constraint model for tracked vehicles is established, and the OD output under the special motion conditions is compensated. The difference between position increments under the carrier coordinate system is used as the observation and the error output of OD is detected and isolated by the two-layer fault detection scheme. The simulation results show that the SINS/OD integrated navigation algorithm based on tracked vehicles kinematic constraints can isolate the fault information of OD in time and greatly improve the accuracy of positioning. It has guiding significance for the engineering practice.

Reference

1. H. S. Zhao, L. J. Miao, J. Shen. High Accuracy Algorithm for SINS /Odometer Integrated Navigation System[J]. Acta Armamentarii, 2014,35（4）：433-420.
2. P D. Groves. Principle of GNSS, inertial and multi-sensor integrated navigation system [M]. Boston: Artech House.2008.
3. X. An. Research on Design and Information Fusion Technology of Ship Integrated Navigation System[D].DaLian: Dalian Maritime University, 2017.
4. J. Gong, Y. Cheng , Y. Y. Qin. Application of vehicle constraints in SINS/OD system's fault detection[J]. Journal of Chinese Inertial Technology, 2013，21(3): 406-410.
5. J. W. Tang, W. P. Liu, D. G. Liu. An Analysis of Steering Track for Tracked Vehicle at Inaccurate Steering[J]. Acta Armamentarii,2006,27(5):779-783
6. H. J. Song, L. H. Gao, J. Li. Correction and Experiment of Steering Performance Index of Tracked Vehicles[J]. Journal of The Academy of Armored Forces Engineering,2008,22(6)：65-68
7. Z. W. Wang, J. Q. Qin, Z. Y. Shi. SINS / OD Error Compensation During Cornering[J]. Fire command and control,2018,43(1):57-61.
8. X. Xiao, Q. Y. Wang, Y. Cheng. High Accuracy Navigation Algorithm for Tightly Coupled INS /Odometer[J].Acta Armamentarii, 2012, 33(4): 395-400.

Research on Contact Pressure Distribution of Pad

Dong Chen, Guofeng Zou

School of Mechanical and Automotive Engineering, South China University of Technology, Wushan Road No.381, Guangzhou, P.R.China.

Abstract. A simplified thermo-structural coupling finite element model of disc brake was developed. The contact pressure distribution of new pad was simulated under an emergency braking condition. The result shows that the contact pressure distribution of pad is not uniform when the new pad is in normal working condition. The inhomogeneity of the contact pressure is a critical cause of tangential partial wear. Compared the contact pressure distribution of new pad, uniform wear pad and tangential partial wear pad, the influence of wear amount and tangential partial wear on the contact pressure of pad was investigated. The results prove that the influence of wear amount on the contact pressure of pad is very small, while tangential partial wear has a great influence on the contact pressure of pad. The wear and force characteristics of pad are analyzed. The wear of its life cycle of pad is obtained, and the reliability and accuracy of the simulation are verified.

1 Introduction

Disc brakes are the most common type of brakes on a car. When braking, the pads firmly hold the disc under the braking force, and the vehicle is decelerated or even stopped by the friction between them. During braking, brake disc and pad are elastic deformed, coupled with the friction, so that the contact pressure distribution is uneven between them. It is likely to cause partial wear pad, braking vibration, braking noise, then seriously affect the braking performance of disc brake. Therefore, the study of brake contact pressure distribution on friction interface is an indispensable part in the design and manufacture of brake.

In recent years, domestic and foreign scholars conducted a large number of detailed researches [1], [2]. Huang et al. [3] proposed a contact pressure distribution model of a disc brake that has a variable effect of moving heat source, the elastic deformation of the brake disc and pad, and the frictional heat flux of the interface. Zhang et al. [4] established a three-dimensional transient thermo-mechanical coupled finite element model, then calculated and analyzed the contact pressure of the brake disc-pad and the thermo-mechanical coupling characteristic of the brake pad under emergency braking conditions. Anders Söderberg et al. [5] conducted a simulation analysis of the wear and contact stress distribution of pad. Gao et al. [6] studied the variation of the contact pressure distribution characteristics of disc brakes under different working conditions by numerical method. Bakar RA et al. [7] obtained the static contact pressure distribution through the pressure indicating film, and obtained the brake contact pressure distribution by establishing the finite element model of disc brake, and studied the influence of the friction surface morphology on the brake contact pressure distribution.

The uneven contact pressure distribution on the frictional interface is the consensus of the above studies, but they rarely involve the contact pressure distribution of the wear and partial wear pads. In this paper, the contact pressure distribution of new pad under emergency braking condition is firstly simulated and analyzed. The contact pressure distribution of wear pad and partial wear pad are then analyzed. The differences between them were compared and the influence of wear amount and partial wear on the contact pressure of pad was investigated. Finally, the wear and force characteristics of pad are analyzed. The wear of life cycle of pad is obtained, and the reliability and accuracy of the simulation are verified. This creative method provides a completely new idea for the study of partial wear of pad. It has certain reference value for the future research of brake design, optimization and vibration noise.

2 Thermo-Structure Coupling Model of Disc Brake

2.1 Disc Brake

A disc brake model is shown in Fig.1, including brake disc, pads, brake caliper, carrier, piston and other components. The brake disc has a thickness of 26 mm and a diameter of 281 mm. There are a total of 36 ventilation slots in the circumferential direction of brake disc. A pair of pads is divided into the inner pad (near the piston side) and the outer pad (near the caliper finger side). It has a thickness of 19 mm, a wrap angle of 60 °, an outer diameter of 140 mm and an inner diameter of 100 mm.

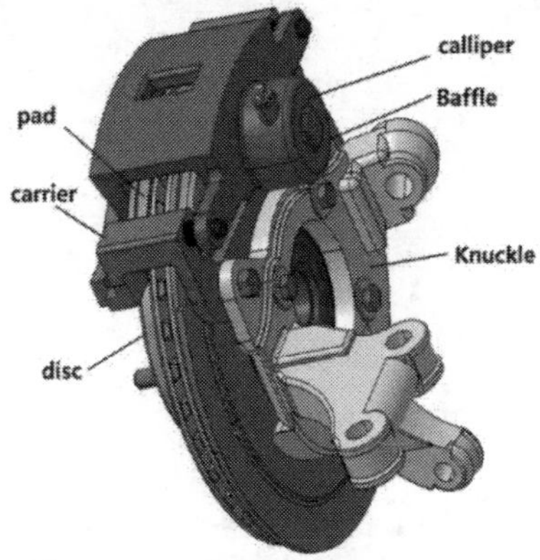

Figure 1. The structure of disc brake.

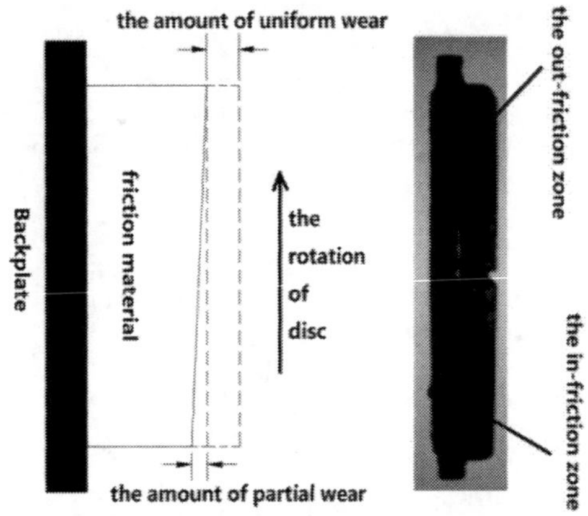

Figure 2. Tangential partial wear design.

It has been described above that the thickness of new pad is 19 mm, whereas the wear pad discussed in this paper is a pad with a uniform wear amount of 4 mm, namely have a thickness of 15 mm. The partial wear pad has a uniform wear amount of 4 mm and a tangential partial wear amount of 1.5 mm. Tangential partial wear pad shown in Fig.2, the in-friction zone has the maximum amount of partial wear.

2.2 Finite element model

In order to ensure that the model can accurately simulate the real friction interface contact pressure distribution under the emergency braking condition and ignoring minor details to simplify the model, the modeling of this paper is based on the following assumptions: (1) The materials of components are uniformly distributed and isotropic. (2) The material of brake disc is HT250.It is a linear material and its material properties do not change with temperature. The material of pad is a resin-reinforced composite material, which is a non-linear material. The elastic modulus and thermal expansion coefficient change with temperature. (3) There is elastic contact between brake disc and pad. (4) Consider only thermal conduction and convection, ignoring thermal radiation.⑤ Discard brake disc wear during simulation analysis.

Because this paper mainly analyzes the brake friction phenomenon between brake disc and pad, the structural components such as the caliper body and the hub connecting part which have little relation with the research are not considered in the modeling. In order to obtain better convergence of analysis and calculation, remove the ventilation slots and bolt holes in brake disc, but also remove chip straight flute in the middle of pad. Moreover, the shape of pad is also properly simplified to a regular shape. Finally, a simplified finite element model based on ABAQUS software, including brake discs and friction plates, is established as shown in Fig.3. Model is composed of 5950 hexahedron thermal-mechanical coupling units.

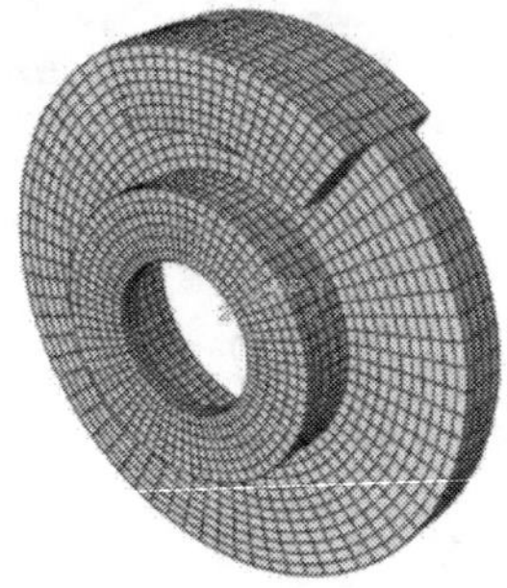

Figure 3. Finite element model of brake disc-pad.

2.3 The relevant settings of simulation

(1) Contact relation definition: Contact between brake disc and pads.

(2) The boundary conditions are set as follows: the pressure of brake oil is 8MPa, which is converted to a uniform brake pressure of 3.36MPa applied to the upper surfaces of pads; Constrained X and Y translational degrees of freedom of pad as well as Z rotation degree of freedom of pad; Define a reference point somewhere in the axis of rotation to control the movement of brake disc. The initial angular velocity is 80 rad / s and the corresponding initial braking speed is 72 km / h. Brake disc slow down for uniform deceleration. The duration of the entire emergency braking process is 2 s; The convection heat exchange coefficient between pad and surrounding environment and the convective heat dissipation coefficient between brake disc and surrounding environment are known from the literature [8].

(3) Analysis step settings: Simulation is divided into two analysis steps. The first analysis step takes 1 s, applying a uniform brake pressure slowly to pad to improve the convergence of the model calculation. The second analysis step introduces the rotation of brake rotor and it takes 2 s.

(4) Parameter definition: The material properties of each part in the simulation model are shown in Tab.1 and Tab.2. The influences of temperature on the material properties and the friction coefficient of the composite material can be seen in Tab.3 and Tab.4.

 https://doi.org/10.1051/matecconf/201819801002

Table 1. Material properties of HT250 (brake disc)

Density (kg/m3)	Young's modulus (Pa)	Thermal conductivity (W•(m•℃)-1)
7220	1.05E11	48
Poisson's ratio	Thermal expansion (℃-1)	Specific heat (J•(kg•℃)-1)
0.3	4.39E-6	419

Table 2. Properties of composites material (pad)

Density (kg/m3)	Thermal conductivity (W•(m•℃)-1)
1550	0.9
Poisson's ratio	Specific heat (J•(kg•℃)-1)
0.25	1200

Table 3. Effect of temperature on composites material properties

Temperature (℃)	20	100	200	300
Young's modulus (Pa)	2.2E9	1.3E9	5.3E8	3.2E8
Thermal expansion (℃-1)	1E-5		3E-5	

Table 4. Effect of temperature on friction coefficient

Temperature (℃)	20	100	200	300	400
Friction coefficient	0.37	0.38	0.41	0.39	0.24

3 Contact pressure analysis of pads

Since a pairs of pads are symmetrically arranged with respect to the brake disc, only one pad needs to be analyzed. The pad studied in this paper is the inner pad (near the piston side).

The finite element simulation of the braking process was performed to obtain the contact pressure distribution of new pad at different times, as shown in Fig.4. In order to more intuitively observe the contact pressure changes of new pad during braking process, the contact pressure values of 10 points as shown in Fig. 5 are extracted. The contact pressure curves at circumferential and radial points of new pad were obtained as showed in Fig. 6.

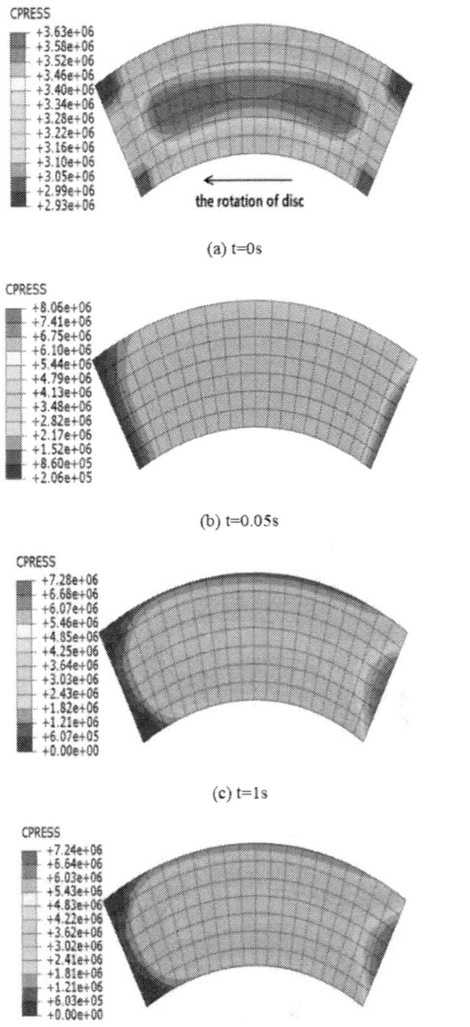

(a) t=0s

(b) t=0.05s

(c) t=1s

(d) t=2s

Figure 4. Contact pressure distribution of new pad at different times.

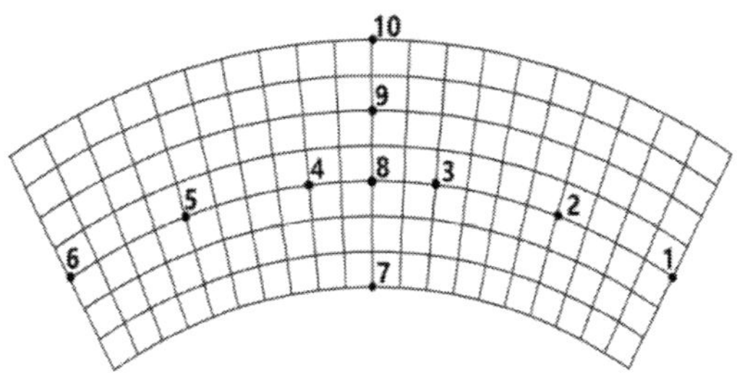

Figure 5. Distribution of measuring points on pad.

https://doi.org/10.1051/matecconf/201819801002

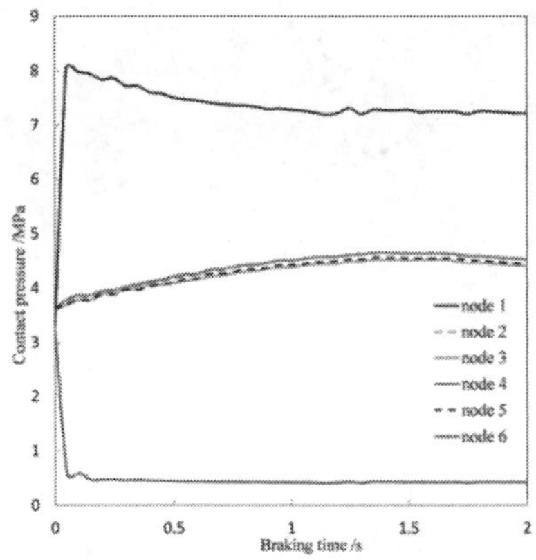

(a) Circumferential

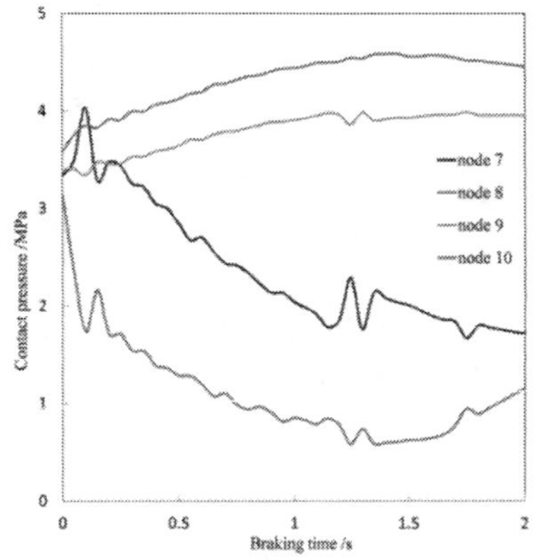

(b) Radial

Figure 6. Contact pressure curves of each point on new pad.

3.1 Contact pressure distribution of new pad

Static (brake disc has not yet rotated) contact pressure distribution of new pad can be seen from Fig.4 (a). It can be seen that when the friction force is not considered, the contact stress distribution of new pad is not uniform. Distribution trend is that the middle is bigger than the around, and the maximum pressure appears in the middle of pad with a value of 3.63MPa. This uneven distribution of contact pressure is caused by the elastic deformation of brake disc and pad.

As can be seen from Fig.4 and Fig.6 (a): (1) When brake disc just started to rotate, the contact pressure distribution of new pad changed obviously under the combined influences of positive braking force and friction force. The circumferential distribution more uneven, the maximum contact pressure appears in the in-friction zone, and its value increased from 3.63MPa to 8MPa. The minimum contact pressure appears in the out-friction zone with a value of 0.2MPa. The contact pressure of the central part did not change much, has been slowly increasing. (2) The contact pressure in the central part of new pad is almost equal.

3.2 Effect of wear amount and partial wear on contact pressure of pad

The contact pressure values of six nodes in the wear pad and partial wear pad were extracted. The contact pressure curve is shown in Fig.7. Comparing Fig.6 (a) and Fig.7, we can see that the trend of the contact pressure curve of new pad, wear pad and partial wear pad is the same, but the values is quite different. Since the contact pressures of the four nodes in the middle are almost the same, the contact pressure curves of the nodes 1, 4, and 6 on the three pads are compared to obtain Fig. 8.

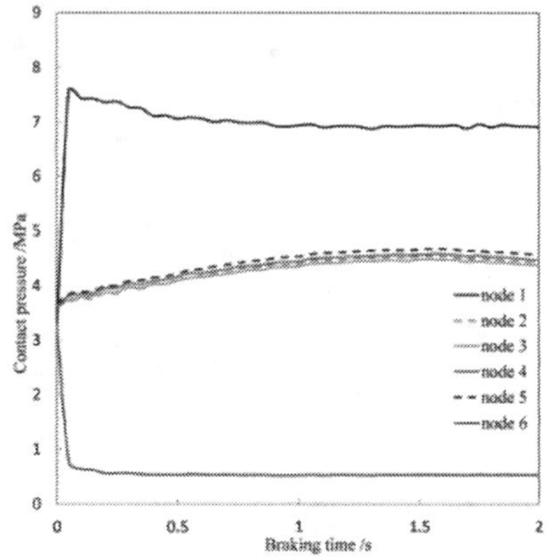

(a) wear pad

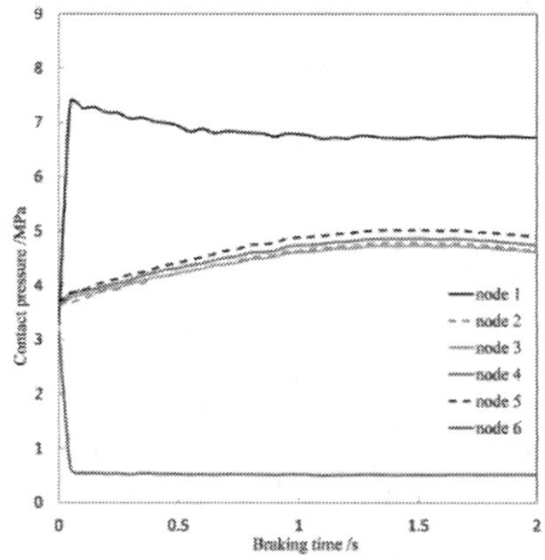

(b) partial wear pad

Figure 7. Contact pressure curves of circumferential points on wear pad and partial wear pad.

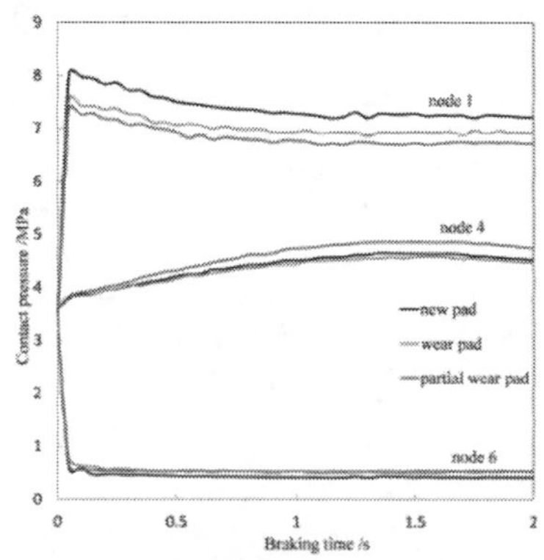

Figure 8. Contact pressure curves of node1, 4, 6 on different pad.

From Fig. 8 we can see: In the in-friction zone (the area with the highest contact pressure), the contact pressure of partial wear pad is obviously lower than that of the new pad and wear pad. In the middle part and the out-friction zone, the contact pressure of partial wear pad is obviously higher than the latter two, namely the contact pressure distribution of partial wear pad is more even. The above shows that wear amount has little effect on the contact pressure distribution of pad, while partial wear has significant effect on the contact pressure distribution of pad.

4 Force characteristics and wear characteristics of pad

By analyzing the force characteristics of pad in different stages, the wear characteristics of pad over the whole life cycle are explored.

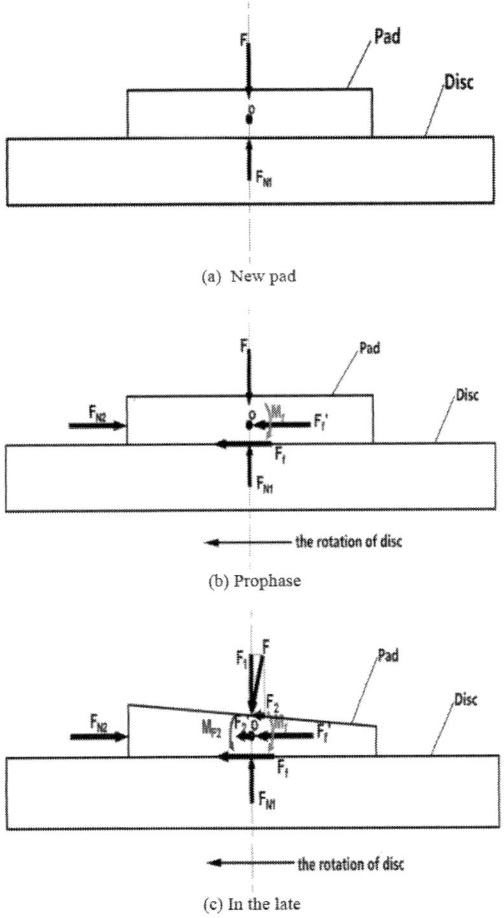

(a) New pad

(b) Prophase

(c) In the late

Figure 9. Force analysis of pad at different Stages.

Fig. 9 (a) shows the force characteristics of new pad when it is not working, meanwhile the brake disc has not been rotated yet. Brake pressure F makes pad and brake disc pressed tightly, then brake disc produce a reverse support force FN1 on pad, so that the forces on pad are balanced in the vertical direction.

Fig. 9 (b) shows the force characteristics of pad in its prophase of life cycle, meanwhile brake disc began to rotate and wear has not been produced yet. So that friction force Ff is generated on the friction interface, which is equivalent to the force Ff' and the moment Mf on the center of mass of pad. And the support force FN2 of carrier to pad makes the forces balance in the horizontal direction. Therefore, the only moment Mf will make pad rotate around its center of mass, resulting in the in-friction zone compact and the out-friction zone relax. This is consistent with the contact pressure distribution of new pad in the previous section, which verifies the correctness of the simulation.

According to Archard's law of wear [9] [10]:

$$\Delta h = k_d p_c \Delta s \tag{1}$$

where Δh is the depth of wear in time Δt , Δs is the sliding displacement of brake disc relative to pad, k_d is a constant, and p_c is the contact pressure at each point on the contact interface of pad.

According to equation (1) shows that the wear of pad is directly related to the contact pressure. Therefore, according to the contact pressure distribution of new pad, it can be seen that new pad begins partial wearing in the early stage.

Fig. 9 (c) shows the force characteristics of pad after partial wear. At this time, the brake pressure F is still perpendicular to the upper surface of pad but at an angle to the vertical. F1 and F2 are the horizontal and vertical components of the force F respectively. F2 is equivalent to the center of mass of pad, and obtain the force F2' and the torque MF2. The direction of MF2 is the opposite of Mf. When the amount of partial wear reaches a certain amount, the two moments are equal size and offset each other. Then the contact pressure of pad tends to be uniform, the wear of pad tends to be uniform. This is consistent with the more even contact pressure distribution of partial wear pad in the previous section, and verifies the correctness of the simulation again.

The above analysis shows that due to the friction moment Mf, the contact pressure distribution of new pad is uneven, resulting in tangential partial wear. As the amount of partial wear increases, the moment MF2 generated by the horizontal component of brake pressure gradually increases, then the contact pressure of pad becomes more and more uniform. When these two moments are equal in size, the contact pressure of pad tends to be uniform and the amount of wear tends to be uniform, then the pad has been uniformly worn since then.

5 Conclusion

(1) During the whole braking process, the temperature distribution and the contact pressure distribution of new pad are not uniform, and the maximum contact pressure appears in the in-friction zone. The contact pressure of each point on pad varies and the contact pressure at each point in the middle part keeps rising slowly. In addition, the contact pressure distribution of pad has changed a lot

when brake disc is rotated, which shows that the friction force is the crucial factor that affects the contact pressure distribution of pad, but the temperature and thermal stress are the very small factor.

(2) By comparing the contact pressure distribution of wear pad, tangential partial wear pad and new pad, we can see that: the wear amount had little effect on the contact pressure of pad, while the tangential partial wear had a great influence on the contact pressure of pad.

(3) The force and wear characteristics of pad are analyzed. The wear of life cycle of pad is obtained, and the reliability and accuracy of the simulation are verified. The wear of new pad is tangential partial wear in the early stage, and the wear tends to be uniform when the amount of partial wear reaches a certain moment. Namely, the pad wear has an adaptive adjustment process. In general, the tangential partial wear of pad is difficult to avoid, but the amount of partial wear can be controlled in a small range.

References

1. Ouyang Huajiang. A combined analysis of heat conduction, contact pressure and transient vibration of a disk brake [J]. Drive System Technique, 2003; 51.1/2: 190-206
2. Zhao Bo, Fan Pingqing. Analyzing in braking efficiency and contact stress of disc brake [J]. Machinery Design & Manufacture, 2011(9): 134-136
3. Huang Jianmeng, Gao Chenghui, Lin Xiezhao, et al. Study on the contact pressure distribution of a disc brake [J]. Chinese Journal of Solid Mechanics, 2007，28 (3)：297-302
4. Zhang Lijun, Chen Yuan, Diao Kun, et al. Computational investigation into disc-pads pressure distribution and thermo-mechanical coupling characteristics of brake pads in disc brake [J]. Journal of Tongji University(Natural Science), 2013， 41(10): 1554-1561
5. Söderberg A, Andersson S. Simulation of wear and contact pressure distribution at the pad-to-rotor interface in a disc brake using general purpose finite element analysis software[J]. Wear, 2009，267(12): 2243-2251
6. Gao Hongxiang, Wang Xiaofei. Research of pressure distribution state of disc brake [J]. Journal of Mechanical Transmission, 2014(9): 39-42
7. Bakar A R A, Ouyang H, Siegel J E. Brake pad surface topography part I: contact pressure distribution [J]. Automotive Sector, 2005
8. Zhao Tenglun. ABAQUS 6.6 application in mechanical engineering [M]. China Water & Power Press, 2007.

Research on Reducer Power Measurement System of Bridge Crane

Zhihao Ge[1],Yifei Tong[1],MengZhong[1] and Xiangdong Li[2]

[1]*Nanjing University of Science and Technology,Nanjing,China*
[2]*Jiangsu Province Special Equipment Safety Supervision Inspection Institute, Nanjing, China*

Abstract. In this paper, the reducer is taken as the research object for power measurement. First, the demand of reducer power measurement is analyzed, and the method for measuring power is determined. Then, in view of the power measurement method adopted by the reducer, the overall framework of the measuring system is carried out, and the host system is designed.

1 Introduction

With the development of bridge crane, energy consumption of crane transmission has attracted much attention [1].As an important transmission part, the reducer plays an important role in the hoisting mechanism. Therefore, measuring the efficiency of the reducer in service is essential for studying the energy consumption of the hoisting mechanism for bridge crane. The efficiency of reducer is traditionally measured in open and closed manners. The parameters measured in both manners to calculate the transmission efficiency of reducer are the same: torques and speeds of both high speed and low speed shafts. Currently, because of the structure of torque sensor, it is hard to measure a working reducer. Therefore, it is extremely rewarding if the efficiency can be measured when the reducer is in service. This paper investigates a measurement system that can measure torque and speed of a working reducer and transmit the data to a host computer to calculate the power of both low-speed and high-speed shafts of the reducer.

2 Demand Analysis of Reducer Power Measurement

The hoisting mechanism is one of the most important transmission mechanisms of the bridge crane and the transmission efficiency is an important indicator for its energy consumption[2]. The input energy for entire hoisting mechanism is electricity which is applied onto both the motor and the brake. However, the mechanical energy converted from the electricity by the brake acts as the braking energy and is not transferred to transmission parts.The mechanical energy converted from the electricity by the motor during non-braking process and transferred to each transmission part in serial way, as shown in Figure 1. If the entire mechanism or a single part is taken as the object to study the energy consumption, it is hard to find the solution since a lot of variables have to be studied. A method is adopted to break down and simplify the complicated problem so as to easily solve a difficult or even dissolvable problem. So, the hoisting mechanism under non-braking condition is broken down into a reducer, a coupling and a suspension system, which is composed of a drum, wire rope, a plurality of pulleys and a lifting hook[3]. This method enables the study of mechanical transmission efficiency to be performed on each subsystem instead of the entirety. As the intermediate part of the energy transmission, the input power and output power of the reducer can be used to calculate the efficiency of other transmission parts. Therefore, it is necessary and meaningful to measure the power of the reducer of the bridge crane.

Actually, very few measuring devices can be used to effectively measure the energy consumption of reducer on-site due to inappropriate on-site conditions. Although the reducer power cannot be directly measured by a device, two parameters: torque and speed, which are related to the power, can be measured to calculate the power and the efficiency of reducer. And then Equation(1) and Equation(2) can be used to calculate the power and real-time efficiency of reducer.

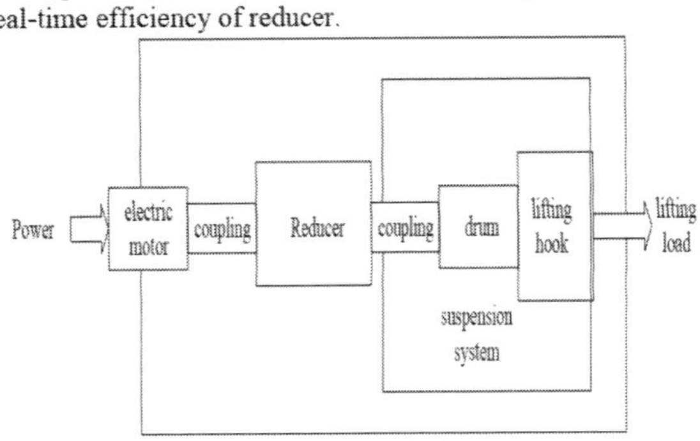

Figure 1. Transferring process of mechanical energy

$$P=\frac{n\times T}{9500} \quad (1)$$

$$\eta_{Ji}=\frac{P_{JOi}}{P_{JIi}}\times 100\% \quad (2)$$

Where

T is the torque to be measured in $N{\cdot}m$;

n is the speed in r/min;

P is the power in kW;

P_{JIi} is the high speed shaft power of the reducer, namely the input power;

P_{JOi} is the low speed shaft power of the reducer, namely the output power.

3 Overall Design of System Framework

As described above, it is of need to measure the torque and speed of both low-speed and high-speed shafts of the reducer. Chuck type power meter comprises two sections: the torque measuring and the speed measuring section. It utilizes a synchronized method to measure the toque and speed of high or low speed shaft on the reducer. Then the measured values are packed and transmitted to the host computer to calculate the power.

The block diagram of measuring system is shown in Figure 2, in which the measuring object, namely the transmission mechanism, is located in the leftmost block; the slave, namely the measuring circuit, is located in the second block; the communication section including data transmission mode and communication protocol is located in the third block[4]; and the host program which is developed by Visual C++ is located in the fourth block. The host computer can real-time display present torque and speed, calculate and display present power, and save the data in a SQL database. Some parameters of the measuring system like shaft material parameters, the acquisition frequency of system and the output format are all set in the host computer[5].

The chuck type power meter located in the leftmost block is shown in Figure 3. The chuck is consisted of an upper part and a lower part to clamp both internal and external metallic guiderails into the groove and then the guiderails are fixed onto the chuck by screws. An extension spring is arranged inside the moveable slider for connection to ensure reliable contact of the high speed metallic sheave with both internal and external metallic guiderails. The light source for speed measuring is arranged inside the slider and located right above the orbital groove on the guiderail plate. A plurality of light receivers is arranged in the orbital groove. The slider rotates relatively to the guiderail plate and the relative rotation speed is considered to be the shaft rotation speed. The external power is supplied into the chuck via the contact between metallic sheave and metallic guiderails.

A plurality of strain gauges is attached on the shaft to constitute a Wheatstone bridge [6]. The chuck is bolted onto the shaft to avoid relative rotation. The power supply inside the chuck provides stable working voltage for the Wheatstone bridge. When the shaft deforms under torque, the bridge outputs the voltage variation.

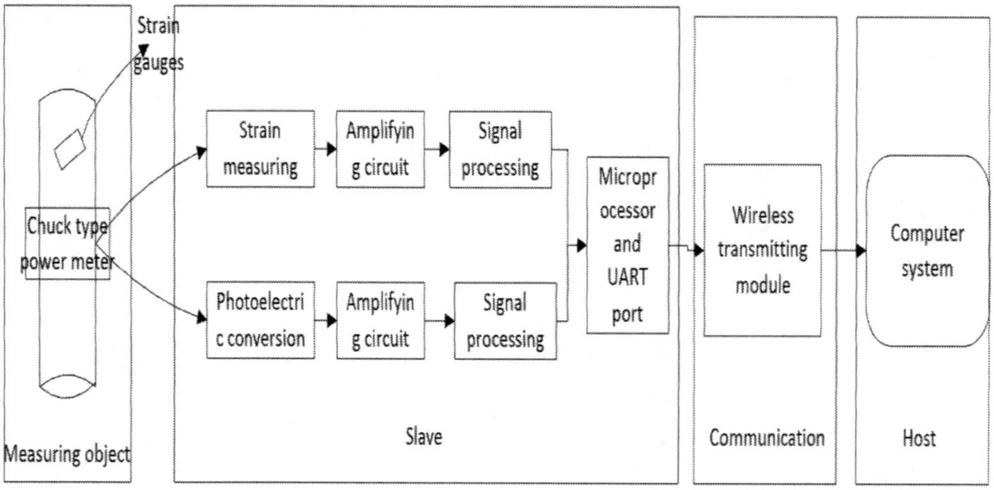

Figure 2. Block diagram of measuring system

The overall hardware structure of slave is divided into three sections based on their functions: a torque measuring section, a speed measuring section and a signal transmission section. The torque measuring section comprises a strain gage bridge, an amplifier circuit, a strain voltage processing circuit and a microprocessor computing circuit. The speed measuring section is designed on the basis of photoelectric measurement and the signal transmission section is physically a wireless transmission module.

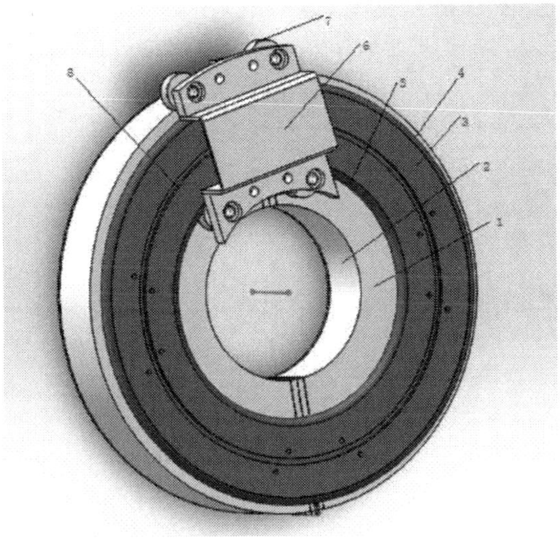

Figure 3. Chuck structure

1. Chuck 2. Internal diameter of chuck 3. High speed guiderail plate 4. External metallic guide ring 5. Internal metallic guide ring 6. Moveable slider 7. High speed metallic sheave 8. Orbital groove on guiderail plate

4 Design of System Parameters

Design parameters of each component are correlated to each other in hardware design. The voltage variation range of strain bridge is dependent on the shaft strain and the strain bridge. The amplification factor of amplifier is determined by the input voltage of V/F converter. The selection of microprocessor is based on above parameters and the speed acquisition frequency. The calibration test is performed at last to determine the maximum measuring range and the accuracy. Their correlation is shown in Figure 4.

During hardware design and selection of microprocessor, parameters have to be reasonably set for successful hardware design. In principle, the strain variation range on the measured shaft and the resulting voltage variation range have to be determined first. Then use the input voltage of V/F converter to determine the amplification factor of amplifier, after which the crystal oscillator and the microprocessor can be selected. Thetransmitting module and the power module are designed at last.

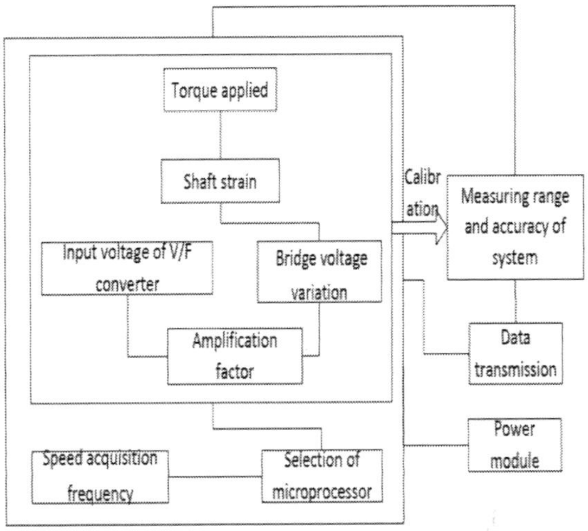

Figure 4. Correlation among parameters

5 Host System Design

The host program system is constituted by a four-layer architecture, as shown in Figure 5. The first layer is a user layer, which is used as the human-machine interface of system for visual display of data, setting of measuring parameters, selection of communication channel, setting of data save and identification for graphic data.

The second layer is an interaction layer, which is mainly used for data receiving, data calculation, data storage and data monitoring. The data receiving refers to utilizing the Wi-Fi module and the IP communication protocol on the laptop to transmit the torque and speed values measured by the chuck type power meter to the host program system[7]. The data calculation refers to calculating the data which has been processed to obtain the torque, speed and power values of measured object. The data monitoring refers to monitoring the data which has been processed and the host computer shall raise an alarm once a value exceeds the threshold. The processed and calculated data shall be stored for data playback and postanalysis.

The third layer is a core layer which is mainly used for processing data and performing compensation algorithm. The received data will be filtered and the graph display will be processed for curve-smoothing on this layer in order to make the data curve smoother. The curve is fitted by the least square method in order to obtain a curve model with minimum error to the true value. The curve fitting is a very important function during system initialization since it can make the measured data graph closer to the true data graph.

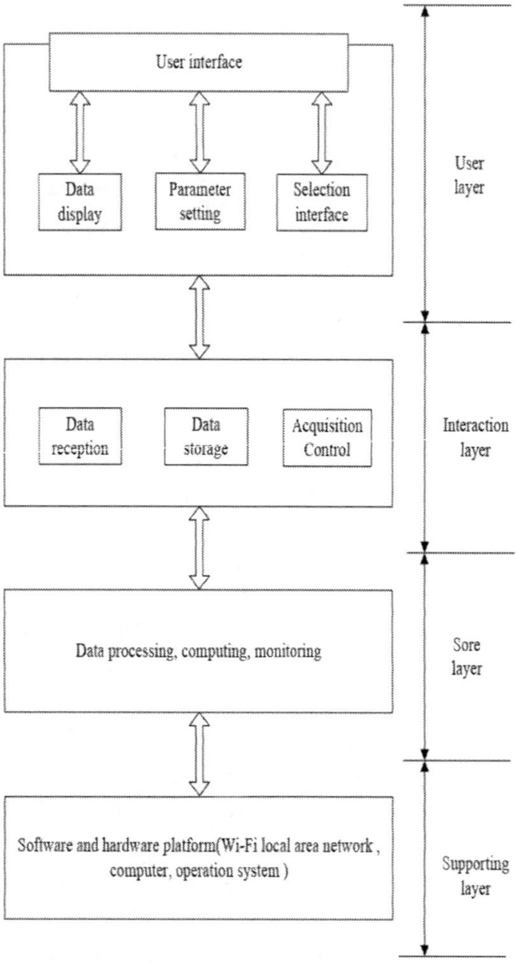

Figure 5. Host program system

The fourth layer is a supporting layer, which is used as a software and hardware platform, on which various elements like Wi-Fi local area network (called as LAN hereinafter), computer, operation system and database interact with each other.

For a measuring system which has been calibrated, be sure the chuck type power meter is in service if the host system is used for data measuring. Then use the host system to check whether the Wi-Fi module is online, the slave is in service, and the sensors and the communication work normally. Then measuring parameters related to input such as the diameter of measured shaft and the shear modulus into the host system. After that, balance channels and remove null points in order to start the measuring.

6 Conclusion

This paper investigates a measuring method that can measure torque and speed of the reducer, and carry out framework of the system. And, the host system is designed according to the requirement of measurement method. In further study, the hardware circuit that is an important part of the power measuring system is needed to design.

7 Acknowledgment

This work was financially supported by the National Key Research and Development Program of China (2017YFF0209804). The supports are gratefully acknowledged.

References

1. Wei Ye,YifeiTong,DongboLi,Xiangdong Li, XuchengYuan,Research on bridge cranes energy-saving index reduction based on rough set theory.J. Telkomnika Indonesian Journal of Electrical Engineering, 11(2013): 7269-7276.
2. Liu Yang, Crane hoisting mechanism of variablepolylinereformationHelical tube. D. Northeastern University, (2010).
3. Wei He, ZhekunLi, ZhenyuWang,An Liu, The Mathematical Model of Crane Hoisting Mechanism.J. Applied Mechanics & Materials,477-478(2013):315-320.
4. Hao Zhang, Vibration data acquisition system based on wireless transmission technology.J. China Science and technology information, 7(2016): 50-51.
5. Cheng Yang, Ping Song, WenjiaPeng, Gaoshou Deng, Xiongjun Liu, Design of host computer platform for armored vehicle integrated test system, J. Journal of Jilin University (Engineering and Technology Edition), 47(2017) :1796-1803.
6. DariuszKacprzak, IoanTuleasca, Wheatstone bridge approach to the inspection of

https://doi.org/10.1051/matecconf/201819801003

composite material, J.2017 Eleventh International Conference on Sensing Technology (ICST), IEEE, (2017), 1-6.

7. Zhijun Zhang, XiaoliXu, Guoxin Wu, Wireless Data Acquisition System Based on LabVIEW and Wi-Fi Technology, J. Control & Instruments in Chemical Industry (2013).

Finite element analysis of Bridge Crane Metal Structure Based on ABAQUS

Zhong Meng[1], Tong Yifei [1] and LI Xiangdong [2]

[1]*Nanjing University of Science and Technology, Nanjing, China*
[2]*Jiangsu Province Special Equipment Safety Supervision Inspection Institute, Nanjing, China*

Abstract. Taking the QD-type (50/10t-31.5) bridge crane as the research object and the finite element method as theoretical basis, the metal structure using ABAQUS software is analyzed. Firstly, the three-dimensional model of the metal structure is established according to the design parameters. Second, the bridge crane under three typical operation conditions is analyzed and the stress and displacement analysis results are presented. And, the strength and stiffness of the crane are verified based on the analysis results. Moreover, the maximum stress and displacement under dangerous conditions of the crane are calculated by theoretical calculation. By comparing the results of the theoretical method and the finite element method. The research can favor to increase the design efficiency.

1 Introduction

With the continuous development of the crane industry, the demand for design quality has increased [1]. In order to reduce the design cost, an effective design method is inevitably required. Traditional method is for engineers to estimate the strength at the initial stage of design [2]. Although it is safe and reliable, it will cause waste of resources due to the high safety factor. The modern design generally uses the finite element method as the theoretical basis to comprehensively use CAE analysis software to analyze the structure of the crane [3]. The analysis process of finite element method can be summarized as: discretization of continuum, unit analysis, global analysis, constraint conditions determination, solution to finite element equations, and analysis and discussion of results [4]. The key to this analysis method is to establish a reasonable and effective model. This article provides a more accurate analysis for the static structural analysis of the bridge crane metal structure.

2 The FEA modeling of metal structure

2.1 Three-dimensional modeling

As the main bearing component, the metal structure of the bridge crane is composed of the main girder, end beam and trolley rail. The metal structure of crane is shown in Fig 1[5].

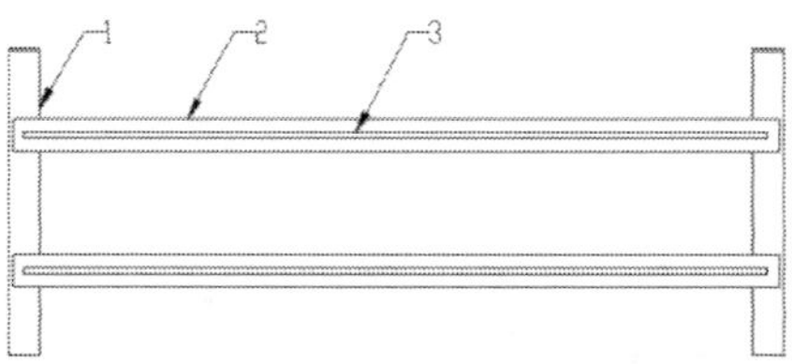

1. End beam 2. Main girder 3. Trolley rail
Figure 1. Metal structure of the crane.

The establishment of the three-dimensional finite element model is the basis for structural analysis. The accuracy of the model directly affects the analysis results. Select the shell element model according to the structural characteristics of the welded box beam, and the minimum aspect ratio a=371/8=46.25>10 in all steel plates. Firstly, using ABAQUS/CAE module to establish the three parts of the rail, main girder and end beam [6]. Then using Assembly module to obtain a three-dimensional finite element model. In order to improve the analysis efficiency, the model can be simplified, such as removing fillets, small holes, etc. Three-dimensional model is shown in Fig 2

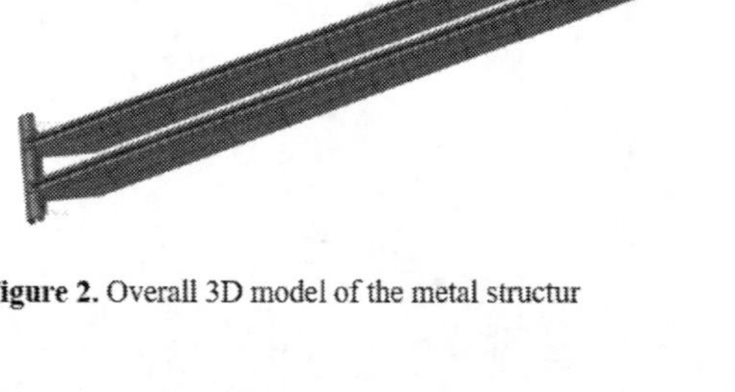

Figure 2. Overall 3D model of the metal structur

https://doi.org/10.1051/matecconf/201819801004

2.2 Setting material

The girder is welded using Q235 steel plate. The main mechanical properties of the material are shown in Table 1.

Table 1. Material properties of the girder.

Type	Young's Modulus (GPa)	Poisson's Ratio	Density (g/cm^3)	Tensile Strength (MPa)	Yield Strength (MPa)
Q235	210	0.3	785	375	235

2.3 Interactions definition

The analysis model is assembled from the main girder, end beam and trolley rail. The contact between the cover plate of main girder and the lower surface of the rail has little impact on the entire model and is always in close contact with the entire analysis process. Therefore, use Tie to define the relationship between them. This relationship not only eliminates rigid body displacement of the model, but also greatly reduces the iterations for calculation of the contact state. Set the position tolerance as 0.01 to eliminate contact surface gap due to dimensional deviation.

2.4 Loads definition

This paper analyzes the stress and displacement of the structure under three working conditions: fully loaded travelling crab is located in the middle of the girder, 1/4 and the end of the girder. According to the actual working conditions, the structure can be treated as simply supported beam. Fig 3 shows the loads of the girder when the travelling crab is in the middle, where G is the evenly distributed weight of the structure , Q1 and Q2 are the concentrated loads caused by the lifting weights at the two wheels, and P is the weight of the travelling crab.

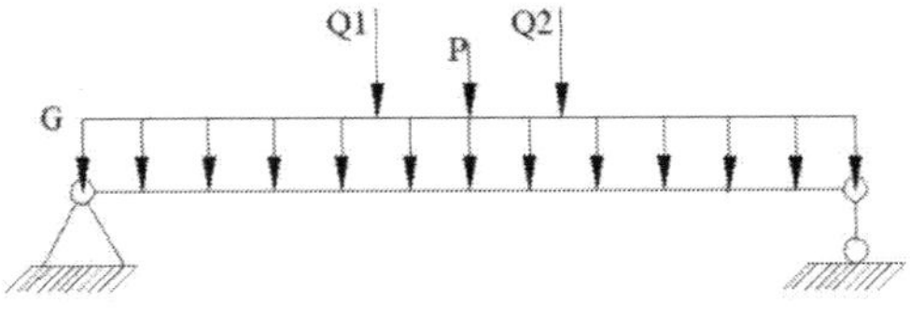

Figure 3. Load condition of the girder.

Gravity load. Simply set the material density $\rho = 7.85E-9t/mm^3$ and gravitational acceleration $g = 9800\phi_1 = 10780mm/s^2$ in Abaqus software. Introduce hoisting impact factor $\phi_1 = 1.1$ to take into account the influence of crane's own weight vibration load.

Working load. The weight of travelling crab P=153860N, and lifting load Q=490000N. Average working load on four wheels of the travelling crab and get $F = \frac{P + Q\phi_2}{4} = 178115N$, where the lifting dynamic load factor $\phi_2 = 1.15$ is set to take into account the dynamic effect of the increased load due to inertial forces.

2.5 Boundary conditions

Set the reference points at the wheel center of the end beam, and couple these points to the end faces of the beam, as shown in Fig 4 The boundary condition is imposed on reference points: constrained movement of the x, y, z direction and rotational freedom in the x, z direction of one fixed end; and at another hinged end, constrained movement in the y, z direction and rotational freedom in the x, z direction. The reference points are shown in Fig 5.

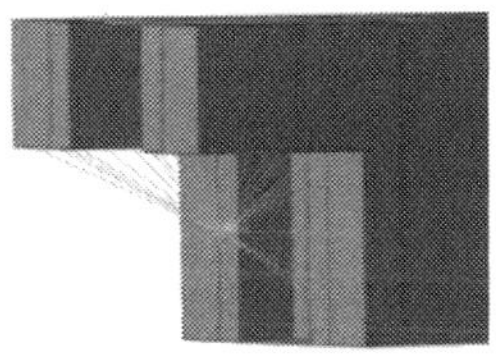

Figure 4. End faces of the beam.

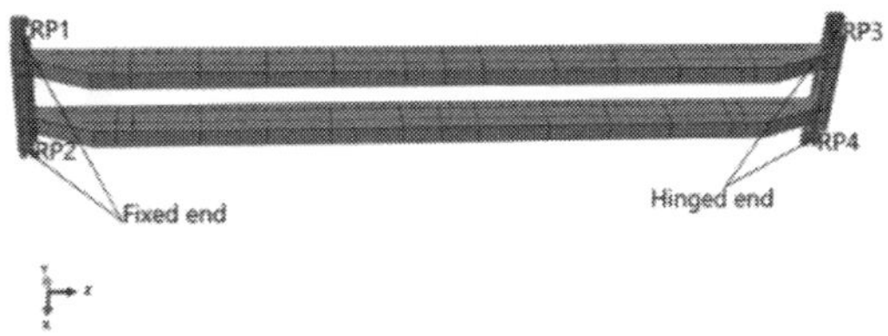

Figure 5. Location of reference point.

2.6 Mesh

The part rail is a three-dimensional solid model, for which the seeds with an approximate global size of 50 are arranged, the hexahedral grid is divided by the sweep technology, and the cell type is C3D8R (linear reduction of the eight-node three-dimensional solids). The mesh is verified without errors or warnings. The main girder is a shell element model, for which the seeds with an approximate global size of 50 are arranged. The free technology is used to divide the quadrilateral mesh. The element type is S4R (linear reduction of the four-node shell). There are three highlight warnings in the model. Solve these problems through geometric clean function module. The mesh of end beam is set in the same way as the main girder.

The pre-processing of the finite element analysis has been completed, and next, the model will be submitted to the module of Abaqus/Standard for calculation.

3 Analysis result

The stress nephogram when the travelling crab is fully loaded in the middle of the girder is shown in Fig 6, and Fig 7 shows the displacement nephogram of trolley with full loaded in 1/2 cross girder.The results of finite element calculations show that the maximum stress and displacement and their main hotspot areas of the crane metal structure under three typical operating conditions are shown in Table 2.

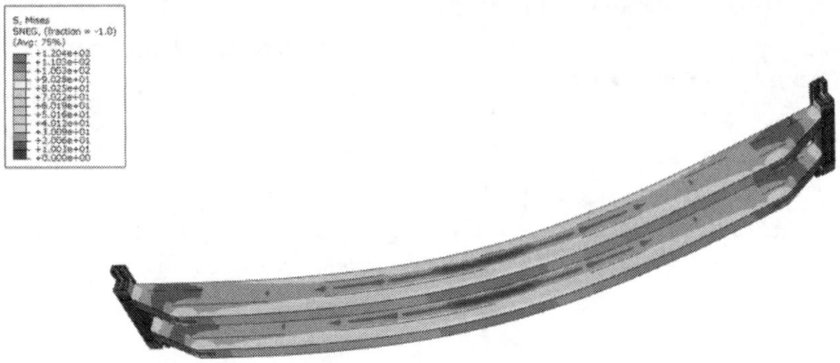

Figure 6. Equivalent stress nephogram of trolley with full loaded in 1/2 cross girder.

Figure 7. Equivalent displacement nephogram of trolley with full loaded in 1/2 cross girder.

Table 2. Stress and displacement of the girder.

Load conditions	Maximum stress [MPa]	Maximum stress position	Maximum displacement [mm]	Maximum displacement position
In the middle of the girder	120.4	Where the load is applied and the middle of lower cover	43.3	The middle of the web
At 1/4 of the girder	112.4	Where the load is applied and the middle of lower cover	33.2	The middle of the web
At the end ofthe girder	119	The end of the upper cover	17.5	The middle of the web

Calibrate the strength and stiffness of the girder when the travelling crab is in the mid-span position.

The girder material is Q235 steel, and its Yield strength $\sigma_s = 235MPa$. Take the safety factor as $n_s = 1.33$, and $\sigma_{\max} = 120.4MPa$.

Allowable stress:

$$[\sigma] = \frac{\sigma_s}{n_s} = 176.7MPa \quad (1)$$

Strength conditions:

$$\sigma_{\max} = 120.4 < [\sigma] = 176.7 \quad (2)$$

Therefore, the strength of the crane main girder meets the design requirement

https://doi.org/10.1051/matecconf/201819801004

The span of the crane is 31500mm, and the maximum displacement $u_{max} = 43.3(mm)$.

$$u_{max} = 43.3 \le \frac{1}{700}S = 45 \quad (3)$$

Therefore, the stiffness of the crane main girder meets the design requirements.

4 Theoretical calculation

In order to verify the accuracy of the finite element analysis results, the following theoretical calculations are performed on the stress and displacement of the main beam.

4.1 Main girder loading situation

Loads in the vertical plane:

Main beam weight $G = 163856(N)$, and the weight evenly distributed to the girder is $q = \frac{G}{31.5} = 5202(N/m)$.

trolley weight $P = 153860(N)$

Working load $Q = 490000(N)$

Calculated load $Q_1 = Q_2 = \frac{P + Q\phi_2}{4} = 178115N$

Loads in the horizontal plane: Inertia force during crane operation $q_0 = \frac{G}{g} \cdot a = 4846(N)$,Where acceleration $a = 0.32(m/s^2)$.

The loading of the main girder in the vertical and horizontal planes is shown in Fig 8 and Fig 9.

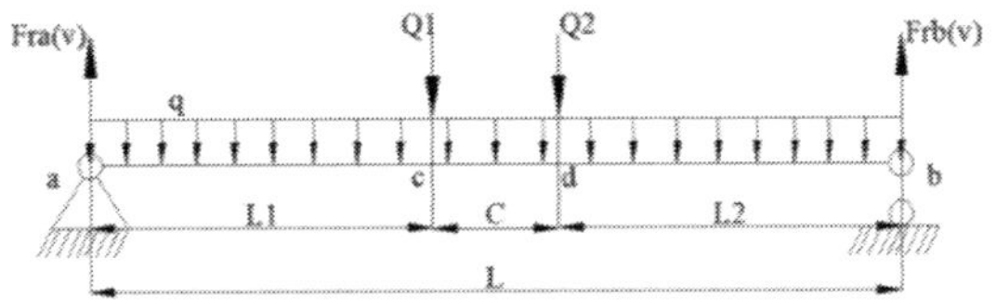

Figure 8. Loads in the vertical plane.

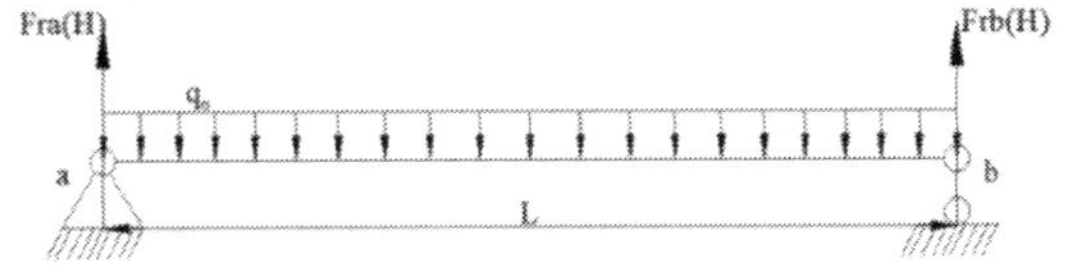

Figure 9. Loads in the horizontal plane.

4.2 Main girder section

For the convenience of calculation, the main girder is treated as an equal section, and its structural dimensions are shown in Fig 10.

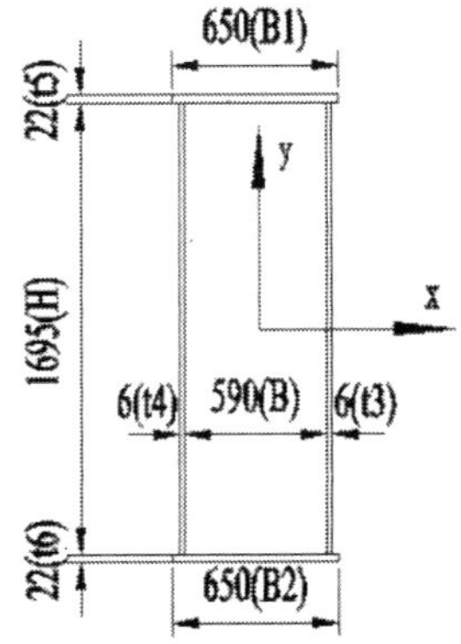

Figure 10. Main girder section.

4.3 The shear and bending moments of the vertical plane

Inertia moments I_x, I_y.

$$I_{x\text{上}} = I_{x\text{下}} = \frac{B_1 t_5^{\,3}}{12} + \left(\frac{H}{2}\right)^2 \cdot B_1 t_5$$

$$I_{x\text{左}} = I_{x\text{右}} = \frac{t_4 H^3}{12}$$

$$I_{y\text{上}} = I_{y\text{下}} = \frac{t_4 H^3}{12}$$

$$I_{y\text{左}} = I_{y\text{右}} = \frac{H t_4^{\,3}}{12} + \left(\frac{B}{2}\right)^2 \cdot t_4 H$$

$$I_x = 2I_{x\text{上}} + 2I_{x\text{左}} = 28.1 \times 10^{-3}(m^4)$$

$$I_y = 2I_{y\text{下}} + 2I_{y\text{右}} = 2.91 \times 10^{-3}(m^4)$$

Bending section coefficient W_x, W_y.

$$W_x = \frac{I_x}{y_{max}} = 33.16 \times 10^{-3}(m^3)$$

$$W_y = \frac{I_y}{x_{max}} = 9.86 \times 10^{-3}(m^3)$$

Vertical surface reaction force $F_{ra}^{\,V}$, $F_{rb}^{\,V}$.

Balance condition:

$$\sum M_B = F_{ra}^{\,V} L - \frac{1}{2}qL^2 - Q_1(C + L_2) - Q_2 L_2 = 0$$

$$\sum Y = F_{ra}^{\,V} + F_{rb}^{\,V} - qL - Q_1 - Q_2 = 0$$

Where $L = 31.5(m)$, $C = 3.58(m)$, $L_2 = 13.96(m)$, $Q_1 = Q_2 = 177237.5(N)$, $q = 5202(N/m)$, and get $F_{ra}^{\,V} = F_{rb}^{\,V} = 259169(N)$.

Shear force $F(x)$.

Part a-c, $F(x) = F_{ra}^{\,V} - qx$, $0 \le x \le 13.96$

Part b-d, $F(x) = F_{ra}^{\,V} - qx - Q_1$, $13.96 \le x \le 17.54$

Part c-b, $F(x) = F_{ra}^{\,V} - qx - Q_1 - Q_2$, $17.54 \le x \le 31.5$

Maximum shear $F_{max} = 259165.72(N)$.

 https://doi.org/10.1051/matecconf/201819801004

Fig 11 shows the main girder vertical shear diagram.

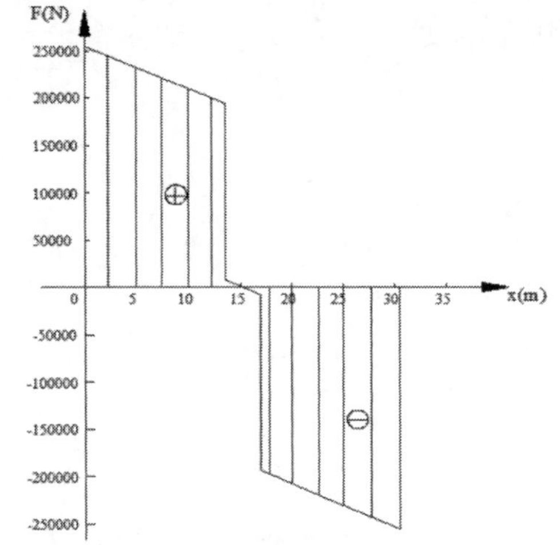

Figure 11. Main girder vertical shear diagram.

Bending moment $M(x)$.

Part a-c, $M(x)=F_{ra}{}^{V}x-\frac{1}{2}qx^2, 0\leq x\leq 13.96$

Part b-d,

$M(x)=F_{ra}{}^{V}x-\frac{1}{2}qx^2-Q_1(x-13.96), 13.96\leq x\leq 17.54$

Part c-b,

$M(x)=F_{ra}{}^{V}x-\frac{1}{2}qx^2-Q_1(x-13.96)-Q_2(x-17.54)$,
$17.54\leq x\leq 31.5$

Maximum bending moment

$M_{max}{}^{V}=3111086.9(N\cdot m)$.

Fig 12 shows the main girder vertical bending moment diagram.

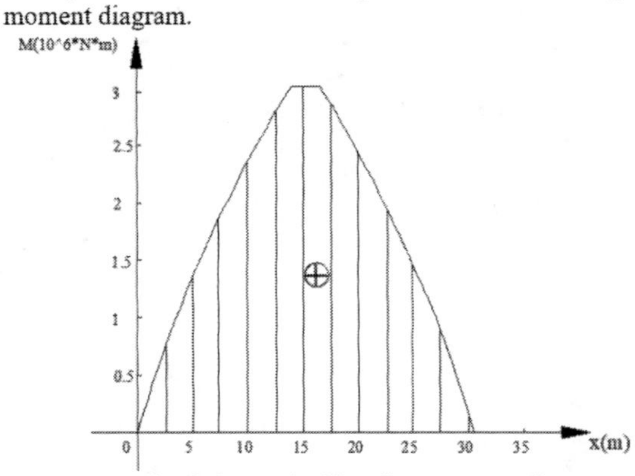

Figure 12. Main girder vertical bending moment diagram.

4.4 The shear and bending moments of the horizontal plane

Maximum shear and bending moments can be obtained in the same way as in Section 4.3.

Maximum shear $F_{max}=2423(N)$

Maximum bending moment $M_{max}{}^{H}=19081.3(N\cdot m)$

4.5 Calculation of maximum stress and displacement

Maximum bending normal stress

$$\sigma=\frac{M_{max}{}^{V}}{W_x}+\frac{M_{max}{}^{H}}{W_y}=116.2(MPa)$$

Maximum shear stress $\tau=\frac{F_{max}S^*}{2I_x t_4}=13.62(MPa)$

Comprehensive stress $\sigma_{max}=\sqrt{\sigma^2+3\tau^2}=118.6(MPa)$

Displacement of the main beam in the vertical direction $f=\frac{(1.1G_{小车}+1.14Q)L^3}{2\times 48EI_x}=42.3(\text{mm})$.

5 Summary

In this paper, the finite element analysis of QD-type (50/10t-31.5) hooked bridge crane metal structure was carried out using ABAQUS software. Through the comparison of three different working conditions, it can be seen that the stress and displacement of the girder is maximum when trolley with full loaded in 1/2 cross girder. The maximum stress $\sigma_{max}=120.4MPa$, and the maximum displacement $u_{max}=43.3mm$. It is obvious that it meets the design requirements and has a certain amount of margin. Moreover, the maximum stress and displacement under dangerous conditions of the crane are calculated by theoretical calculation. By comparing the results of the theoretical method and the finite element method, we find that the finite element calculation model has good precision.

References

1. Tang R, Huang J, Control of bridge cranes with distributed-mass payloads under windy conditions, Mechanical Systems & Signal Processing, **72-73**(2016):409-419.
2. Liu B, Zhang Q, Xing J X, Design of Bridge Crane Metal Structure Based on Limit State Design Method, Mechanical Engineering & Automation, **3**(2011):031.
3. Jiong G, Weichen Y, Research on crane rapid customization design system, Hoisting and Conveying Machinery, **2**(2014): 015.
4. Tong Y, Tang Z, Ye W, Yang Z, Research on energy-saving optimization design of bridge crane, Polskie Naukowo-Techniczne Towarzystwo Eksploatacyjne, **15**(2013):449-457.
5. GUAN Z, ZHANG Q, GUO J, The Finite Element Analysis of QD250/50/10t-25.5 Bridge Crane, Journal of Shanghai Electric Technology, **2**(2011): 004.
6. Kloosterman G, Bouwman V, Automatic remeshing and rezoning for the simulation of 3D glass bottle forming with Abaqus/CAE, Abaqus/Standard and Abaqus/Explicit, International Journal of Material Forming, 2(2009):105-108.

Mass Imbalance Compensation Control for Stabilized Platform Based on UKF Identification

Zhang WeiMing[1,2,3,4], SHI ZeLin[1,3,4]

[1] *Shenyang Institute of Automation, Chinese Academy of Sciences, Shenyang, 110016, China*

[2] *University of Chinese Academy of Sciences, Beijing. 100049, China*

[3] *Key Laboratory of Opto-Electronic Information Processing CAS, Shenyang, 110016, China*

[4] *The Key Lab of Image Understanding and Computer Vision, Liaoning Province, Shenyang, 110016, China*

Abstract. Due to the mass imbalance about the center of rotation, the stability of stabilized platform system degrades with carrier's disturbances. Various feed-forward control methods are provided by reaserchers to solve this problem, however these methods are not well applied because the eccentricity of stabilized platform could not be measured directly. The dynamics model of a typical 2-axis stabilized platform is given. The eccentricity vector is identified through Unscented Kalman Filter(UKF) algorithm. Imbalance torque is precisely observed so that the real-time nonlinear compensation for mass imbalance is achieved through a feed-forward loop. The simulation result indicates that the Root Mean Squared Error (RMSE) of parameters estimation is 0.024 after convergence. the LOS stabilization with carrier's 2.5Hz vibration is 0.04 rad/s, which improves 78% compared to conventional feed-back control.

1 Introduction

There are several effects inherent to the stabilized platform that could have a marked influence on overall pointing and tracking performance, such as noise, random error, and disturbance torque. Among these factors, the most complex nonlinear phenomenon is the effective torque due to mass imbalance and gimbal friction[1], [2], [3], [6].

When the center of gravity (CG) of the gimbal frame deviates from the rotation center, the imbalance torque is generated due to the acceleration in the direction perpendicular to the eccentricity. The line-of-sight (LOS) stabilization degrades seriously when the vibration of carrier is reinforced by high speed airflow or other environment factors.

A conventional approach is to increase the stabilizing loop gain or insert proportional-integral cascaded compensator[7]. However, these methods may not work well for dynamic, highly nonlinear disturbances, or even increase loop sensitivity to sensor noise. If the imbalance torque is observed, feed-forward controller could be introduced to compensate the imbalance effects[8], [9].

2 Dynamics model of stabilized platform

Two processes should be considered in the dynamics model of the stabilized platform system, the first is the disturbance transmitting process from the carrier to the payload, the second is the controlling process. A typical 2-axis stabilized platform is given (see fig.1). $OX_BY_BZ_B$ is the carrier coordinate system in which X_B is the spin axis, Y_B is the yaw axis and Z_B is the pitch axis. $OX_OY_OZ_O$ is defined when the carrier coordinate system rotate the angle ψ with Z_B. $OX_IY_IZ_I$ is defined when $OX_OY_OZ_O$ rotate the angle θ with Y_O. The origin of these coordinate system coincides with the intersection of the gimbal axis system.

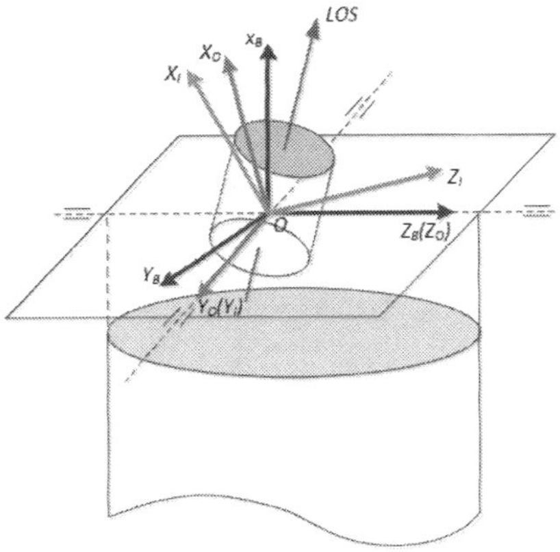

Figure 1. 2-axis stabilized platform.

According to kinematics relationship the inertial LOS rate is given by:

$$\boldsymbol{\omega}_I = \begin{bmatrix} \omega_{Ox}\cos\theta - \omega_{Oz}\sin\theta \\ \omega_{Oy} + \dot{\theta} \\ \omega_{Ox}\sin\theta + \omega_{Oz}\cos\theta \end{bmatrix} \quad (1)$$

The torque of the gimbal axis is:

$$\boldsymbol{M}_r = \frac{\mathrm{d}\boldsymbol{H}}{\mathrm{d}t} + \boldsymbol{\Omega}\times\boldsymbol{H} + \boldsymbol{\rho}\times m\boldsymbol{a} \quad (2)$$

In equation (2), $\boldsymbol{H} = \boldsymbol{J\omega}$ is the angular momentum of the frame, ρ is the eccentricity vector, m is the frame mass, and a is the acceleration of the frame. The torque $\boldsymbol{M}_r$ contains the control torque and other torque disturbances such as coulomb and viscous friction and cable damping effects.

Take the inner frame's situation as example, according to kinematics relationship the coupled acceleration from the carrier is given by:

$$\begin{bmatrix} a_{Ix} \\ a_{Iy} \\ a_{Iz} \end{bmatrix} = \begin{bmatrix} a_{Bx}\cos\psi\cos\theta + a_{By}\sin\psi\cos\theta - a_{Bz}\sin\theta \\ -a_{Bx}\sin\psi + a_{By}\cos\psi \\ a_{Bx}\cos\psi\sin\theta + a_{By}\sin\psi\sin\theta + a_{Bz}\cos\theta \end{bmatrix} \quad (3)$$

The imbalance torque of inner frame is given by:

$$\begin{aligned} M_{imb} = {} & \rho_{Iz} m_I (a_{Bx}\cos\psi\cos\theta + a_{By}\sin\psi\cos\theta - a_{Bz}\sin\theta) \\ & -\rho_{Ix} m_I (a_{Bx}\cos\psi\sin\theta + a_{By}\sin\psi\sin\theta + a_{Bz}\cos\theta) \end{aligned} \quad (4)$$

3 Feed-forward compensation design

If the imbalance torque is observed, feed-forward compensation could be possible to achieve (see fig.2), and the feed-forward controller is designed as: $G_N = -M_{I\_imb} / G_{cl}C_m$ .It is hard to precisely measure the eccentricity vector of the platform system, for the reason that the most important part of feed-forward compensation designing is the eccentricity identification.

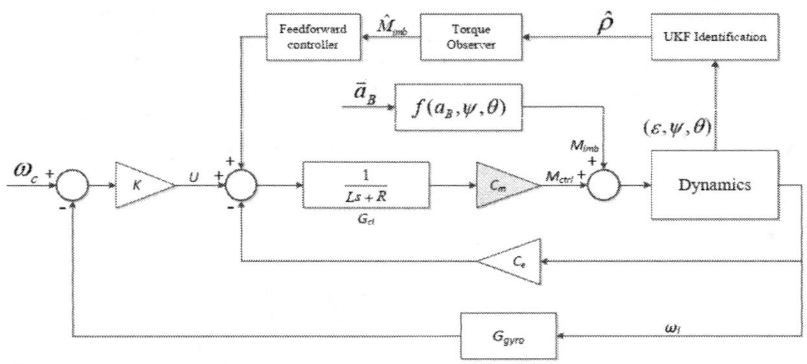

Figure 2. Principle of feed-forward compensation.

Because the relationship between the states and the observations of platform system is nonlinear, it is suitable for using UKF algorithm to estimate the eccentricity [4], [5]. The relationship between the angle acceleration and the eccentricity is:

$$\begin{aligned} \varepsilon_I = {} & \rho_{Ix} m_I (a_{Bx}\cos\psi\sin\theta + a_{By}\sin\psi\sin\theta + a_{Bz}\cos\theta)/J_I \\ & -\rho_{Iz} m_I (a_{Bx}\cos\psi\cos\theta + a_{By}\sin\psi\cos\theta - a_{Bz}\sin\theta)/J_I \end{aligned} \quad (5)$$

The state vector is given by:

$$\boldsymbol{X} = [\rho_{Ix} \quad \rho_{Iz} \quad a_{Bx} \quad a_{By} \quad a_{Bz} \quad \psi \quad \theta]^{\mathrm{T}} \quad (6)$$

The measurement vector is given by:

$$\boldsymbol{Z} = \left[\varepsilon_I \quad a_{Bx} \quad a_{By} \quad a_{Bz} \quad \psi \quad \theta\right]^{\mathrm{T}} \quad (7)$$

The initialization of UKF is:

$$\begin{cases} \hat{X}_0 = \mathrm{E}[X_0] \\ P_0 = \mathrm{E}\left[(X_0 - \hat{X}_0)(X_0 - \hat{X}_0)^{\mathrm{T}}\right] \end{cases} \quad (8)$$

And the sigma points are calculated as:

$$\chi_{k-1}^a = \left[\hat{X}_{k-1}^a \quad \hat{X}_{k-1}^a + \gamma\sqrt{P_{k-1}^a} \quad \hat{X}_{k-1}^a - \gamma\sqrt{P_{k-1}^a}\right] \quad (9)$$

The time-update equations are:

$$\begin{cases} \chi_{k|k-1}^x = \mathbf{F}\left(\chi_{k-1}^x, u_{k-1}, \chi_{k-1}^v\right) \\ \hat{X}_k^- = \sum_{i=0}^{2L} W_i^{(m)} \chi_{i,k|k-1}^x \\ P_k^- = \sum_{i=0}^{2L} W_i^{(c)} \left(\chi_{i,k|k-1}^x - \hat{X}_k^-\right)\left(\chi_{i,k|k-1}^x - \hat{X}_k^-\right)^{\mathrm{T}} \\ Z_{k|k-1} = \mathbf{H}\left(\chi_{k|k-1}^x, \chi_{k-1}^n\right) \\ \hat{Z}_k^- = \sum_{i=0}^{2L} W_i^{(m)} Z_{i,k|k-1} \end{cases} \quad (10)$$

And the measurement-update equations are:

$$\begin{cases} P_{Z_kZ_k} = \sum_{i=0}^{2L} W_i^{(c)} \left(Z_{i,k|k-1} - \hat{X}_k^-\right)\left(Z_{i,k|k-1} - \hat{Z}_k^-\right)^{\mathrm{T}} \\ P_{X_kZ_k} = \sum_{i=0}^{2L} W_i^{(c)} \left(\chi_{i,k|k-1} - \hat{Z}_k^-\right)\left(Z_{i,k|k-1} - \hat{Z}_k^-\right)^{\mathrm{T}} \\ K_k = P_{X_kZ_k} P_{Z_kZ_k}^{-1} \\ \hat{X}_k = \hat{X}_k^- + K_k\left(Z_k - \hat{Z}_k^-\right) \\ P_k = P_k^- - K_k P_{Z_kZ_k} K_k^T \end{cases} \tag{11}$$

where $X^a = \left[X^{\mathrm{T}} \quad v^{\mathrm{T}} \quad n^{\mathrm{T}}\right]^{\mathrm{T}}$, $\chi^a = \left[(\chi^x)^{\mathrm{T}} \quad (\chi^v)^{\mathrm{T}} \quad (\chi^n)^{\mathrm{T}}\right]^{\mathrm{T}}$, $\gamma = \sqrt{L+\lambda}$.

The eccentricity is extracted from state estimation vector after every filter step, and is used to calculate the feed-forward control command.

4 Simulation analysis

According to the system dynamics the simulation model is founded. And the control parameters are given: the motor inductance is 2.8mH, the coil resistance is 5.1 Ω, the torque coefficient is 0.21, the feed-back control gain is 34.5, the filter step is 0.01s, and the vibration of the carrier forms a sinusoidal signal with amplitude of 0.5g and frequency of 2.5Hz.

The simulation result indicates that the convergence time of UKF algorithm is <0.4s, and the Root Mean Squared Error (RMSE) of parameters estimation is 0.024 after convergence(see Fig.3-4). the LOS stabilized accuracy with carrier's vibration is 0.04 rad/s, which improves 78% compared to feedback control (see Fig.5-6).

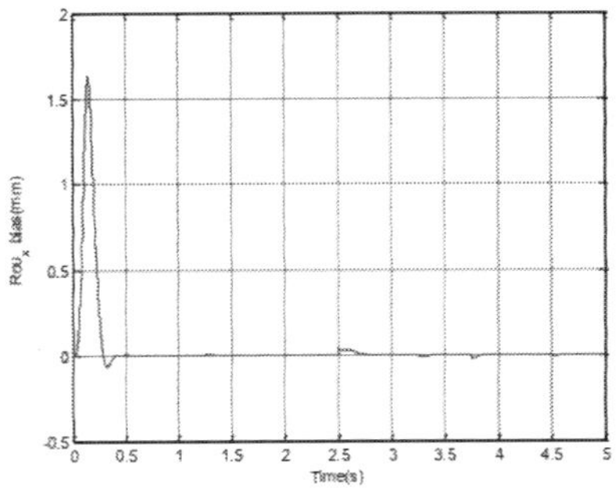

Figure 3. Estimation error of eccentricity in x.

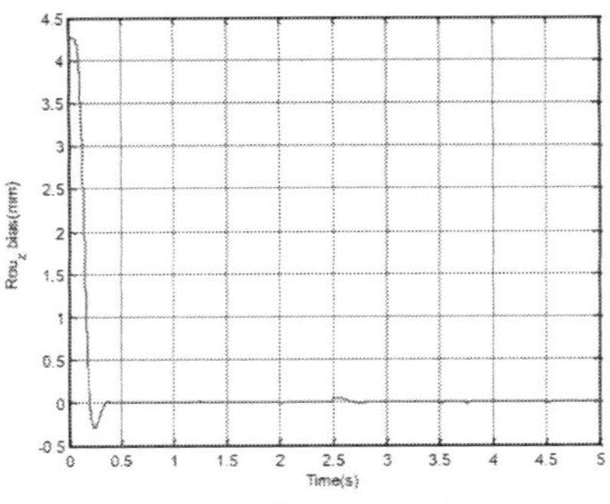

Figure 4. Estimation error of eccentricity in z.

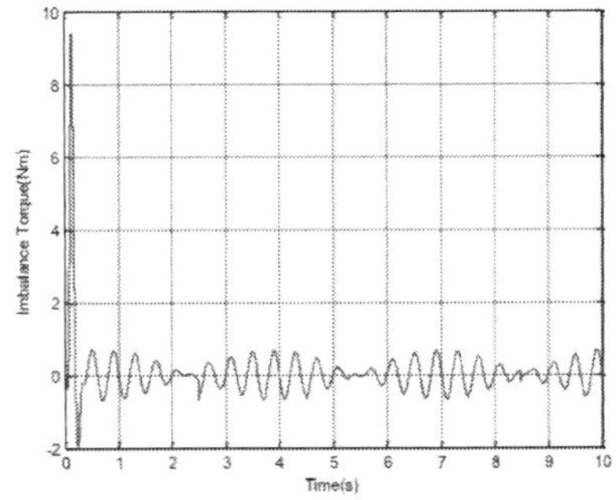

Figure 5. Observed imbalance torque.

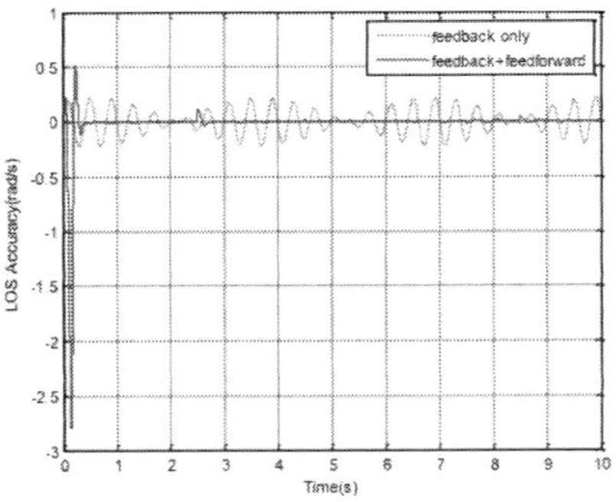

Figure 6. LOS stability comparison.

5 Summary

Because measuring a stabilized platform's eccentricity by conventional instruments is almost impossible, an real-time method of identification is provided. The eccentricity vector is precisely estimated through UKF algorithm, so that the imbalance torque could be completely observed. The convergence time of UKF is very short, and the algorithm consumes few hardware resources. The feed-forward compensation is designed which is effective to suppress the disturbances from the carrier, and better LOS stabilization of platform system is achieved compared to the traditional simple feed-back control method.

6 References

1. C L Lin, Y H Hsiao, Adaptive feedforward control for disturbance torque rejection in seeker stabilizing loop. IEEE Transaction on Control Systems Technology, 2001. 9(1):108-111.
2. Gong W, Z Cai, J Yang, X Li, L Jian. Parameter identification of an SOFC model with an efficient adaptive differential evolution algorithm. International Journal of Hydrogen Energy, 2014. 39(10): 5083-5096
3. Lu W, L Liu, J Sun, Y Zhou, Y Wu. Analysis of complex axis control loop in satellite laser communications. Optik 2012. 123: 458-461
4. YU H, WANG G H, SUN Y, Particle filtering algorithm of state estimation on fusion of UKF and

EKF. Systems Engineering and Electronics. 1001-506X(2003)07-1375-05.
5. Andreas S, Hans A. Bridging the ensemble Kalman filter and particle filters: the adaptive Gaussian mixture filter. Computational Geosciences, 2011, 15(2): 293-305.
6. Zhang Y P, Wang W X. Analysis on the stability of visual axis of airborne high precision stabilized platform. Foreign electronic measurement technology, 2015, 34(11):40-44.
7. Peter J Kennedy, Rhonda L Kennedy. Direct versus indirect line of sight (LOS) stabilization[J]. IEEE Transactions on Control Systems Technology, Vol. 11, NO. 1, January 2003:1063-1066.
8. Wei W, Dai M, Li J Q. ADRC control system for airborne opto-electronic platform. Optics and Precision Engineering. 2015,23（8）:2296-2299.
9. Han Z J. Adaptive control. Beijing: Tsinghua University Press, 2002: 151-158.

Research on Vibration Characteristics of Typical Pipe

Shaoming Yu, Tian Lu, Guo Wei, Yanping Hu

Beijing Institute of Structure and Environment Engineering, 100076, Beijing, China

Abstract. Through the study on the vibration characteristics of the typical pipe, it tries to provide the basis for the pipe design and the pipe ground test method. A typical pipe is selected. First, modal analysis is carried out through theoretical analysis and finite element simulation. Then, pipe vibration is studied by finite element simulation and test. The results show that the theoretical results in modal analysis coincide well with those in finite element simulation. The finite element simulation and experimental results are basically consistent in vibration analysis. The reason for the difference is mainly the ideal boundary of simulation. The combination of finite element simulation and test is an important method for the research of pipe reliability and environmental adaptability.

1 Introduction

The launch vehicle, especially the liquid launch vehicle, is equipped with a large number of pipe systems for the realization ofl lecific functions. As the "blood vessel" of the spacecraft, the reliability and environmental adaptability of the pipe system play an important role in the whole model launch. Because of much failure caused by the pipe's damage, the deep study on the vibration characteristics of the pipe is of great significance to the success of the spacecraft launch. The pipe system is different according to the different use environment. The ground pipe mainly concerns the influence of corrosion and flow induced vibration on the pipe. [1-2]The main part of the spacecraft pipe is the external tension, sound and vibration environment, which affects the pipe. [3-4]

In the study of the vibration characteristics of the pipe, there are mainly theoretical analysis, finite element simulation, test verification and other means.

The theoretical analysis of pipe vibration is based on the differential equation of vibration to solve the natural frequency and main vibration mode of pipe system. In the case of low accuracy, the approximate solution is done by Rayleigh-Ritz method and Galerkin method.

Numerical simulation based on finite element method to calculate the pipe system compared to the theoretical analysis more complex, especially relates to damping and forced vibration (random vibration) complex parameters, numerical simulation has great advantages, the simulation results can provide support for the optimization design. [5-6]

Pipe vibration test plays an important role in the design of spacecraft. It not only verifies the results of theory and simulation, but also plays an irreplaceable role in the assessment of pipe reliability and adaptability.

2 Typical pipe

In order to carry on the theory, simulation, experimental research and comparison verification of the vibration characteristics of the pipe, a typical pipe is selected as the research object. As shown in Figure 1, the material of pipe is stainless steel, the inner diameter of the pipe is 32mm, the wall thickness is 1.5mm, the length is 800mm, and the A terminal of pipe is the vibration end (the direction of vibration is Y), and the B end is a fixed end.

The vibration test conditions are shown in Table 1.

Two position paste acceleration sensor in test, respectively in the two three spots line, as shown the position of ①、② in Figure 1. The positions of the three strain gage are respectively at the two ends and the middle pipe pipe, as shown in Figure 1, 1~8, 10, 12 for the axial strain and 9, 11 for the tangential strain.

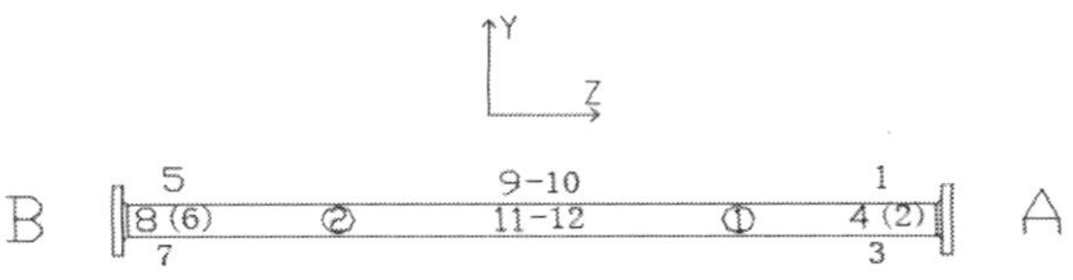

Figure 1 Typical pipe schematic diagram

Table 1. Test conditions for random vibration

Frequency range	PSD	RMS
20~100Hz	3dB/oct	$19.74g_{rms}$
100~2000Hz	$0.2g^2$/Hz	

3 Modal analysis

The damage of pipe system is mostly caused by the resonance between pipe and structure. When the pipe is designed, the resonant frequency can not be coupled with the resonant frequency of the whole rocket and the whole machine, so the modal information is very important for pipe system.

For the typical pipe studied in this paper, the modal analysis is carried out by means of theoretical research and finite element simulation.

3.1 Theoretical analysis

For the pipe in Figure 1, the simplified beam is analyzed. Differential equation for radial vibration of pipe:

$$EI\frac{\partial^4 y}{\partial x^4} + \rho A\frac{\partial^2 y}{\partial t^2} = p(x,t) - \frac{\partial}{\partial x}m(x,t) \tag{1}$$

In which, $y(x,t)$ is radial displacement of tube microelement, $p(x,t)$ is the external force acting on the microelement of the pipe, $m(x,t)$ is external torque acting on pipe microelement.

To get the natural frequency and the main mode of the pipe, $p(x,t)=0$, $m(x,t)=0$, and we can get the general solution of vibration mode of pipe $Y(x)$:

$$Y(x) = C_1\sin\beta x + C_2\cos\beta x + C_3 sh\beta x + C_4 ch\beta x \tag{2}$$

While $\beta^4 = \frac{p^2\rho A}{EI}$

As shown in Figure 1, the B end of the pipe is fixed, and the A end is vibratory (free), so the boundary condition is:

$$\left.\begin{aligned} Y(x)\big|_{x=0} = 0, \left.\frac{dY(x)}{dx}\right|_{x=0} = 0 \\ \left.\frac{d^2Y(x)}{dx^2}\right|_{x=l} = 0, \left.\frac{d^3Y(x)}{dx^3}\right|_{x=l} = 0 \end{aligned}\right\} \tag{3}$$

By taking the boundary condition (3) into the general solution equation (2), the frequency equation of a fixed and free equal section straight pipe with one end and one end is obtained:

$$\cos\beta l \bullet ch\beta l = -1 \tag{4}$$

The first three roots of the pipe frequency equation are $\beta_1 l = 1.875$, $\beta_2 l = 4.694$, $\beta_3 l = 7.855$.

Therefore, the natural frequency of the pipe, the main vibration modes are as follows:

$$p_i = (\beta_i l)^2\sqrt{\frac{EI}{\rho A l^4}} \tag{5}$$

$$Y_i(x) = C_i[\cos\beta_i x - ch\beta_i x + r_i(\sin\beta_i x - sh\beta_i x)] \tag{6}$$

While $r_i = \frac{\cos\beta_i l + ch\beta_i l}{\sin\beta_i l + sh\beta_i l}$

The first three natural frequencies of the pipe in Figure 1 are:

$p_1 = 329.8rad/s$, $p_2 = 2066.8rad/s$, $p_3 = 5787.6rad/s$

That is $f_1 = 52.5Hz$, $f_2 = 329.1Hz$, $f_3 = 921.6Hz$

For the pipe in Figure 1, whose B end is fixed and A end undergo the radial vibration, the natural frequency and the main vibration can be obtained results from the theory. Related to damping and forced vibration (random vibration), It is difficult to get the analytical solution in theory and can only be simulated by the finite element method.

3.2 Finite element simulation

First, the finite element method is used for modal analysis, which is compared with the theory results. In order to correspond with the theoretical calculation, the finite element method does not consider the flange in modal analysis. The finite element simulation uses the commercial software ABAQUS.

The first three modes (in plane) of the finite element simulation of modal analysis are respectively 53Hz, 327Hz and 897Hz. The simulation results are shown in Figure 2.

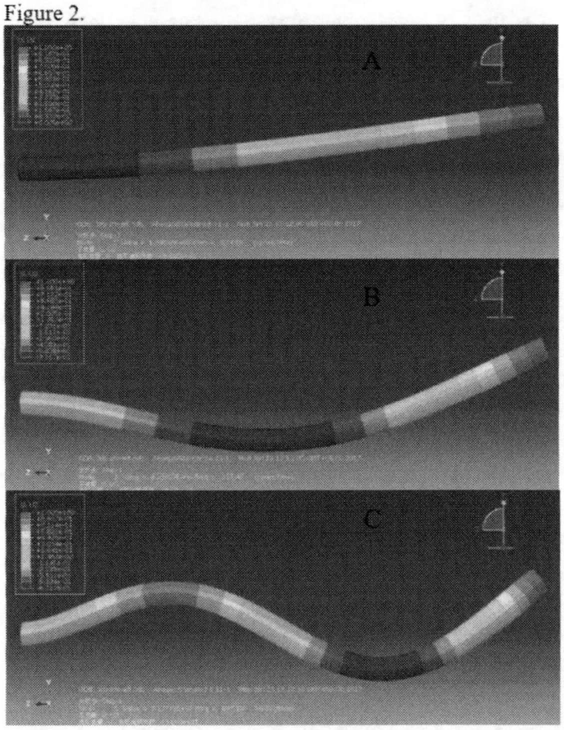

Figure 2 Finite element simulation result diagram of modal analysis

A) First order mode B) Second order mode C) Third order mode

The finite element results of the pipe modal analysis are in good agreement with the theory. The first order frequency deviation is 0.38%, the second order frequency deviation is 0.49%, and the third order frequency deviation is 2.66%. It can be proved from the results that it is advisable to analyze the pipe with beam model theory, and it also proves the accuracy of the finite element modal analysis.

For the finite element method of pipe vibration analysis, ABAQUS adopts the modal superposition method, which is based on the finite element modal analysis results. The accuracy of finite element modal analysis provides the foundation for the accuracy of vibration analysis.

4 Vibration analysis

4.1 Finite element simulation

The ABAQUS finite element software is used to analyze the vibration of the typical pipe. The vibration analysis is based on the modal superposition method, and the finite element simulation of the random vibration environment is carried out on the basis of the previous modal analysis results.

The boundary condition of simulation is the vibration environment at the A end (the loading condition is shown in Table 1) and B terminal is fixed. While random vibration simulation in ABAQUS software can't directly apply the fixed boundary, we add a large mass spot at the fixed end, and tie the fixed boundary with the mass particle. The unidirectional loading method is used for the vibration loading, and the direction is Y direction in Figure 1.

The result of the cloud is as shown in Figure 3.

Figure 3 Finite element simulation results of typical pipe vibration

4.2 Vibration test

A vibration test of a typical pipe is carried out, and a schematic diagram is shown as shown in Figure 4. B end of the pipe is fixed by the ingot; A end of pipe is exerted vibration; the pipe is connected with the slide by fixture; the loading conditions is shown in Table 1; the loading direction for Y in Figure 1. For the convenience of comparison, the test is consistent with the finite element simulation boundary in the previous section.

The mounting of the strain gauge and the installation of the acceleration spot are carried out in accordance with figure 1.

Figure 4 Abridged general view of typical pipe vibration test

4.3 Comparison between finite element simulation and vibration test

The results of the finite element simulation are compared with the experimental results.

The root mean square (RMS) comparison of acceleration spots is shown in Table 2; the spot 1's PSD of acceleration comparison is shown in Figure 5; the root mean square strain comparison of each strain measurement spot is shown in Figure 6.

Through comparison, we can get that:

In Table 2, the test results of root mean square acceleration response are larger than that of simulation results. This is because there is a spike in the high frequency section due to interference and other factors in the test process. When the root mean square is calculated, the experimental value is obviously greater than that of the finite element simulation.

In Figure 5, simulation of resonant frequency is high, this is because the simulation boundary is more ideal, which cannot be strictly fixed in test. The finite element simulation has lower peak, this is because the ideal damping and there are no additional mass sensor in finite element simulation. The simulation results are smooth in high frequency, which is due to no circuit interference, no control iteration, no system other structural signal interference (such as screw loosening, shaking table percussion, etc.).

In Figure 6, the root mean square strain ratio data of simulation is smaller than that of the test. The strain of the vibration end is larger than that of the fixed end, and the simulation results are more obvious. The strain on the upper and lower surface of the pipe is larger than that of the neutral layer, and the strain at the neutral layer is small, which is consistent with the bending theory, and the simulation results are consistent.

Table 2. Acceleration response contrast

No.	RMS in the test(g_{rsm})	RMS in the simulation(g_{rsm})
1	49.656	32.224
2	43.218	31.724

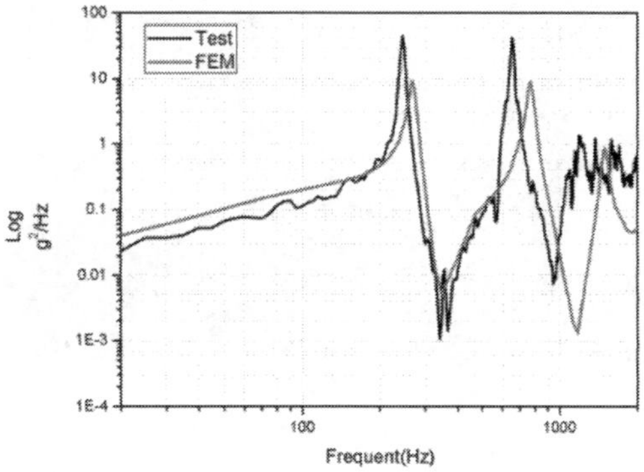

Figure 5 Acceleration PSD contrast of spot 1

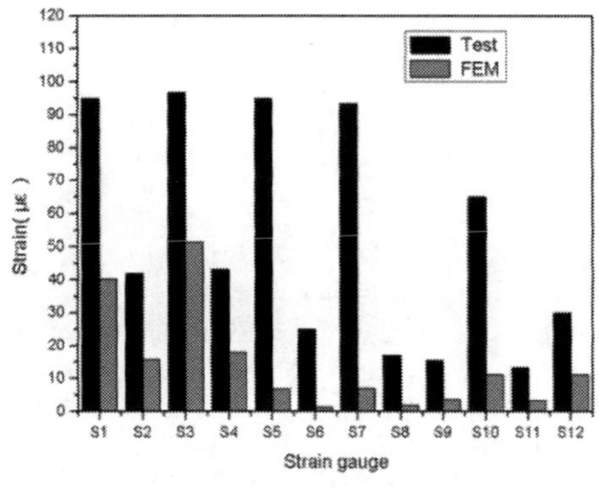

Figure 6 Root mean square strain contrast

5 Summary

In the recent model launching process, the pipe abnormality appears many times. The reliability of the pipe and the adaptability of the environment are concerned. The research on the pipe of the spacecraft is of great significance and is facing the urgent need.

Pipe failure is mostly caused by resonance. At present, a number of rockets have begun to carry out frequency management, requiring pipe design frequency to avoid the resonance frequency of whole machine and whole rocket.

Finite element simulation is an important method of pipe research. In actual work, we should strengthen combination of pipe test and finite element simulation, improve efficiency of test through finite element simulation, and verify the accuracy of simulation and repair model through test.

References

1. M. E. Stevenson, H. C. Iwand, J. L. McDougall, et al. Failure Analysis of an Aluminum Chiller Pipe by Experimental Simulation and Stress Analysis[J]. Journal of Failure Analysis and Prevention. 2017,17(5):1090-1098.
2. Polynikis Vazouras, Spyros A. Karamanos. Structural Behavior of Buried Pipe Bends and Their Effect on Pipe Response in Fault Crossing Areas[J] Bulletin of Earthquake Engineering , 2017,15(11):4999-5024.
3. Yong Bai, Ting Liu, Weidong Ruan, et al. Mechanical Behavior of Metallic Strip Flexible Pipe Subjected to Tension[J] Composite Structures. 2017,170:1-10.
4. P. Dunst, T. Hemsel, W. Sextro. Analysis of Pipe Vibration in An Ultrasonic Powder Transportation System[J] Sensors and Actuators A: Physical. 2017,263:733-736.
5. Wei G, Yu Shaoming, Lu Tian, et al. Environmental Testing Technology of Pipe System[J] Missiles and Space Vehicles, 2017, 3(5):93-96.
6. Yu Shaoming, Lu Tian, Wei G, et al. Research on Vibration of Pipe Based on Strain-analysis[J] Journal of Dynamics and Control , 2017 , 15(6):512-517.

Vibrodiagnostics Used for Evaluating the Technical Condition of a Mechanical Gearbox

Jan Furch[1], Jiří Stodola[2] and Josef Glos[3]

[1]*University of Defence, Faculty of Military Technology, Kounicova 65, 662 10 Brno, Czech Republic*
[2]*University of Defence, Faculty of Military Technology, Kounicova 65, 662 10 Brno, Czech Republic*
[3]*University of Defence, Faculty of Military Technology, Kounicova 65, 662 10 Brno, Czech Republic*

Abstract. The aim of this article is to evaluate the technical condition of a mechanical four-speed gearbox tightly connected to a mechanical two-speed auxiliary gearbox placed in an off-road vehicle. When observing the technical condition of the mechanical gearbox, we used one of technical diagnostics methods, namely vibrodiagnostics. The mechanical gearbox was monitored during all its life cycle up to failure occurrence. In the article there is a detailed description of the stand where the gearbox technical condition was monitored. When observing vibration signals, we used four tri-axial sensors placed in four selected spots depending on the design arrangement of antifriction bearings. In order to evaluate the measured results, two methods were applied, namely the root mean square gRMS based on the measured acceleration in three axes. The other method used for monitoring the technical condition was the Crest factor. The aim of the experiment was to monitor the technical condition of the mechanical gearbox, mainly its antifriction bearings, therefore the values were measured when the gear was put into fourth position and into overdrive in the auxiliary gearbox. When the gear is put into fourth position, the torque is transmitted from the motor through the main gearbox and the auxiliary gearbox directly. Single gearbox wheels are engaged but not loaded (they do not transmit the torque). At the end of our article we introduced the measured values along with the oral evaluation of the technical condition of the mechanical gearbox and the auxiliary gearbox.

1 Introduction

Vibrodiagnostics is one of the most important methods used for identifying a technical condition. With the use of vibration diagnostics we are able to detect an incipient failure, locate the place of an incipient failure and predict the length of time during which a device is going to work before a failure occurs or a preventive action is performed. The proper application of vibrodiagnostics can prevent undesirable damage to machines, thereby preventing costs associated with repair and wasting money when a device is out of service.

Experimental vibrodiagnostic measurements are performed while obtaining primary signals. Primary results of measurements are systematized and analysed. Operating characteristics of rotary systems with bearings of sliding friction and those of roll are compared. Generalization of research results is done and conclusions are formulated [1], [2], [3]. Advanced computational models suitable for the development of a modern powertrain in the field of noise and vibration are presented in [4].

A lot of companies spend large amounts of money on the preventive maintenance of machines and technical equipment despite the fact that there is no need to take a preventive action at that particular moment. This leads to massive financial losses, mainly during the time the machine is out of service. Therefore, it is desirable to perform preventive maintenance at the time when it is really necessary. A decision on taking a preventive action or replacing a part which is based on the real technical condition is proactive maintenance. This is made possible by using a mathematical model describing the technical condition of a machine during its life. The aim of our paper is to describe the technical condition of the vehicle mechanical gearbox by means of vibrodiagnostics [5].

2 Possibilities of determining the technical condition of rolling bearings and gearbox wheels

Vibrations occur as a result of rotating or straight-line moving bodies. The course of vibrations is influenced mainly by the technical condition of individual machine components such as shafts, gear boxes, crank mechanisms, cam mechanisms, antifriction bearings, and also by the imbalance of rotating parts, backlash in friction bearings, wear, material fatigue, cracking phenomenon, corrosion, and other factors affecting the smooth running of a machine. The vibration itself is defined then as a dynamic phenomenon when particles or solid bodies move around a zero equilibrium position. Vibrations are given by a combination of six movements, namely by a shift in an orthogonal coordinate system x, y, z and by the rotation about these axes. We can describe them by the amplitude and phase in a certain period of time. Depending on the time variations of values, vibrations are of a periodical, non-periodical, or random character. As for periodical vibrations, the course of vibrodiagnostic values keeps repeating over time. Harmonic vibration, which has a sinusoidal waveform, is based on these vibrations. For harmonic vibrations we need to set only one determining value and the others can be calculated.

The basic way of describing oscillations is to determine their displacement x, velocity v, acceleration a, maximum value A_{max}, the root of mean square RMS, and the average value A_{ave} [6].

The International Standards Organization (ISO), which establishes internationally acceptable units for the measurement of machinery vibration, suggested the root mean square (RMS) as the standard unit of measurement. This was decided in an attempt to establish criteria that would determine an effective value for the varying function of acceleration. Acceleration tends to provide the energy content in the vibration signal, whereas the acceleration peak is correlated better with the intensity of vibration [7].

A calculation of the mean value A_{ave} and root of mean square RMS is based on acceleration, so called effective value $gRMS$. The following formulas were used for the calculation.

Acceleration can be expressed as the characteristics of motion which shows the way the velocity of a body (particle) changes in time. Acceleration is a vector quantity since it gives both the magnitude and direction of the change. It is possible to calculate the instantaneous acceleration and the average acceleration. Acceleration may also be determined as the time derivation of velocity using the formula below [7]:

$$a = \frac{dv}{dt} = -X_{max}\omega^2 \sin \omega t = X_{max}\omega^2 \sin(\omega t + \pi). \quad (1)$$

If acceleration is related to a change in velocity in terms of a decreasing speed, it is called deceleration and has a minus sign. The mean absolute value A_{ave} can be expressed as follows:

$$A_{ave} = \frac{1}{T}\int_0^T |a| dt. \quad (2)$$

The root of mean square can be calculated by the equation below [3]:

$$gRMS = \sqrt{\frac{1}{T}\int_0^T a^2 dt}. \quad (3)$$

Another parameter used to identify the technical condition of rolling bearings is the Crest factor. The principle of the method is to measure the effective and peak vibration values and their calculation according to the following formula [8]

$$C = \frac{|A_{max}|}{gRMS}. \quad (4)$$

Generally speaking, as any signal exhibits sharpness, its C increases because it would indicate that the peak value is much larger than the RMS value. C is typically used to monitor the change of the signal pattern due to a localized fault such as tooth breakage of a rotating gear which will theoretically generate pulses during meshing. C is also very useful for detecting discrete impulses above the background of the signal, which do not occur frequently enough or have a very short duration to significantly increase the $gRMS$ value. The value of C is in the range of 2.5 to 3.5 is not uncommon for many machines, but in many cases, if its value is 5 or more, it may indicate a problem [8].

3 Description of the experiment

The measurement was performed on a testing stand shown in Figure 1. The main parts of the stand are a frame for anchoring measured equipment, a notebook with software DEWESOFT X2 (1), control unit for break AHB12 (2), measuring equipment DEWE-50-USB2-8 (3), a frequency converter MOVITRAC MCLTE B0040-5A3-4-00 (4), an asynchronous motor with engine speed control in 10 – 3000 RPM (5), a propeller shaft with a flange (6), gearbox with transfer box including four tri-axial acceleration sensors KS943B.100 (7), and air cooled brake AHB/12 (8).

For the measurement itself, we used the DEWE-50-USB2-8 device from the DEWETRON firm. This is an 8-channel device intended for measuring vibrations of equipment. The vibration signal was measured by four KS943B.100 tri-axial accelerometers by which it was possible to measure the vibrations of four places at the same time. Because 12 measuring channels are needed, we use two devices connected in series. These accelerometers were mounted on the gearbox by means of a magnet.

Legend: 1 - Laptop with DEWE software; 2 - Control unit for

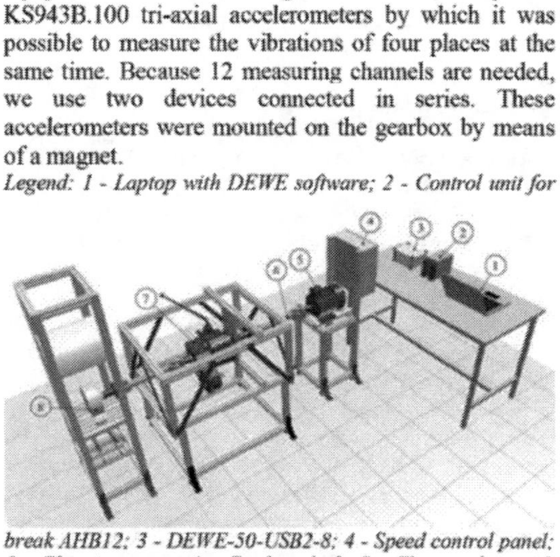

break AHB12; 3 - DEWE-50-USB2-8; 4 - Speed control panel; 5 - Electric motor; 6 - Cardan shaft; 7 - The gearbox with transfer box; 8 - Air-cooled brake AHB-12.

Figure 1. Testing stand for vibration measurements

The aim of this article is to determine the predicted time (detection of potential failure), it is necessary to determine the cause of incipient failure. This is presented in Figure 1. Only then will it be determined whether the failure occurred in a roller element, the inner, the outer ring and the cage of roller bearing or in the gear tooth. This is based on the assumption that each part listed above is of a different relative speed to the shaft and of a different frequency at which the failures occur.

4 Analysis of measurement results

The experiment was conducted to measure mechanical gearbox (a gear box and a transfer box) durability at input revolutions 2000 rpm. During this experiment the mechanical gearbox operated for 3613 hours which is equivalent to 160000 km of a given vehicle. During all this experiment the gears were changed as illustrated in Table 1. The gears were changed continuously between single measurements which means all speed gears were always selected. The mechanical gearbox was equally worn during the experiment which was one of the basic measuring rules. When performing the experiment, different running resistances were simulated using the Air-cooled brake AHB-12 and a Control unit for the brake AHB12.

When calculating the root mean square values calculated from *gRMS* and crest factor *C* acceleration, the input revolutions were 2000 rpm, the gear was put into fourth position in the gearbox and into overdrive in the transfer box.

The root mean square *gRMS* values were calculated in all common axes dimensions (x, y, z). After, the *gRMS* values were summarized in respective axes, and then an average *gRMS* value was calculated and put in the resulting graph, see Fig. 2. The *gRMS* values were calculated for all four sensors in single measuring spots as illustrated in Fig. 2.

Table 1. The operating time for each speed gear of the gearbox and transfer box

Speed gear in transfer box	Operating time (%)	Speed gear in gearbox	Operating time (%)	Operating time (hour)	Operating time (km)
Overdrive	90	1	5	162.6	7200
		2	10	325.2	14400
		3	35	1138.1	50400
		4	50	1625.9	72000
Reduction	10	1	5	18.1	800
		2	20	73.3	3200
		3	35	126.5	5600
		4	40	144.5	6400
Total time (T)				3613.0	160000

Fig. 2 shows that up to about 40000 km the bearing run-out is likely to occur. The *gRMS* values gradually increase up to this mileage. This is followed by a period of routine operation up to about 140000 km during which the *gRMS* values are more or less constant. After, the technical condition of the mechanical gearbox and single antifriction bearings and gearbox wheels rapidly decreases. When the operation reaches 150000 - 160000 km, bearings no. 4 and 2 are damaged. Bearings no. 1 and 3 are in a state of near failure occurrence.

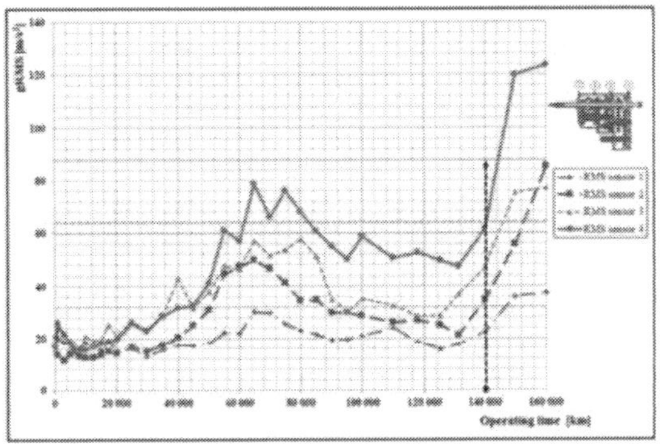

Figure 2. Expressing the dependence of gRMS on mileage the mechanical gearbox

The experiment clearly shows that the measurement can be performed on one box only using more sensors. This applies only when the sensors are properly placed above bearings. First we were afraid that the noise of the box might spread which would result in distorting the measuring results, but fortunately this did not occur

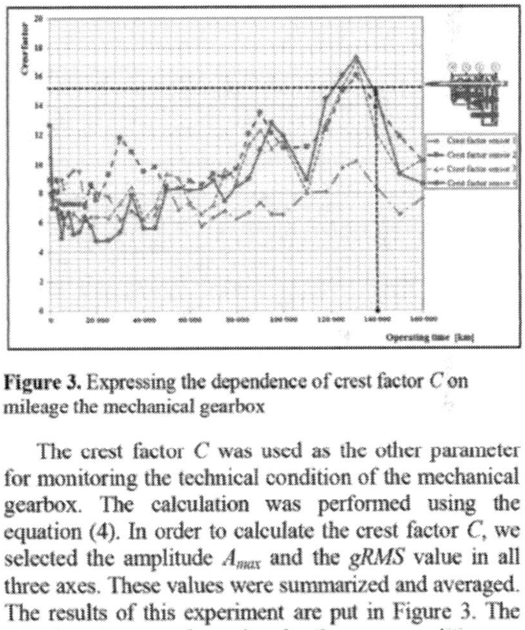

Figure 3. Expressing the dependence of crest factor *C* on mileage the mechanical gearbox

The crest factor *C* was used as the other parameter for monitoring the technical condition of the mechanical gearbox. The calculation was performed using the equation (4). In order to calculate the crest factor *C*, we selected the amplitude A_{max} and the *gRMS* value in all three axes. These values were summarized and averaged. The results of this experiment are put in Figure 3. The experiment was performed under the same conditions as introduced in the previous part. The measured crest factor *C* values show that when the operation of a mechanical gearbox reaches about 140000 km, the abnormal wear and then the failure of antifriction bearings occur. The course of crest factor values show that the introduced value 140000 km is limiting for all four detection spots.

Based on both methods we have come to the conclusion that when the operation of the mechanical gearbox reaches about 140000 km, it is advisable to finish its operation and take appropriate technical measures.

5 Conclusion

The aim of the experiment was to verify that vibrodiagnostics is the proper method used for identifying the technical condition of rotating parts in vehicle mechanical gearboxes. The experiment primarily focused on monitoring the wear of antifriction bearings which are one of the most important parts of a vehicle transmission gear.

The experiment shows us that the vibration signal is an important signal used for identifying the technical condition of rotating parts. When processing the signal, it is essential to select a proper place for sensors, choose appropriate sensors, adjust properly software used for evaluation, introduce the measuring results in the right way, etc. When measuring and analysing, we used Dewesoft hardware and software which turned out to be very suitable. When evaluating the durability of the mechanical gearbox, we used two suitable methods which enabled us to describe the technical condition of antifriction bearings. The first method used for evaluating the technical condition of the mechanical gearbox is the root mean square which helped us to calculate single values put in Figure 2. These values show us that it is possible to determine maximal *gRMS* values acceptable for single measured spots for a specific gearbox. Based on the introduced results we can conclude that when reaching about 140000 km, the technical condition of antifriction bearings dramatically decreases, and at about 150000 km the bearings are damaged which was proved after the mechanical gearbox was dismantled. The results achieved by using the other method calculations (crest factor) support the conclusion made when applying the root mean square gRMS method.

Acknowledgment

Presented work has been prepared with the support of the Ministry of Defence of the Czech Republic, Partial Project for Institutional Development and Specific Research, Department of Combat and Special Vehicles, University of Defence, Brno.

References

1. F. Cong, J. Chen, G. Dong, M. Pecht, Vibration model of rolling element bearings in a rotor-bearing system for fault diagnosis. Journal of Sound and Vibration [online]. vol. **332**, no. 8, p. 2081-2097, (2013)
2. S. Tsyfanskii, V. Beresnevich, Vibrodiagnostics of fatigue cracks in flexible, geometrically nonlinear, beam-type structural elements. Russian Journal of Nondestructive Testing. pp 427-433, (1994)
3. V. Vekteris, M. Jurevicius, A. Cereska, Vibrodiagnostics of the components of technological machines. *8th International Conference on Vibration Measurements by Laser Techniques Location*. Ancona, (2008)
4. P. Novotný, A. Prokop, M. Zubík, K. Řehák, Investigating the influence of computational model complexity on noise and vibration modelling of powertrain. Journal of Vibroengineering. vol. **22,** no. 4, p. 277–392, (2016).
5. J. Furch, J. Glos, J. Blecha, Identifying the Technical Condition of Rotating Parts by Means of Vibrodiagnostics. Transaction of Famena. vol. **39**, no. 2, pp. 75 – 86, (2015)
6. J. Furch, J. Glos, T. T. Nguyen, Identification of the technical condition of roller bearings by means of vibrodiagnostics and tribodiagnostics. Transaction of Famena. vol. **41**, no. 3, pp. 99 – 113, (2017)
7. P. Girdhar, C. Scheffer, *Practical machinery vibration analysis and predictive maintenance*. Burlington: Newnes, 255 p., (2004)
8. M. P. Norton, D. Karczub, *Fundamentals of Noise and Vibration Analysis for Engineers*. Cambridge University Press, 631 p., (2003)

Comparison of Three Methods of Length Compensation in a Parallel Kinematic and Their Equivalence Conditions

Shucen Du[1], Josef Schlattmann[1], Stefan Schulz[1] and Arthur Seibel[1]

[1]Workgroup on System Technologies and Engineering Design Methodology, Hamburg University of Technology, 21073 Hamburg, Germany

Abstract. This paper is a further development of a previous work (PAMM 16, 51–52). In this paper, we derive an analytical formulation of the linear actuators' length variations for a 2-1-2 DOF joint designed parallel hexapod based on Riebe and Ulbrich's method (Arch. Appl. Mech. 72, 817–829) and the Euler angles based matrix method. The equivalence conditions as well as the computation errors are given and compared with a quaternion based method.

1 Introduction

The 2-1-2 DOF (degrees-of-freedom) joint designed parallel hexapod (six universal joints mounted on both the moving and the fixed platform and six linear actuators connecting these two platforms) is a 6-DOF spatial mechanism. If the length variation of a linear actuator is realized by a threaded spindle, a length compensation is necessary because the relative rotation between the upper and the lower part of the linear actuator is coupled with its length variation.

In our previous work [1], we developed an analytical formulation of this passive rotation based on the rotation matrix of the Euler angles of the moving part of the hexapod. Further, for substituting the rotation matrix, a quaternion based method was used, which shows several advantages such as simple geometrical interpretation, compact mathematical representation, coordinate system independency, no gimbal lock, and higher computational efficiency and accuracy [2, 3]. In this paper, we compare these two methods with the classical representation of the passive rotation, which shall be denoted as Riebe and Ulbrich's method [4], and provide their equivalence conditions.

The paper is organized as follows. In Section 2, we derive Riebe and Ulbrich's method and provide an analytical expression for the angle of passive rotation. In Section 3, we do the same with the Euler angles based matrix method. Section 4 compares these two methods with each other by deriving their equivalence conditions. As a further alternative, the quaternion based method is reviewed in Section 5. In Section 6, the computational performance of Riebe and Ulbrich's method and the Euler angles based matrix method as well as the Euler angles based matrix method and the quaternion based method is compared. Finally, Section 7 concludes the work.

The rotation matrix ${}^{*}\boldsymbol{R}_{**}$ used in this work represents the rotation of frame $*$ referred to frame $**$. The operator $\boldsymbol{R}(*,**)$, on the other hand, defines a rotation around the $*$ axis with an angle of $**$. In the analytical operations, the rotation operator $\boldsymbol{R}(*,**)$ is also a substitution of the corresponding rotation matrix.

2 Riebe and Ulbrich's method

In the work of Riebe and Ulbrich [4], the relative rotation ${}^{C_1}\boldsymbol{R}_{C_2}$ between the upper $\{C_2\}$ and the lower cylinder $\{C_1\}$ of the linear actuator is computed in a typical way. The two cross shaft universal joints on both ends of the linear actuator are handled separately, and the rotation of the linear actuator is described by the output angles of the input and bending angles, α_*, β_*, $\gamma_{*,*}$, see Fig. 1. In order to build a relation between Riebe and Ulbrich's method and the Euler angles based matrix method described in Section 3, we use the following assumptions.

1) The orientation of the linear actuators' z axis is constant or fixed to the z axis of the inertial coordinate system $\{0\}$,

2) The centers of the universal joints are fixed in space; note that the lengths of the linear actuators are then constant.

2.1 Upper platform's and linear actuator's upper part's motion

The rotation of the upper platform $\{2\}$ is composed of three rotations: 1) the platform's tilt motion around the inertial x axis, β_2, 2) the platform's rotation around its local z axis with angle $\gamma_{1,2}$, 3) the simultaneous rotation of the platform and the upper part of the linear actuator around the inertial z axis with angle α_2. Hence, this rotation can be expressed as follows:

$$ {}^{0}\boldsymbol{R}_2 = \boldsymbol{R}(z, \alpha_2)\boldsymbol{R}(x, \beta_2)\boldsymbol{R}(z, \gamma_{1,2}) \quad (1) $$

with respect to the inertial coordinate system $\{0\}$.

The corresponding rotation of the upper cylinder of the linear actuator $\{C_2\}$ is composed of: 1) no rotation around the inertial x axis, 2) the passive rotation $\gamma_{2,2}$ around the inertial z axis caused by the platform's local rotation, 3) the simultaneous rotation with platform $\{2\}$ around the inertial z axis with angle α_2. Consequently, this rotation can be expressed by

$$ {}^{0}\boldsymbol{R}_{C_2} = \boldsymbol{R}(z, \alpha_2)\boldsymbol{R}(x, 0)\boldsymbol{R}(z, \gamma_{2,2}) = \boldsymbol{R}(z, \alpha_2 + \gamma_{2,2}) \quad (2) $$

with respect to the inertial coordinate system $\{0\}$.

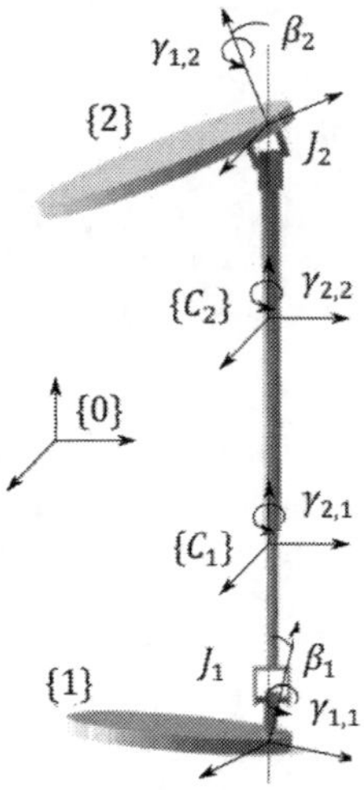

Figure 1. Coordinate system of one link chain in a hexapod.

2.2 Lower platform's and linear actuator's lower part's motion

Similarly to the upper platform, the lower platform's ($\{1\}$) rotation comprises 1) the platform's tilt motion around the inertial x axis, β_1, 2) the platform's rotation around its local z axis with angle $\gamma_{1,1}$. The rotation matrix of platform $\{1\}$ can be thus written as

$$ {}^{0}\boldsymbol{R}_1 = \boldsymbol{R}(x, \beta_1)\boldsymbol{R}(z, \gamma_{1,1}) \quad (3) $$

with respect to the inertial coordinate system $\{0\}$.

The corresponding rotation of the lower cylinder of the linear actuator $\{C_1\}$ is composed of: 1) no rotation around the inertial x axis, 2) the passive rotation $\gamma_{2,1}$ around the inertial z axis caused by the platform's local rotation. Consequently, this rotation can be expressed by

$$ {}^{0}\boldsymbol{R}_{C_1} = \boldsymbol{R}(x, 0)\boldsymbol{R}(z, \gamma_{2,1}) = \boldsymbol{R}(z, \gamma_{2,1}) \quad (4) $$

with respect to the inertial coordinate system $\{0\}$.

2.3 Angular relation between the upper and the lower part of the linear actuator

The rotation inside a linear actuator affects its length. This rotation is described as follows:

$$ {}^{C_1}\boldsymbol{R}_{C_2} = {}^{C_1}\boldsymbol{R}_0 \, {}^{0}\boldsymbol{R}_{C_2} = \boldsymbol{R}(z, \alpha_2 + \gamma_{2,2} - \gamma_{2,1}). \quad (5) $$

In [4], α_2 is 0 because this angle is not *passive* and independent of other pose variables such as β_* and $\gamma_{1,*}$. However, in order to completely obtain the rotation affecting the linear actuator's length, we keep this variable in consideration. The rotation angle ρ between the cylinders $\{C_1\}$ and $\{C_2\}$ and thus the passive rotation of the linear actuator can be then determined from

$$ \tan\rho = \tan(\alpha_2 + \gamma_{2,2} - \gamma_{2,1}) $$
$$ = \frac{\tan(\alpha_2 + \gamma_{2,2}) - \tan\gamma_{2,1}}{1 + \tan(\alpha_2 + \gamma_{2,2})\tan\gamma_{2,1}} $$
$$ = \frac{\tan\alpha_2 + \tan\gamma_{2,2} - \tan\gamma_{2,1} + \tan\alpha_2 \tan\gamma_{2,2} \tan\gamma_{2,1}}{1 - \tan\alpha_2 \tan\gamma_{2,2} + \tan\alpha_2 \tan\gamma_{2,1} + \tan\gamma_{2,2} \tan\gamma_{2,1}} \quad (6) $$

where $\tan\gamma_{2,2} = \tan\gamma_{1,2}/\cos\beta_2$ and $\tan\gamma_{2,1} = \tan\gamma_{1,1}/\cos\beta_1$.

3 Euler angles based rotation matrix method

The main disadvantage of Riebe and Ulbrich's method is that, in most cases, the angles related to the rotation ${}^{C_1}\boldsymbol{R}_{C_2}$ cannot be explicitly given. For this reason, the pose of the hexapod's moving platform is better described by a set of, e. g., Euler angles. In this paper, we use the z-x-z Euler angle-set ψ-φ-θ to describe the hexapod's moving platform's pose. The rotation between the upper platform $\{2\}$ and the lower platform $\{1\}$ can be then written as

$$ {}^{1}\boldsymbol{R}_2 = \boldsymbol{R}(z, \psi)\boldsymbol{R}(x, \varphi)\boldsymbol{R}(z, \theta) \quad (7) $$

In order to formulate the linear actuator's length compensation in an analytical way, the pose of the moving platform $\{2\}$ is described by a set of proper z-x-z Euler angles for the rotation and the vector (x, y, z) for the linear displacement of the coordinates center. Since the passive rotation is independent of the linear actuator's active length (the length variation caused by the motor's rotation), we define the hexapod's moving platform's pose as the vector $(\psi, \varphi, \theta, \alpha, \beta)$, in which (α, β) represents the axial orientation of the linear actuator, i. e., $\boldsymbol{R}(z, \alpha)\boldsymbol{R}(x, \beta)$.

A rotational movement of the hexapod can be described in the following sequence. 1) The whole structure rotates around the inertial x axis with angle β. 2) The whole structure rotates around the inertial z axis with angle α. 3) The platforms $\{1\}$ and $\{2\}$ rotate back to the initial position inversely to the above motion. 4) The platform $\{2\}$ finally rotates to reach the rotational pose ${}^{1}\boldsymbol{R}_2$.

After this sequence, the moving platform $\{2\}$ and each linear actuator reach their target pose, and the base platform $\{1\}$ has returned to its initial position. Here, the rotation of platform $\{1\}$ is described by

$$ {}^{0}\boldsymbol{R}_1 = \boldsymbol{R}(z, \alpha)\boldsymbol{R}(x, \beta)\boldsymbol{R}(x, -\beta)\boldsymbol{R}(z, -\alpha) = \boldsymbol{R}(0) \quad (8) $$

and the rotation of platform {2} by

$$^{0}\boldsymbol{R}_2 = \boldsymbol{R}(z,\alpha)\boldsymbol{R}(x,\beta)\boldsymbol{R}(x,-\beta)\boldsymbol{R}(z,-\alpha)\boldsymbol{R}(z,\psi)\boldsymbol{R}(x,\varphi)\boldsymbol{R}(z,\theta) = \boldsymbol{R}(z,\psi)\boldsymbol{R}(x,\varphi)\boldsymbol{R}(z,\theta) \quad (9)$$

with respect to the inertial coordinate system {0}. Point 3) is treated as the rotations of platforms {1} and {2} around their own x and z axes, respectively. Consequently, for the individual rotations of the lower cylinder $\{C_1\}$ and the upper cylinder $\{C_2\}$, we get

$$^{0}\boldsymbol{R}_{C_1} = \boldsymbol{R}(z,\alpha)\boldsymbol{R}(x,\beta)\boldsymbol{R}(z,\gamma_1) \quad (10)$$

and

$$^{0}\boldsymbol{R}_{C_2} = \boldsymbol{R}(z,\alpha)\boldsymbol{R}(x,\beta)\boldsymbol{R}(z,\gamma_2) \quad (11)$$

with respect to the inertial coordinate system {0}. Thus, the rotation between $\{C_1\}$ and $\{C_2\}$ to be compensated can be written as

$$^{C_1}\boldsymbol{R}_{C_2} = {}^{C_1}\boldsymbol{R}_0\, {}^{0}\boldsymbol{R}_{C_2} = {}^{C_1}\boldsymbol{R}_0^{-1}\, {}^{0}\boldsymbol{R}_{C_2} = \boldsymbol{R}(z,\gamma_2-\gamma_1). \quad (12)$$

The rotation angle ρ between the lower cylinder $\{C_1\}$ and the upper cylinder $\{C_2\}$ and thus the passive rotation of the linear actuator can be then determined from

$$\tan\rho = \tan(\gamma_2-\gamma_1) = \frac{\tan\gamma_2 - \tan\gamma_1}{1+\tan\gamma_2\tan\gamma_1}. \quad (13)$$

For cylinder $\{C_1\}$, γ_1 is caused by $\boldsymbol{R}(x,-\beta)\boldsymbol{R}(z,-\alpha)$, and $\tan\gamma_1 = -\tan\alpha/\cos\beta$. For cylinder $\{C_2\}$, γ_2 is caused by $\boldsymbol{R}(x,-\beta)\boldsymbol{R}(z,\psi-\alpha)\boldsymbol{R}(x,\varphi)\boldsymbol{R}(z,\theta)$. Based on the disassembly of the rotations of the universal joint's cross shaft in the appendix, the output rotation of the upper cylinder is described by

$$\tan\gamma_2 = N_1/D_1 \quad (14)$$

where $N_1 = \cos(\psi-\alpha)\sin\theta + \sin(\psi-\alpha)\cos\varphi\cos\theta$ and $D_1 = \cos(\psi-\alpha)\cos\varphi\cos\theta\cos\beta - \sin(\psi-\alpha)\sin\theta\cos\beta + \sin\varphi\cos\theta\sin\beta$.

4 Equivalence of Riebe and Ulbrich's method and the Euler angles based matrix method

The analytical expression of the passive rotation with Riebe and Ulbrich's method is described by Eq. (6). In order to demonstrate the equivalence of this method and the Euler angles based matrix method, described by Eq. (13), we have to show the following relation in a non-revolution case:

$$\boldsymbol{R}(z,\alpha_2+\gamma_{2,2}-\gamma_{2,1}) = \boldsymbol{R}(z,\gamma_2-\gamma_1), \quad (15)$$

that is, we have to find the connection between $(\alpha_*,\beta_*,\gamma_{*,*})$ and $(\psi,\varphi,\theta,\alpha,\beta)$. This can be realized by the following three constraints.

Constraint 1: The z axis of the lower cylinder $\{C_1\}$ in both methods is the same. This constraint can be written as

$$^{1}\boldsymbol{R}_{C_1}\begin{bmatrix}0\\0\\1\end{bmatrix} = {}^{1}\boldsymbol{R}_0^{-1}\,{}^{0}\boldsymbol{R}_{C_1}\begin{bmatrix}0\\0\\1\end{bmatrix} = \boldsymbol{R}(z,-\gamma_{1,1})\boldsymbol{R}(x,-\beta_1)\boldsymbol{R}(z,\gamma_{2,1})\begin{bmatrix}0\\0\\1\end{bmatrix} = \boldsymbol{R}(z,\alpha)\boldsymbol{R}(x,\beta)\begin{bmatrix}0\\0\\1\end{bmatrix}, \quad (16)$$

which leads to

$$\begin{bmatrix}\sin\beta_1\sin\gamma_{1,1}\\ \sin\beta_1\cos\gamma_{1,1}\\ \cos\beta_1\end{bmatrix} = \begin{bmatrix}\sin\alpha\sin\beta\\ -\cos\alpha\sin\beta\\ \cos\beta\end{bmatrix} \Rightarrow \begin{cases}\beta_1 = -\beta,\\ \gamma_{1,1} = -\alpha.\end{cases} \quad (17)$$

Further,

$$\tan\gamma_{2,1} = \frac{\tan\gamma_{1,1}}{\cos\beta_1} = -\frac{\tan\alpha}{\cos\beta} = \tan\gamma_1. \quad (18)$$

Constraint 2: The z axis of the upper cylinder $\{C_2\}$ in both methods is the same. This constraint can be written as

$$\boldsymbol{R}(z,-\gamma_{1,2})\boldsymbol{R}(x,-\beta_2)\boldsymbol{R}(z,\gamma_{2,2})\begin{bmatrix}0\\0\\1\end{bmatrix} = \boldsymbol{R}(z,-\theta)\boldsymbol{R}(x,-\varphi)\boldsymbol{R}(z,-\psi)\boldsymbol{R}(z,\alpha)\boldsymbol{R}(x,\beta)\begin{bmatrix}0\\0\\1\end{bmatrix}, \quad (19)$$

which leads to

$$\begin{bmatrix}\sin\gamma_{1,2}\sin\beta_2\\ \cos\gamma_{1,2}\sin\beta_2\\ \cos\beta_2\end{bmatrix} = \begin{bmatrix}E_1\\E_2\\E_3\end{bmatrix} \quad (20)$$

where $E_1 = -\sin(\psi-\alpha)\cos\theta\sin\beta - \cos(\psi-\alpha)\sin\theta\cos\varphi\sin\beta + \sin\varphi\sin\theta\cos\beta$, $E_2 = \sin(\psi-\alpha)\sin\theta\sin\beta - \cos(\psi-\alpha)\cos\theta\cos\varphi\sin\beta + \sin\varphi\cos\theta\cos\beta$, and $E_3 = \cos(\psi-\alpha)\sin\varphi\sin\beta + \cos\varphi\cos\beta$.

The first entry divided by the second one gives us $\tan\gamma_{1,2}$. Thus,

$$\begin{cases}\tan\gamma_{1,2} = N_2/D_2,\\ \cos\beta_2 = \cos(\psi-\alpha)\sin\varphi\sin\beta + \cos\varphi\cos\beta\end{cases} \quad (21)$$

where $N_2 = \sin\varphi\tan\theta - \sin(\psi-\alpha)\tan\beta - \cos(\psi-\alpha)\tan\theta\cos\varphi\tan\beta$ and $D_2 = \sin\varphi + \sin(\psi-\alpha)\tan\theta\tan\beta - \cos(\psi-\alpha)\cos\varphi\tan\beta$.

Constraint 3: The two expressions of the rotation around the inertial z axis are equal (this is the relation we want to show):

$$\boldsymbol{R}(z,\alpha_2+\gamma_{2,2}-\gamma_{2,1}) = \boldsymbol{R}(z,\gamma_2-\gamma_1). \quad (22)$$

Or in matrix form:

$$\begin{bmatrix}\cos(\alpha_2+\gamma_{2,2}-\gamma_{2,1}) & -\sin(\alpha_2+\gamma_{2,2}-\gamma_{2,1}) & 0\\ \sin(\alpha_2+\gamma_{2,2}-\gamma_{2,1}) & \cos(\alpha_2+\gamma_{2,2}-\gamma_{2,1}) & 0\\ 0 & 0 & 1\end{bmatrix}$$

$$=\begin{bmatrix}\cos(\gamma_2-\gamma_1) & -\sin(\gamma_2-\gamma_1) & 0\\ \sin(\gamma_2-\gamma_1) & \cos(\gamma_2-\gamma_1) & 0\\ 0 & 0 & 1\end{bmatrix}. \quad (23)$$

Since the matrices are always unitary, the following relation holds:

$$tan(\alpha_2+\gamma_{2,2}-\gamma_{2,1}) = tan(\gamma_2-\gamma_1)$$

$$\Leftrightarrow \begin{cases} cos(\alpha_2+\gamma_{2,2}-\gamma_{2,1}) = cos(\gamma_2-\gamma_1),\\ sin(\alpha_2+\gamma_{2,2}-\gamma_{2,1}) = sin(\gamma_2-\gamma_1).\end{cases} \quad (24)$$

After inserting Eqs. (6) and (13), the above equation becomes

$$\frac{\tan(\alpha_2+\gamma_{2,2})-\tan\gamma_{2,1}}{1+\tan(\alpha_2+\gamma_{2,2})\tan\gamma_{2,1}} = \frac{\tan\gamma_2-\tan\gamma_1}{1+\tan\gamma_2\tan\gamma_1}. \quad (25)$$

Consequently,

$$\tan(\alpha_2+\gamma_{2,2}) = \frac{\tan\alpha_2+\tan\gamma_{2,2}}{1-\tan\alpha_2\tan\gamma_{2,2}}$$

$$= \frac{\cos\beta_2\tan\alpha_2+\tan\gamma_{1,2}}{\cos\beta_2-\tan\alpha_2\tan\gamma_{1,2}} = \tan\gamma_2 \quad (26)$$

where $\tan\gamma_{2,2} = \tan\gamma_{1,2}/\cos\beta_2$. Rearranging to $\tan\alpha_2$ finally leads to:

$$\tan\alpha_2 = \frac{\cos\beta_2\tan\gamma_2-\tan\gamma_{1,2}}{\cos\beta_2+\tan\gamma_{1,2}\tan\gamma_2}. \quad (27)$$

Combining Eqs. (17), (21), and (27), the *equivalence conditions* are summarized as follows:

$$\begin{cases}\beta_1 = -\beta,\\ \gamma_{1,1} = -\alpha,\\ \tan\gamma_{1,2} = N_2/D_2,\\ \cos\beta_2 = \cos(\psi-\alpha)\sin\varphi\sin\beta+\cos\varphi\cos\beta,\\ \tan\alpha_2 = \frac{\cos\beta_2\tan\gamma_2-\tan\gamma_{1,2}}{\cos\beta_2+\tan\gamma_{1,2}\tan\gamma_2}.\end{cases} \quad (28)$$

5 Passive rotation in quaternion form

As proven in [5], a rotation matrix can be replaced by a set of four numbers, the quaternions. Using quaternions to compute the passive rotation has shown advantages in the reduction of trigonometric computations. In this case, the pose and location of the moving platform {2} is defined as

$$\begin{cases} {}^1\boldsymbol{q}_{12}(t) = [0 \quad x(t) \quad y(t) \quad z(t)],\\ {}^1\boldsymbol{R}_2 = \boldsymbol{R}(\boldsymbol{v},\theta).\end{cases} \quad (29)$$

The corresponding rotation quaternion is thus

$${}^1\boldsymbol{q}_2 = \left(\cos\tfrac{\theta}{2} \quad \boldsymbol{v}\sin\tfrac{\theta}{2}\right) \quad (30)$$

where $\boldsymbol{v}$ is a unitary vector that represents the rotation axis, and θ is the rotational displacement. The inverse kinematics of a hexapod in quaternion form can then be written as

$${}^1\boldsymbol{q}_{J_1J_2} = {}^1\boldsymbol{q}_{12} + {}^1\boldsymbol{q}_2\,{}^2\boldsymbol{q}_{2J_2}\,{}^1\boldsymbol{q}_2^{-1} - {}^1\boldsymbol{q}_{1J_2} \quad (31)$$

and the rotation between the cylinders $\{C_1\}$ and $\{C_2\}$ as

$${}^{C_1}\boldsymbol{q}_{C_2} = {}^1\boldsymbol{q}_{C_1}^{-1}\,{}^1\boldsymbol{q}_2\,{}^2\boldsymbol{q}_{C_2}. \quad (32)$$

In Eq. (32), ${}^1\boldsymbol{q}_{C_1}$, e. g., is determined as follows:

$${}^1\boldsymbol{q}_{C_1} = \boldsymbol{q}(x,\beta_1)\cdot\boldsymbol{q}(y,\alpha_1)$$

$$= S_{\alpha_1/2}C_{\beta_1/2} + C_{\alpha_1/2}S_{\beta_1/2}\boldsymbol{i} + C_{\alpha_1/2}C_{\beta_1/2}\boldsymbol{j} + S_{\alpha_1/2}S_{\beta_1/2}\boldsymbol{k} \quad (33)$$

where

$$S_{\alpha_1/2} = \sin\tfrac{\alpha_1}{2} = \pm\sqrt{(1-\cos\alpha_1)/2}\begin{cases}\geq 0 \text{ for } \sin\beta_1 \geq 0,\\ < 0 \text{ for } \sin\beta_1 < 0,\end{cases}$$

$$C_{\alpha_1/2} = \cos\tfrac{\alpha_1}{2} = \sqrt{(1+\cos\alpha_1)/2},$$

$$S_{\beta_1/2} = \sin\tfrac{\beta_1}{2} = \pm\sqrt{(1-\cos\beta_1)/2}\begin{cases}\geq 0 \text{ for } \sin\alpha_1 \leq 0,\\ < 0 \text{ for } \sin\alpha_1 > 0,\end{cases}$$

$$C_{\beta_1/2} = \cos\tfrac{\beta_1}{2} = \sqrt{(1+\cos\beta_1)/2}.$$

The expression for ${}^2\boldsymbol{q}_{C_2}$ is obtained in the same way.

6 Simulation results

The equivalence conditions of Riebe and Ulbrich's method and the Euler angles based matrix method are given in Section 4. In this section, a simulation of the errors of these two methods is carried out depending on the variables $(\psi,\varphi,\theta,a,b)$. The z values indicate the maximum errors of the pose space formed by (ψ,φ,θ) and how are they distributed on a linear actuator's orientation surface formed by (α,β). Figure 2 indicates that the error between the two algorithms is limited in a co-domain with a magnitude of $10^{-11\circ}$, while the definition domain is $[-89°, 89°]$ for all five variables. However, when we increase the resolution to $10^{-12\circ}$, the error increases to a magnitude of $10^{-6\circ}$.

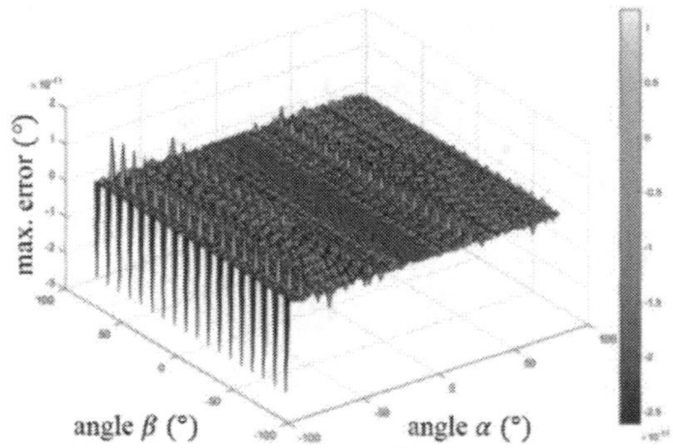

Figure 2. Maximum length compensation computation error distribution between Riebe and Ulbrich's method and the Euler angles based matrix method on an α-β surface.

Similarly, the computation error of the Euler angles based matrix method and the quaternion based method is shown in Fig. 3, where the maximum error remains in a $10^{-4\circ}$ magnitude.

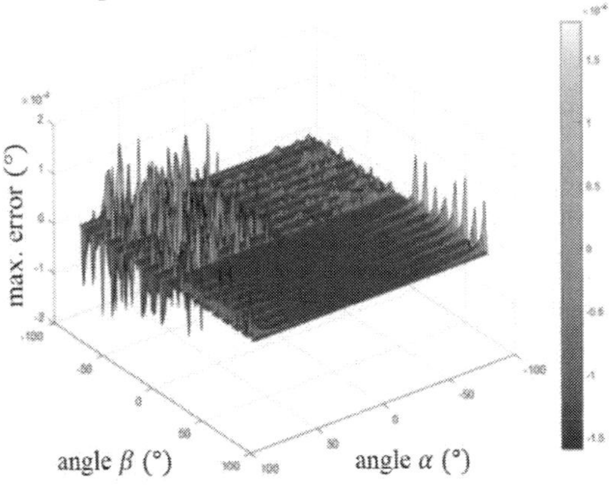

Figure 3. Maximum length compensation computation error distribution between the Euler angles based matrix method and the quaternion based method on an α-β surface.

7 Conclusion

The method for the length compensation problem in the parallel hexapod referred to universal joints provided by Riebe and Ulbrich [4] is inexplicit for the engineering occasion. This inexplicit expression disserves the rotation around the universal joints from the Euler angles of the moving platform. This paper provided two more methods that fit for practical use and proved the equivalence of them by proposing three equivalence constraints. The equivalence is true if the equivalence conditions in Eq. (28) are true. This equivalence is then verified by simulation of the computation error among these three methods. The error of computation remains in a range of roundness error rather than inequality of the computation methods of length compensation.

Acknowledgements

The research described in this paper was supported by the German Research Foundation (DFG) under grant SCHL 275/15-1.

References

1. S. Du, J. Schlattmann, S. Schulz, and A. Seibel, Passive *rotation compensation in parallel kinematics using quaternions*, Proceedings in Applied Mathematics and Mechanics 16, 51–52, (2016).
2. E. B. Dam, M. Koch, and M. Lillholm, *Quaternions, Interpolation and Animation*. Technical Report DIKU-TR-98/5, Department of Computer Science, University of Copenhagen, Copenhagen, Denmark, (1998).
3. K. Shoemake, *Animating rotations with quaternion curves*, In Proceedings of the 12th Annual Conference on Computer Graphics and Interactive Techniques (SIGGRAPH), pp. 245–254, (1985).
4. S. Riebe and H. Ulbrich, *Modelling and online computation of the dynamics of a parallel* kinematic *with six degrees-of-freedom*, Archive of Applied Mechanics 72, 817–829, (2003).
5. J. B. Kuipers, *Quaternions & Rotation Sequences: A Primer with Applications to Orbits*, Aerospace and Virtual Realitz, Princeton University Press, Prinston, New Jersey, USA, (1999)

https://doi.org/10.1051/matecconf/201819802003

Appendix. This appendix illustrates the passive rotation caused by two times bending of a cross-shaft universal joint; $\{I\}$, $\{II\}$, and $\{M\}$ indicate the lower frame, the upper frame, and the cross shaft of the universal joint, respectively.

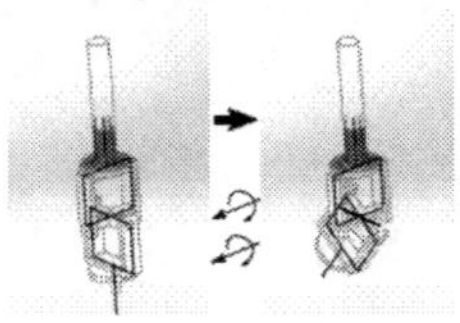	$\{I\}: \boldsymbol{R}(x,A_1),\ {}^0\boldsymbol{R}_{I(1)} = \begin{bmatrix}1&0&0\\0&\cos A_1&-\sin A_1\\0&\sin A_1&\cos A_1\end{bmatrix}$; $\{II\}: \boldsymbol{R}(0,0),\ {}^0\boldsymbol{R}_{II(1)} = \boldsymbol{I}$; $\{\mathrm{M}\}: \boldsymbol{R}(x,A_1)$, ${}^0\boldsymbol{R}_{M(1)} = [\boldsymbol{x}_{II(1)}\ \ \boldsymbol{y}_{I(1)}\ \ \boldsymbol{x}_{II(1)}\times\boldsymbol{y}_{I(1)}] = \begin{bmatrix}1&0&0\\0&\cos A_1&-\sin A_1\\0&\sin A_1&\cos A_1\end{bmatrix}$
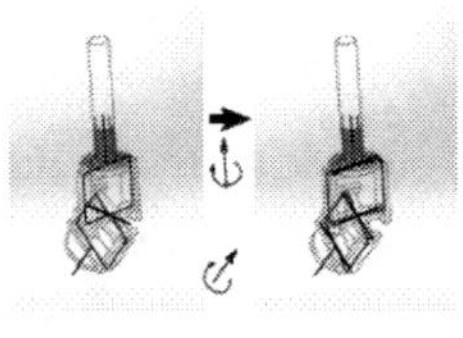	$\{I\}: \boldsymbol{R}(z,A_2),\ {}^0\boldsymbol{R}_{I(2)} = {}^0\boldsymbol{R}_{I(1)}\begin{bmatrix}\cos A_2&-\sin A_2&0\\\sin A_2&\cos A_2&0\\0&0&1\end{bmatrix} =$ $\begin{bmatrix}\cos A_2&-\sin A_2&0\\\cos A_1\sin A_2&\cos A_1\cos A_2&-\sin A_1\\\sin A_1\sin A_2&\sin A_1\cos A_2&\cos A_1\end{bmatrix}$; $\{II\}: R(z,X_1),\ {}^0\boldsymbol{R}_{II(2)} = \begin{bmatrix}\cos X_1&-\sin X_1&0\\\sin X_1&\cos X_1&0\\0&0&1\end{bmatrix}$; $\{M\}:\ {}^0\boldsymbol{R}_{M(2)} = [\boldsymbol{x}_{II(2)}\ \ \boldsymbol{y}_{I(2)}\ \ \boldsymbol{x}_{II(2)}\times\boldsymbol{y}_{I(2)}] =$ $\begin{bmatrix}\cos X_1&-\sin A_2&\sin X_1\sin A_1\cos A_2\\\sin X_1&\cos A_1\cos A_2&-\cos X_1\sin A_1\cos A_2\\0&\sin A_1\cos A_2&\cos X_1\cos A_1\cos A_2+\sin X_1\sin A_2\end{bmatrix}$
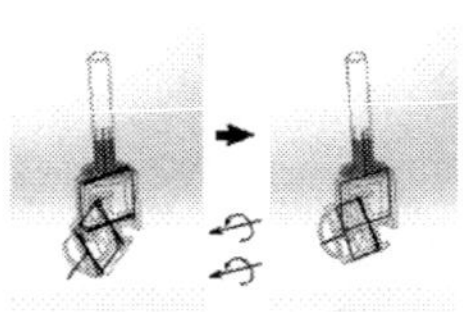	$\{I\}: \boldsymbol{R}(x,A_3),\ {}^0\boldsymbol{R}_{I(3)} = {}^0\boldsymbol{R}_{I(2)}\begin{bmatrix}1&0&0\\0&\cos A_3&-\sin A_3\\0&\sin A_3&\cos A_3\end{bmatrix} =$ $\begin{bmatrix}\cos A_2&-\sin A_2\cos A_3&\sin A_2\sin A_3\\\cos A_1\sin A_2&\cos A_1\cos A_2\cos A_3-\sin A_1\sin A_3&-\cos A_1\cos A_2\sin A_3-\sin A_1\cos A_3\\\sin A_1\sin A_2&\sin A_1\cos A_2\cos A_3+\cos A_1\sin A_3&-\sin A_1\cos A_2\sin A_3+\cos A_1\cos A_3\end{bmatrix}$; $\{II\}: \boldsymbol{R}(0,0),\ {}^0\boldsymbol{R}_{II(3)} = {}^0\boldsymbol{R}_{II(2)}\boldsymbol{I} = \begin{bmatrix}\cos X_1&-\sin X_1&0\\\sin X_1&\cos X_1&0\\0&0&1\end{bmatrix}$; $\{M\}:\ {}^0\boldsymbol{R}_{M(3)} = [\boldsymbol{x}_{II(3)}\ \ \boldsymbol{y}_{I(3)}\ \ \boldsymbol{x}_{II(3)}\times\boldsymbol{y}_{I(3)}] = \begin{bmatrix}\cos X_1&-\sin A_2\cos A_3&\vdots\\\sin X_1&\cos A_1\cos A_2\cos A_3-\sin A_1\sin A_3&\vdots\\0&\sin A_1\cos A_2\cos A_3+\cos A_1\sin A_3&\end{bmatrix}$
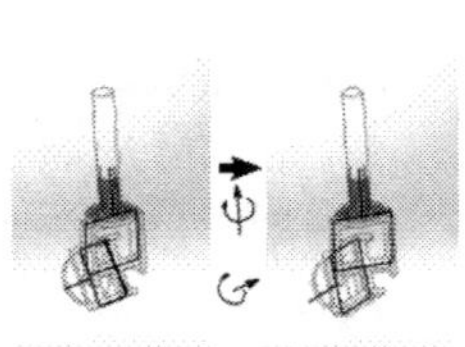	$\{I\}: \boldsymbol{R}(z,A_4),\ {}^0\boldsymbol{R}_{I(4)} = {}^0\boldsymbol{R}_{I(3)}\begin{bmatrix}\cos A_4&-\sin A_4&0\\\sin A_4&\cos A_4&0\\0&0&1\end{bmatrix} =$ $\begin{bmatrix}&-\cos A_2\sin A_4-\sin A_2\cos A_3\cos A_4&\\\vdots&-\cos A_1\sin A_2\sin A_4+\cos A_1\cos A_2\cos A_3\cos A_4-\sin A_1\sin A_3\cos A_4&\vdots\\&-\sin A_1\sin A_2\sin A_4+\sin A_1\cos A_2\cos A_3\cos A_4+\cos A_1\sin A_3\cos A_4&\end{bmatrix}$; $\{II\}: \boldsymbol{R}(z,X_2),\ {}^0\boldsymbol{R}_{II(4)} = {}^0\boldsymbol{R}_{II(3)}\begin{bmatrix}\cos X_2&-\sin X_2&0\\\sin X_2&\cos X_2&0\\0&0&1\end{bmatrix} = \begin{bmatrix}\cos(X_1+X_2)&-\sin(X_1+X_2)&0\\\sin(X_1+X_2)&\cos(X_1+X_2)&0\\0&0&1\end{bmatrix}$; $\{M\}:\ {}^0\boldsymbol{R}_{M(3)} = [\boldsymbol{x}_{II(3)}\ \ \boldsymbol{y}_{I(3)}\ \ \boldsymbol{x}_{II(3)}\times\boldsymbol{y}_{I(3)}] =$ $\begin{bmatrix}\cos(X_1+X_2)&-\cos A_2\sin A_4-\sin A_2\cos A_3\cos A_4&\\\sin(X_1+X_2)&-\cos A_1\sin A_2\sin A_4+\cos A_1\cos A_2\cos A_3\cos A_4-\sin A_1\sin A_3\cos A_4&\vdots\\0&-\sin A_1\sin A_2\sin A_4+\sin A_1\cos A_2\cos A_3\cos A_4+\cos A_1\sin A_3\cos A_4&\end{bmatrix}$

$$\boldsymbol{x}_{II(3)}\perp\boldsymbol{y}_{I(3)} \Rightarrow \boldsymbol{x}_{II(3)}\cdot\boldsymbol{y}_{I(3)} = 0$$
$$\Rightarrow \cos(X_1+X_2)\,(-\cos A_2\sin A_4-\sin A_2\cos A_3\cos A_4)$$
$$+\sin(X_1+X_2)\,(-\cos A_1\sin A_2\sin A_4+\cos A_1\cos A_2\cos A_3\cos A_4-\sin A_1\sin A_3\cos A_4) = 0$$

$$\frac{\sin(X_1+X_2)}{\cos(X_1+X_2)} = \frac{\cos A_2\sin A_4+\sin A_2\cos A_3\cos A_4}{\cos A_1\cos A_2\cos A_3\cos A_4-\cos A_1\sin A_2\sin A_4-\sin A_1\sin A_3\cos A_4}$$

Experimental Study on Grinding Force of Zirconia Ceramics in Dry/Wet Grinding Environment

Zhang Junping[1],Wang Weidong[2],Li Songhua[2],Han Tao[2]

[1]Shanghai Vocational College of Science & Technology,201800,Shanghai,China;
[2]School of Mechanical Engineering, Shenyang Jianzhu University,110168,Shenyang,China;

Abstract. The impacts of different linear speed of grinding wheel, grinding depth and workpiece feed speed with or without grinding fluid on grinding force were studied by plane grinding machining of zirconia ceramics. The impacts of different machining environment and grinding parameter on normal and tangential grinding forceswere studied by testing the grinding force during grinding with a force measuring device. The studies showed that the normal and tangential grinding forces decrease with the increase of the linear speed of grinding wheel and increase with the improvement of grinding depth and workpiece feed speed. The grinding depth has the greatest impacts on the normal and tangential grinding forces in dry grinding environment; while in wet grinding environment, the grinding depth exerts the greatest impacts on the normal grinding force and the linear speed of grinding wheel imposes the greatest impacts on the tangential grinding force. In addition, it was found that the normal grinding force in dry grinding is minor than that in wet grinding, that the tangential grinding force in dry grinding is greater than that in wet grinding, and that the grinding force ratio in dry grinding is lower than that in wet grinding.

1 Introduction

With the development of the times, high-end manufacturing equipment has gradually become the focus of development; and engineering ceramics have been widely applied to the basic components of high-end machine tools by virtue of its excellent performance. Such ceramic materials have such advantages as high hardness, wear resistance, oxidation resistance, corrosion resistance and high temperature resistance [1]. Such materials have broad application prospect in aerospace, national defense and military, industrial and chemical equipment and other fields.

Since engineering ceramics have high hardness and great brittleness, the processing and production of engineering ceramics is a worldwide problem. In practical production, grinding is generally applied to the processing of engineering ceramics [2]. In the process of grinding, the grinding force is a key parameter. The magnitude of grinding force may affect the integrity of the workpiece surface during processing. Therefore, it is very necessary to study the grinding force [3]. For this purpose, this paper focused on the study of the impacts of different grinding parameters on the normal and tangential grinding forces in dry/wet grinding of zirconia ceramics. The factors influencing the normal and tangential grinding forces of zirconia ceramics were studied through a large number of experiments. Besides, this paper tried to find efficient grinding methods suitable for zirconia ceramics.

2 Experimental Conditions

Experiments were carried out on the precision forming grinder. The test piece was 20*20*10mm workblanks of zirconia ceramics. The tool was a resin bond diamond grinding wheel with a particle size of D91. Kistler9257B plane force measuring device was used for real-time detection of the grinding force.

3 EXPERIMENT ON IMPACTS OF GRINDING PARAMETERS ON GRINDING FORCE

The grinding force is composed of the deformation force of smear metal and the mutual friction on its surface. Since zirconia ceramic materials are hard and brittle, it is difficult to cut over abrasive particles during grinding. Such materials are removed mostly in a brittle manner and rarely in a plastic manner. Therefore, when analyzing the grinding of zirconia ceramics, the grinding force mainly is friction [4].

$L9(3^3)$ orthogonal experiment was adopted. The scheme and results of the orthogonal experiment are as shown in Table 1. In order to improve the accuracy of experimental data, repeated experiments were carried out for each group of experiments. The experimental data with large deviation were removed; and the averages of the remaining data were regarded as experimental results.

Table 1 Scheme and Results of Orthogonal Experiment

Number of experiment	Linear speed of grinding wheel/[m/s]	Grinding depth/[mm]	Workpiece feed speed /[mm/min]	Normal grinding force in wet grindin-g/[N]	Tangenti-al grinding force in wet grinding/[N]	Norm-al grindi-ng force in dry grind-ing/[N]	Tangen-tial grinding force in dry grinding/[N]
1	30	0.005	1000	251	31	96	35
2	30	0.010	3000	448	34	245	70
3	30	0.015	5000	653	58	383	85
4	40	0.005	3000	320	30	145	32
5	40	0.010	5000	452	33	295	60
6	40	0.015	1000	459	36	200	40
7	50	0.005	5000	201	20	160	29
8	50	0.010	1000	352	23	135	30
9	50	0.015	3000	495	31	260	65

The response tables for the normal and tangential grinding forces in dry/wet grinding should be established to accurately judge the impacts of grinding parameters on normal and tangential grinding forces. The response tables for the normal and tangential grinding forces in dry/wet grinding are as shown in Tables 2 and 3.

Table 2 Response Table for Fn and Ft in Dry Grinding

Factor / Level	Linear speed of grinding wheel [m/s]		Grinding depth [mm]		Workpiece feed speed [mm/min]	
Grinding force	Fn	Ft	Fn	Ft	Fn	Ft
1	241	63	134	32	143	35
2	213	44	225	53	216	56
3	185	41	281	63	279	58
Range	56	22	147	31	136	23

Table 3 Response Table for Fn and Ft in Wet Grinding

Factor / Level	Linear speed of grinding wheel [m/s]		Grinding depth [mm]		Workpiece feed speed [mm/min]	
Grinding force	Fn	Ft	Fn	Ft	Fn	Ft
1	450	41	257	27	354	30
2	410	29	417	30	421	32
3	349	25	535	41	435	37
Range	101	16	278	14	81	7

In wet grinding environment, grinding depth has the greatest impacts on the normal grinding force, while the linear speed of grinding wheel exerts the greatest impacts on the tangential grinding force. If the range of grinding depth, the linear speed of grinding wheel and workpiece feed speed decreases in order, the impacts on the normal grinding force are gradually weakened; if the range of the linear speed of grinding wheel, grinding depth and workpiece feed speed decreases in turn, the impacts on the tangential grinding force are gradually weakened.

In dry grinding environment, grinding depth imposes the greatest impacts on the tangential and normal grinding force. If the range of grinding depth, workpiece feed speed and the linear speed of grinding wheel decreases in order, the impacts on tangential and normal grinding force are gradually weakened.

4 Analysis of Experimental Results

4.1 Impacts of linear speed of grinding wheel on grinding force

It can be seen from Fig. 1 that in both wet and dry grinding environment, the tangential and normal grinding force in grinding zirconia ceramics decreases with the improvement of the linear speed of grinding wheel. In order the study this case, the theory of the maximum thickness of undeformed smear metal in grinding dmax was introduced [5]. It may be formulated as follows:

$$d_{max}=\{3/c\tan a\}^{1/2}\{v_w/v_s\}^{1/2}\{a_p/d_s\}^{1/4} \quad (1)$$

Where: c is the number of effective grains on the grinding contact area; a is the half-angle of the cross section of undeformed smear metal; aρ is the grinding depth; and ds is the diameter of the grinding wheel [6]. The grinding force of single grain increases with the increase of the maximum undeformed cutting thicknessdmax. It can be seen from Formula (1) that if the linear speed of grinding wheel increases and other parameters remain the same, the value of dmax will decrease; the grinding force decreases with the increase of the linear speed of grinding wheel for the variation trend of the maximum undeformed cutting thickness is the same as that of grinding force; when the linear speed of grinding wheel increases to a certain extent, with the further increase of the linear speed of grinding wheel, the variation of maximum undeformed cutting thickness is no longer obvious, thus the grinding force diagrams show a slow downward trend. In dry grinding, with the rise of the temperature on workpiece surface, the removal method in grinding gradually shifts from brittle removal to plastic removal to increase the tangential grinding force. Therefore, the tangential grinding force in dry grinding is greater than that in wet grinding. Consequently, relative greater linear speed of grinding wheel may be used in grinding process, which can not only reduce the grinding force and improve the surface quality, but also improve the removal rate of materials.

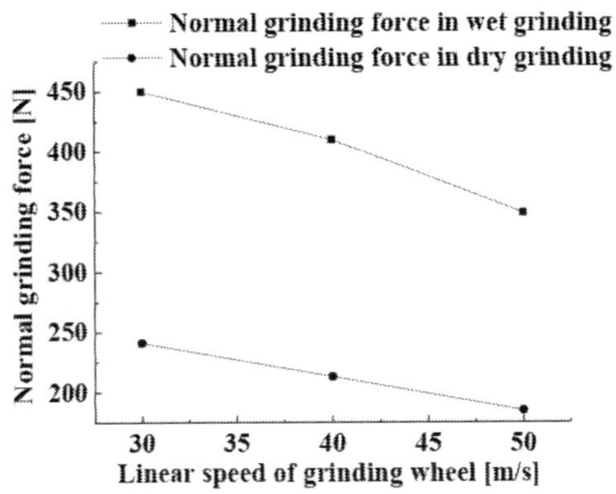

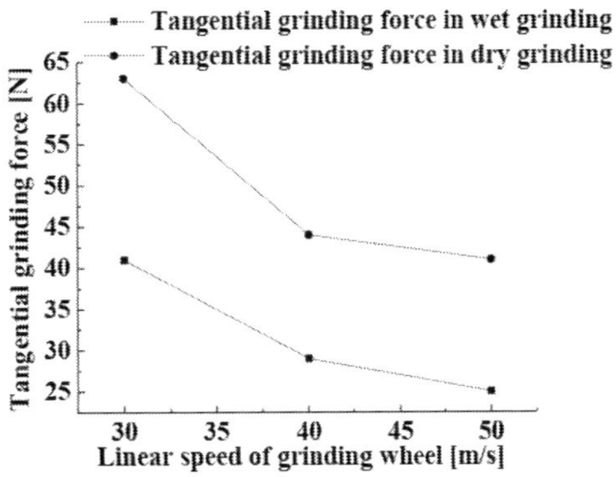

Fig. 1 Impacts of Linear Speed of Grinding Wheel on Normal and Tangential Grinding Forces

4.2 Impacts of grinding depth on grinding force

In both wet and dry grinding environment, with the gradual increase of grinding depth, the normal and tangential grinding forces in grinding zirconia ceramics gradually increase as shown in Fig. 2. When the grinding depth increases, the normal and tangential grinding forces also increase; and the grinding depth has greater impacts on the normal grinding force. This is because when the grinding depth is smaller, the ceramic materials are removed by the slight plastic deformation, thus the grinding force is relatively small. The removal mechanism of zirconia materials in grinding changes with the gradual increase of grinding depth. It can be seen from experimental study that if the grinding depth gradually increases and exceeds a certain critical value, the material removal method in grinding zirconia ceramics gradually shifts from plastic deformation to brittle shedding. This may impose certain impacts on the normal grinding force. The increase of the normal grinding force is because with the gradual increase of grinding depth, the number of effective grains involved in grinding and the contract area between grains and workpiece surface increase accordingly. In dry grinding environment, with the gradual increase of grinding depth and due to the absence of cooling liquid, the grinding temperature rises sharply, which will cause phase change on zirconia surface in grinding process. In such case, the removal method will be gradually converted from brittle removal to brittle-plastic removal, which increases the tangential grinding force. Therefore, the tangential grinding force in dry grinding is greater than that in wet grinding. In order to improve the surface quality of grinding ceramic materials, small grinding depth should be selected. It would be best if the grinding depth is below the critical grinding depth, thus zirconia ceramics can be removed through ideal plastic deformation [7].

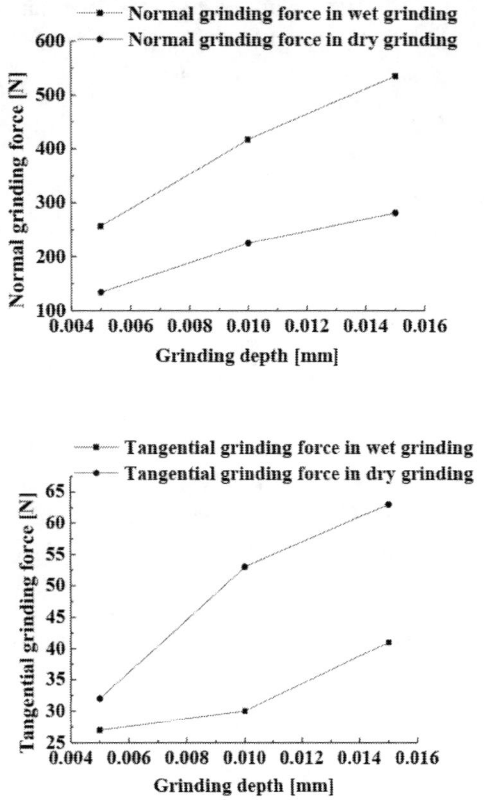

Fig. 2 Impacts of Grinding Depth on Normal and Tangential Grinding Forces

4.3 Impacts of workpiece feed speed on grinding force

In wet and dry grinding environment, the tangential and normal grinding forces in grinding zirconia ceramics increase with the increase of workpiece feed speed as shown in Fig. 3. The normal grinding force in wet grinding of zirconia ceramics is greater than that in dry grinding, but the tangential grinding force in dry grinding of zirconia ceramics is greater than that in wet grinding. This is because when the workpiece feed speed is low and with slow temperature rise, zirconia ceramics are mainly removed by plastic removal. The tangential and normal grinding forces increase with the improvement of workpiece feed speed; and the change is obvious. In dry grinding, with the improvement of workpiece feed speed, the grinding temperature also rises, making the removal method for zirconia ceramics workpieces gradually shift from brittle removal to plastic removal. In this way, the tangential grinding force in dry grinding is greater than that in wet grinding.

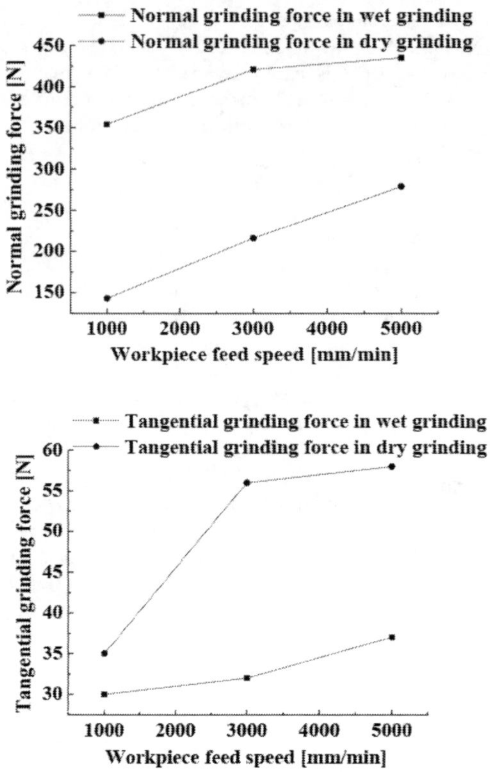

Fig. 3 Impacts of Workpiece Feed Speed on Normal and Tangential Grinding Forces

4.4 Impacts of dry/wet grinding environment on grinding force ratio

The grinding force ratio can reflect the wear of the grinding wheel and the difficulty level during the grinding. The grinding ratio is defined as the ratio of the normal grinding force Fn to the tangential grinding force Ft. The grinding force ratio in dry/wet grinding environment can be calculated by the data in Table 1. As shown in Fig. 4, it can be clearly found that when grinding zirconia ceramics, the grinding force ratio in wet grinding is greater than that in dry grinding. It can be concluded from the correlation between the grinding force ratio and the difficulty level in grinding the material that the grinding effects without grinding fluid in grinding of zirconia ceramics are better that those with grinding fluid. According the analysis above, a lot of heat will be generated in the grinding process due to the continuous friction and cutting between the grinding wheel and ceramic surface. If there is no cooling liquid, the surface of zirconia ceramics will be softened, thus the removal method for such material will change from brittle removal to plastic removal. In such case, the machinability of zirconia ceramics is improved.

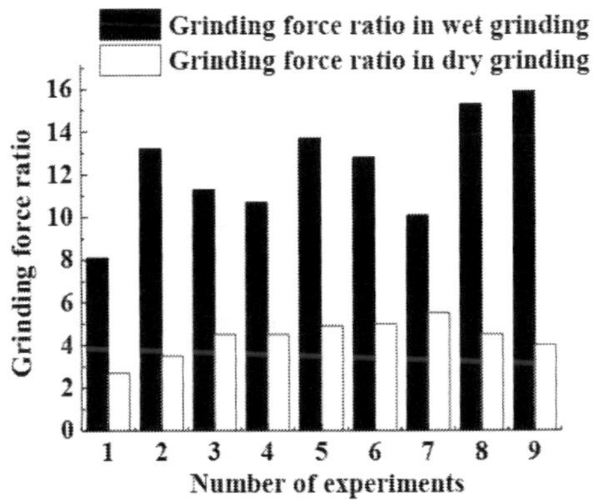

Fig. 4 Comparison of Grinding Force Ratio in Dry/Wet Grinding Environment

5 Conclusions

When grinding zirconia ceramics, the normal and tangential grinding forces are negatively correlated to the linear speed of the grinding wheel and positively correlated to the workpiece feed speed.In wet grinding, the impacts of the grinding depth, the linear speed of the grinding wheel and the workpiece feed speed on the normal grinding force gradually weakens; and the impacts of the linear speed of the grinding wheel, the grinding depth and the workpiece feed speed on the tangential grinding force gradually weakens. In dry grinding, the major and minor order of the impacts of various process parameters on the tangential and normal grinding forces is the grinding depth, the workpiece feed speed and the linear speed of the grinding wheel.The tangential grinding force in dry grinding is greater than that in wet grinding, while the normal grinding force in dry grinding is minor than that in wet grinding. Dry grinding is easier for grinding zirconia ceramics than wet grinding.

References

1. YUAN Jiulong, Zhang feihu, Development Research of Science and Technologies in Ultraprecision Machining Field, J. Journal Of Mechanical Engineering, 46 (2010) 161-176.
2. Hou Yonggai, Tian Jiugen, The Current Research Status of Zirconia Ceramics Grinding, J. China Ceramic, 50 (2014) 6-9.
3. Bahman Azarhoushang, Amir Daneshi, Dal Ho Lee, Evaluation of thermal damages and residual stresses in dry grinding by structured wheels, J. Journal of Cleaner Production, 11 (2016).
4. Zhang Ningju, Zhao Meilin, The Optimization Of Grinding Dosage On Grinding Hardening, J. Journal of Mechanical Strength, 38 (2016) 526-530.
5. WANG Yu, WU Yuhou, Experimental Studdy on Inner Grinding Force of Zirconia Ceramic Bearing Rings, J. Machinery Design & Manufacture, 9 (2015) 159-165.
6. WU Yuhou,WANG Kai. Experimental Study on Grinding Force of HIPSN Engineering Ceramics by Internal Grinding, J. Journal of Shenyang Jianzhu University, Natural Science, 5 (2008) 863-867.
7. Wu Shuan, Zhu Xijing, Modelling Analysis on Ductile Removal of Hardbrittle Material by Single Grain, J. Surface Technology, 45 (2016) 219-223.

Dynamic Response of Aluminum Honeycomb Sandwich Plate Subjected to Impact by Flying Spear

Chun Cheng, Xi Chen , Zhonghua Du, Zhaojun Pang and Jiyue Si

School of Mechanical Engineering, Nanjing University of Science and Technology, 210094Nanjing , China

Abstract. Numerical simulation and experiment were applied to study the dynamic response of aluminum honeycomb sandwich plate subjected to the impact by flying spear as well as the the proper velocity of the flying spear impacting on the plate. The deformation mode and damage form of the aluminum honeycomb sandwich plate were obtained. Moreover, the proper velocity of the flying spear is around 40m/s~50m/s. It is confirmed that The deformation area of the upper and lower panels decreases with the increase of the impact velocity, whereas the buckling angle of the panel increases with the increase of the impact velocity. The damage forms of the aluminum honeycomb cell are collapse and buckling sequentially from the impact point to the surrounding. In addition, The collapse area of the honeycomb cells increases with the increase of the maximum diameter of the flying spear and the range of buckling decreases with the increase of impact velocity. The experimental results are in good agreement with the simulation results.

1 INTRODUCTION

The aluminum honeycomb sandwich plate is widely used in aviation, aerospace, shipbuilding, automobile, wind energy system, rail locomotives and other industrial fields due to its excellent energy absorption function, high specific strength and low density[1-3]. Based on its wide application in various industries, the dynamic response of aluminum honeycomb sandwich plate under different load conditions has been a hot topic in recent years. P Ren[4] considered the dynamic response and impact resistance of aluminum alloy honeycomb sandwich plate subjected to the underwater explosion shock wave loading using the underwater explosion shock wave loading device. He found that when compared with mono-layer plate with the same surface density, the compress of aluminum honeycomb core subjected to underwater shock loading can effectively reduce the plastic deformation of the back panel and improve the overall impact resistance of sandwich structures. Y Song[5]found the deflection of the back plate of the clamped square after impacted by foam metal bullet reaches the maximum at the center and minimizes around the edge, the overall deformation of the sandwich panels is quaquaversal. J Hang[6] carried out hyper-velocity impact test and numerical simulation for honeycomb sandwich structure with insulation layer. The results showed that honeycomb core restricts radial expansion of debris cloud. M Hozhar[7] analyzed the dynamic response of aluminum honeycomb sandwich plate with different foam filling materials under different impact loads using finite element method, and carried out the crash-worthiness design of the structure. The dynamic response and deformation of aluminum honeycomb sandwich plates with different thickness under the explosive load were studied by S Li, and the numerical simulation was carried out with LS-DYNA[8]. H Zhou[9] studied the dynamic response and energy absorption of the honeycomb sandwich structures with the same mass and thickness but different densities under high speed impact.

The good energy absorption function of aluminum honeycomb sandwich plate was conformed by the above studies, therefore the aluminum honeycomb sandwich plate is widely used in aviation and spaceflight[1]. The Earth orbit is in a serious predicament caused by millions of space debris. Operational satellite vital for mankind infrastructure are threatened to be destroyed by space debris. These space debris produced by invalid satellites or a rocket fragments. Many enabling space debris capturing and removal methods have been proposed in the past decade and several methods have been tested on ground and/or in parabolic flight experiments[10-12]. The sketch map of a capture system suing flying spears is shown in Figure 1, which is designed to capture and remove the invalid satellites and space debris or remove from orbit. Based on the flying spear capture system, in this paper we studied the dynamic response of aluminum honeycomb sandwich plate which is applied to satellite under the impact of flying spear. Through experiment and numerical simulation, The deformation mode and damage form of aluminum honeycomb sandwich plate and the impact velocity of the correct action of flying spear were obtained, which providing theoretical and experimental basis for further improvement of the flying spear capture system.

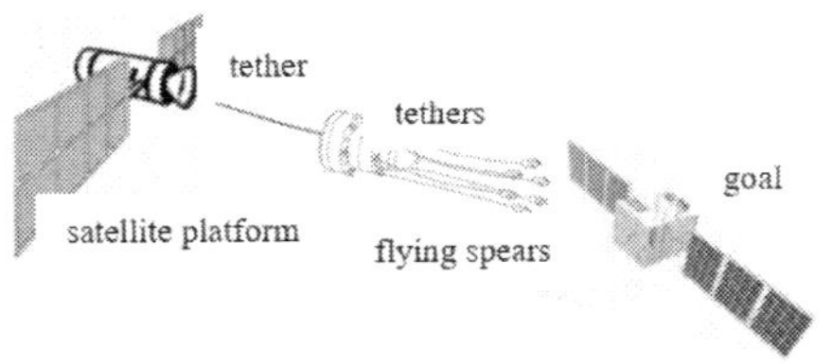

Figure1. Design scheme of a capture system using flying spears

2 Numerical simulation

2.1 Numerical simulation model

The numerical simulation of a flying spear impacting on aluminum honeycomb sandwich plate was carried out by the LS-DYNA. The thickness of the upper and lower panels on the aluminum honeycomb sandwich plate is 1mm. The section of a single honeycomb cell is regular hexagon, the thickness and length of the cell wall are 0.1mm and 8mm respectively, and the height of the cell is 8mm. In order to avoid the termination of the calculation, the grids of the head of the spear and the center of the aluminum honeycomb sandwich plate were refined by the hexahedron mesh. Johnson-Cook material model and Gruneisen equation of state were used for both flying spear and aluminum honeycomb sandwich plate. 45# steel was chose for the spear. There is no plastic deformation and failure of the flying spear in the process of impact. Therefore, the Johnson-Cook constitutive relation of 45# steel was only considered in numerical calculation, and the failure of material was not considered. The main parameters of 45# steel are shown in Table1[13]. 2A12 aluminum was used for aluminum honeycomb sandwich plate and honeycomb cells. The plastic deformation and damage of the sandwich plate under the impact condition are large, so the modified Johnson-Cook model was used for it.

The expression of the Johnson-Cook strength model is as follows,

$$\sigma_{eq} = \left(A + B\varepsilon^{n}\right)\left(1 + C\ln\frac{\dot{\varepsilon}}{\dot{\varepsilon}_0}\right)\left(1 - T^{*m}\right) \quad (1)$$

Where σ_{eq} denotes equivalent stress, A is the yield strength of material at reference strain rate and reference temperature, the coefficients B, C and m denote strain-strengthening coefficient, the coefficient of strain-rate sensitivity and coefficient of softing due to temperature respectively, the exponent n denotes strain-strengthening exponent, $\dot{\varepsilon}_0$ is user - defined reference strain rate which is usually set to 1.

$$T^{*} = \left(T - T_{room}\right)/\left(T_{melt} - T_{room}\right) \quad (2)$$

Where T , T_{room} and T_{melt} are absolute temperature, room temperature and melting temperature of the material respectively.

The failure of 2A12 aluminum can be calculated by Johnson-Cook failure model,

$$\varepsilon_f = \left[D_1 + D_2\exp\left(D_3\sigma^{*}\right)\right]\cdot\left[1 + D_4\ln\frac{\dot{\varepsilon}}{\dot{\varepsilon}_0}\right]\cdot\left[1 + D_5T^{*}\right] \quad (3)$$

Where σ^{*} is stress triaxiality, $D_1 \sim D_5$ all are material parameters. The modified Johnson-Cook failure model was obtained by modifying the temperature softening term into an exponential form because the fracture strain obtained by tensile test at different temperatures does not have a linear relationship with the temperature[14],

$$\varepsilon_f = \left[D_1 + D_2\exp\left(D_3\sigma^{*}\right)\right]\cdot\left[1 + D_4\ln\frac{\dot{\varepsilon}}{\dot{\varepsilon}_0}\right]\cdot\left[1 + D_5\left(1 - \exp\left(D_6T^{*}\right)\right)\right] \quad (4)$$

Where $D_1 \sim D_6$ all are material parameters. The values of each parameter in the Johnson-Cook strength model and the failure model are shown in Table 2. The contact between the flying spear and the aluminum honeycomb sandwich plate adopted the erosion contact algorithm. The cell model of aluminum honeycomb sandwich plate and 1/4 calculation model are shown in Figure 2 and Figure 3 respectively.

Table.1 Main material parameters of 45# steel

Density/g·cm^{-3}	Poisson's ratio	Young's modulus/GPa	*A*/MPa	*B*/MPa	*n*	C_1	C_2	*m*
7.83	0.3	210	497.75	647.15	0.393	2.76×10^{-3}	0.4	0.626

Table.2 Johnson-Cook strength model parameters of 2A12 aluminum

Young's modulus/GPa	Poisson's ratio	Density /g·cm^{-3}	β[a]	T_{room}/K	T_{melt}/K	*m*	c_1	c_2 /MPa	σ_n [b] /MPa	ε_n [c]
71.7	0.33	2770	0.9	293	863	1.426	0.071	288	635	0.126
Yield strength/MPa	Specific heat /J·(kgK)$^{-1}$	strain rate	*ω*	*C*	D_1	D_2	D_3	D_4	D_5	D_6
400	921	0.00111	0	0.001	0.116	0.211	-2.172	0.012	0.0126	13.04

[a] β- Conversion coefficient of plastic work

[b] σ_n -Necking stress

[c] ε_n -Necking strain

Figure 2. Cell model

Figure 3. 1/4 calculation model

2.2 Numerical simulation results and analysis

The dynamic response of aluminum honeycomb sandwich plate under different impulse was studied.The impact velocity was set at 50m/s, 60m/s, 70m/s, 80m/s and 90m/s respectively. The 1/2 frame diagram and 1/2 open view of the damage of the sandwich plate were shown in Figure 4 as the spear impacted on the aluminum honeycomb sandwich plate at different velocities.

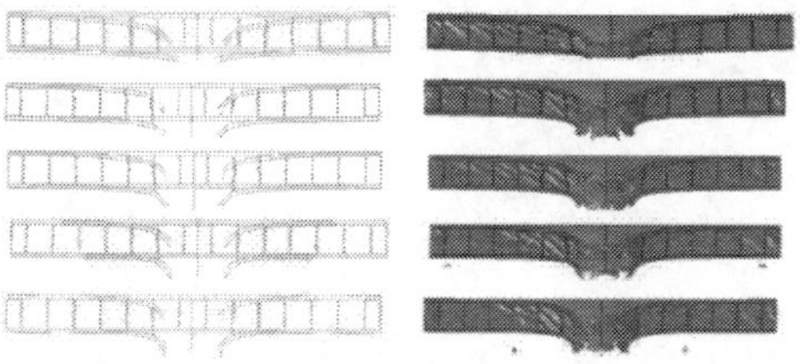

Figure 4 1/2 frame diagram and 1/2 open view of the plates

From Figure 4, it can be seen that the main damage forms of the upper and lower panels of the sandwich plate are shear tear and buckling deformation. When the impact velocity is lower, the buckling deformation area of the upper and lower panels is larger, and the flexion angle of the panel is smaller. Whereas, when the impact velocity is higher, the diameter of the buckling deformation area of the upper and lower panels is smaller, and the flexion angle of the panel is larger. However, the buckling angle of the upper panel is always less than the buckling angle of the lower panel, which is related to the aluminum honeycomb cell. The aluminum honeycomb cell hindered shear tear and buckling deformation of the upper panel. The failure form of aluminum honeycomb cell appears to be collapse and buckling successively from the impact center outwards. The collapse area of the aluminum honeycomb cells is relevant to the maximum diameter of the flying spear, but it has nothing to do with the impact velocity. Whereas, the bucking area decreases with the increase of impact velocity.

The velocity of the correct action of the spear refers to the velocity of the flying spear to capture the satellite correctly, which means the main flying spear embeds in the sandwich plate and the barbs on the flying spear open after the flying spear penetrates the aluminum honeycomb sandwich plate on the satellite. We define this velocity as the proper velocity of the flying spear which is an important parameter for the the flying spear capture system.

It can be seen from the results of numerical simulation that when the velocity of the flying spear was 40m/s, the spear penetrated the aluminum honeycomb sandwich plate only with the tip of the head, and the rest failed to penetrate the plate as seen in Figure 5 (on the left). Whereas when the impact velocity was 50m/s, the most part of the flying spear penetrated the aluminum honeycomb sandwich plate but the thicker part on the back embedded in it as shown in Figure 5 (on the right) and the velocity curve of the flying spear in the process of penetrating the aluminum honeycomb sandwich plate is shown in Figure 6. It can be seen from Figure 6, the velocity of the spear was slowly reduced to about 32m/s, and it penetrated into the aluminum honeycomb sandwich plate at a speed of 32m/s for a period of time, then it sharply reduced to zero, and then increased to 5m/s in reverse direction. The reason for the reverse acceleration of the flying spear is that when the velocity of the spear was reduced to zero, a slight rebound occurred on the aluminum honeycomb sandwich plate. It is can be conjectured from the results of numerical simulation that when the impacting velocity is between 40m/s and 50m/s, the flying spear will penetrate the aluminum honeycomb sandwich plate and embed therein, and the barbs on the spear open. In other words the proper velocity of the flying spear is around 40m/s~50m/s.

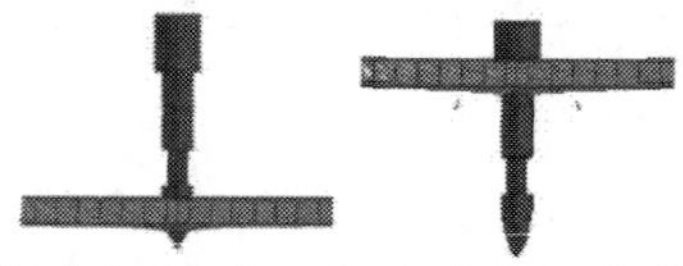

Figure 5. Simulation results under the impact velocities of 40m/s and 50m/s

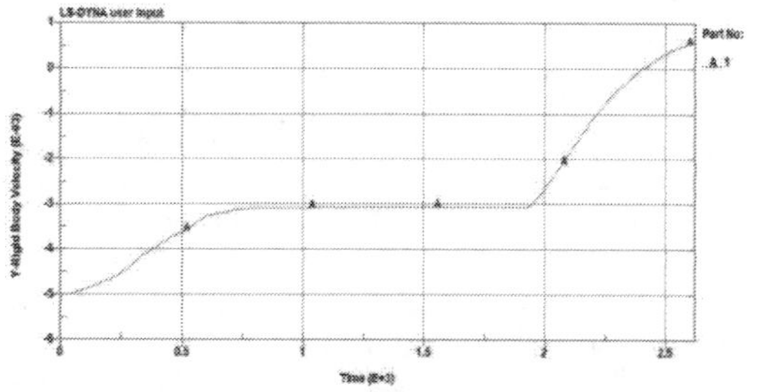

Figure 6. Velocity of flying spear during the perforation under the impact velocity of 50m/s

3 Experiment and experimental results

3.1 Experimental condition

The physical picture of the flying spear with its launcher are shown in Figure 7. The launcher mainly consist of a shell, an end cap, a storage tube for tether, a seal cover and some tether. The end cap was in fixed connection with the shell by screw and the cavity formed between the end cap and the shell was used to install the propellant. Thread connection was used between the storage tube for tether and the end cap. One end of the tether was tied with the spear. A conical slot was machined in the shell which was used to plant a nylon conical device. The nylon conical device connected with the end of the spear by thread was used to seal up gunpowder gas and push the spear forwards. The head of the spear and the cylindrical part of the spear were threaded through a thread, and the barbs were set between the head of the spear and the cylindrical part of the spear.

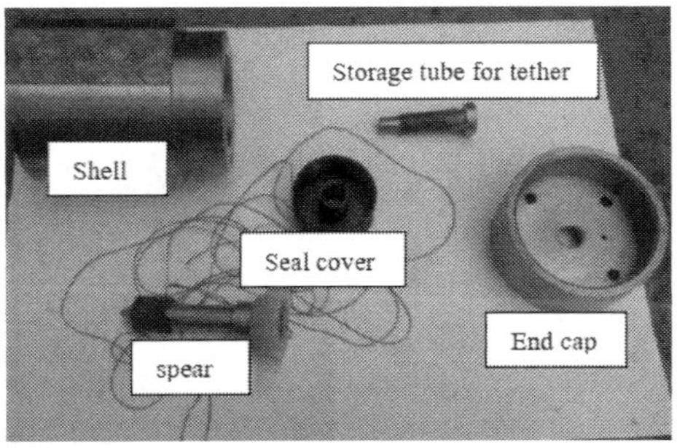

Figure 7. Physical picture of the flying spear and its launcher

The experimental layout is shown in Figure 8. A bench vice was used to fix the launching device. The aluminum honeycomb sandwich plate was fixed at one end of the target frame. Two pieces of aluminum foil paper for speed measurement was used in one experiment, one placed before the aluminum honeycomb sandwich plate and the other one set on one end of the target frame near the bench vice. The other end of the wire used to measure velocity placed at insecurity area. The propellant was ignited by electrical trigger primer.The thickness of the upper and lower panels made by 2A12 aluminum on the aluminum honeycomb sandwich plate is 1mm. The section of the a single honeycomb cell is regular hexagon, the thickness and length of the cell wall are 0.1mm and 8mm respectively, and the height of the cell is 8mm. The overall size of an aluminum honeycomb sandwich plate is 200mm×200mm×10mm. In the experiment, the impact velocity of the spear was controlled by adjusting the charge amount of the propellant.

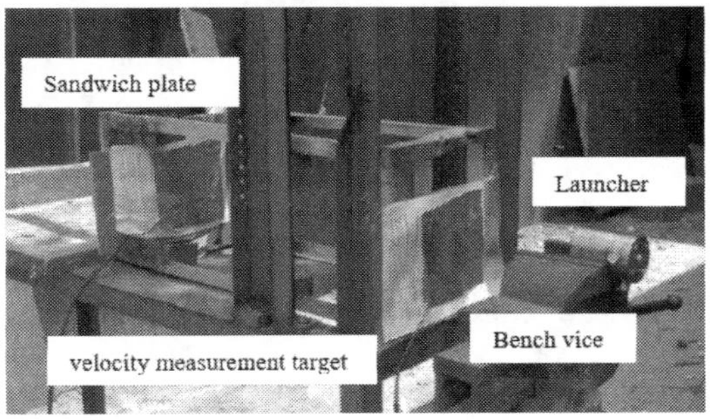

Figure 8. Experimental setup

3.2 Experimental results and analysis

There were 5 spears were launched in the experiment, of which two flying spears failed to impact the aluminum honeycomb sandwich plate correctly due to the poor working accuracy of the spear. The velocities of the other three spears impacting the aluminum honeycomb sandwich panels vertically were 47m/s, 60m/s, 92m/s respectively. When the impact velocity were 60m/s and 92m/s, the spears penetrated the aluminum honeycomb sandwich plates. The perforation and open view of aluminum honeycomb sandwich plates are shown in Figure 9 and Figure 10.

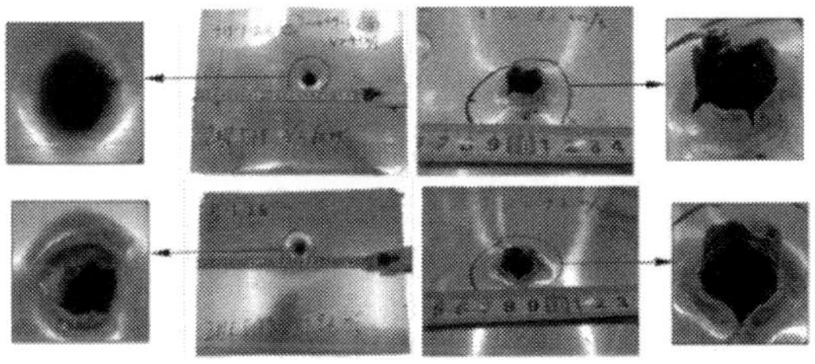

Figure 9. Experimental results under the impact velocities of 60m/s and 92m/s

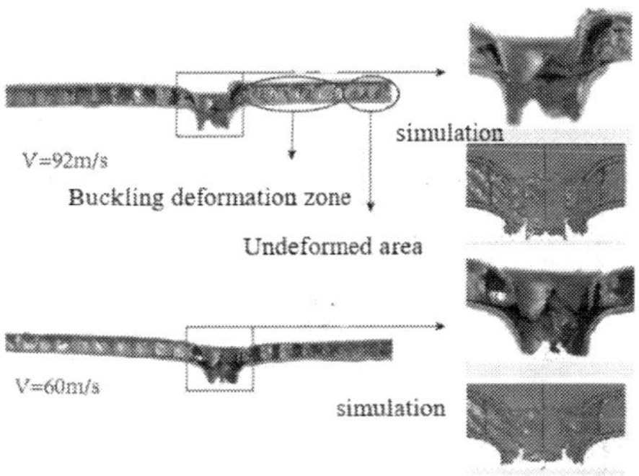

Figure 10. Open view of aluminum honeycomb sandwich plates

It can be seen from Figure 9, there were perforations formed by shear tear at the impact point on the upper panels of the aluminum honeycomb sandwich plates (the panels near the launcher and on the left in the figure), the plastic deformation occurred around the perforation, and the panels flexed inside the sandwich plates. Petal shaped perforations were produced on the lower panels of the aluminum honeycomb sandwich plates (on the right in the figure), When the impact velocity is 60m/s, the diameter of buckling plastic deformation zone on the upper panel is 34mm and that on the lower panel is 26mm; whereas when the impact velocity is 90m/s, the diameter of buckling plastic deformation zone on the upper panel is 32mm, that on the lower panel is 24mm; buckling deformation area on the upper and lower panel shows a decreasing trend as the impact velocity increases. It can be clearly seen from Figure 10, the honeycomb cells collapsed in and around the perforations, buckled near the perforations and did not deform obviously at areas far from the perforations. the range of buckling decreases with the increase of impact velocity. It can be seen from the perforation magnification view that the buckling angle of the upper and lower panels increases with the increase of the impact velocity. The damage and deformation of the upper and lower panels and the honeycomb cells obtained in the experiment are in agreement with the simulation results.

In the experiment, the flying spear with the impact velocity of 47 m/s penetrated the aluminum honeycomb sandwich plate and embedded in it as seen in Figure 11. It is proved that the conjecture from the numerical simulation is correct. That is the proper velocity of the flying spear is around 40m/s~50m/s.

https://doi.org/10.1051/matecconf/201819802005

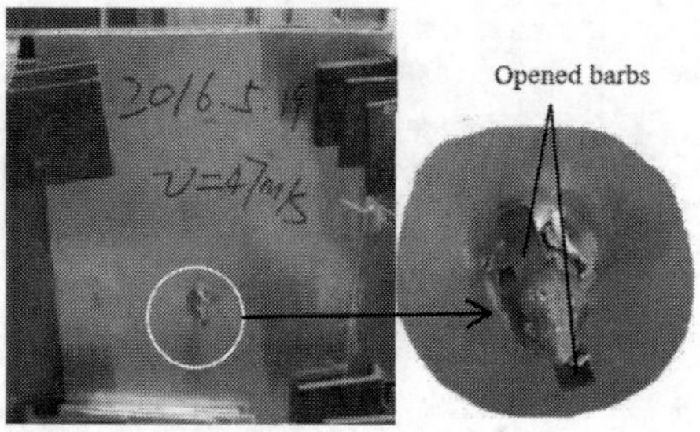

Figure 11. Experimental results under the impact velocity of 47m/s

4 Conclusions

The dynamic response of the aluminum honeycomb sandwich plate under the impact of flying spear was studied by numerical simulation and experiment, and the following conclusions were obtained.

a.The main damage and failure modes of the upper and lower panels of the sandwich plate are shear tear and buckling deformation. The buckling deformation area of the upper and lower panels decreases with the increase of the impact velocity, whereas the buckling angle of the panel increases with the increase of the impact velocity.

b.The damage forms of the aluminum honeycomb cell are collapse and buckling sequentially from the impact center to the surrounding. The collapse area of the honeycomb cells increases with the increase of the maximum diameter of the flying spear and the range of buckling decreases with the increase of impact velocity.

c.The flying spear with the impact velocity of 40m/s~50m/s can penetrate the aluminum honeycomb sandwich plate and embed in it, and the barbs open.

References

1. A Ashab, R Dong, G Lu, et al. Quasi-static and dynamic experiments of aluminum honeycombs under combined compression-shear loading[J]. Material and Design, 2016(97):183-194.
2. N Michihata, H Matsui, K Fushimi. An inverse parallel genetic algorithm for the identification of skin/core debonding in honeycomb aluminium panels[J]. Structural Control & Health Monitoring, 2015, 22(12):1426-1439.
3. V Crupi, G EpastoE, Guglielmino. Comparison of aluminium sandwiches for lightweight ship structures: honeycomb vs. foam[J]. Marine Structures, 2013, 30: 74-96.
4. P Ren, W Zhang, JH Liu et al. Dynamic analysis of aluminium alloy honeycomb core sandwich panels subjected to underwater shock loading[J]. Journal of Vibration & Shock, 2016, 35(2): 7-11.
5. Y Song, Z Wang, T Zhao, et al. Dynamic response of foam sandwich plates subjected to impact loading[J]. Explosion & Shock Waves, 2010, 30(3):301-307.
6. J Huang, Z Ma, S Lan, et al. Study on Hypervelocity Impact Characteristics for Honeycomb Sandwich with Multi-Layer Insulation[J]. Journal of Astronautics, 2010,31(8):2043-2049.
7. M Hozhar, K Sotoush, M Habinollah , et al. Finite element analysis of foam-filled honeycomb structures under impact loading and crash-worthiness design[J]. Taylor & Francis, 2015, 21(2): 2043-2049.
8. S Li, X Li, Z Wang, et al. Finite element analysis of sandwich panels with stepwise graded aluminum honeycomb cores under blast loading[J]. Composites Part A, 2016, 80:1-12.
9. H Zhou, X Wang, Z Zhao. High velocity impact mitigation with gradient cellular solids[J]. Composites Part B: Engineering, 2016, 85: 93-101.
10. M Shan, J Guo, E Gill. Review and comparison of active space debris capturing and removal methods[J]. Progress in Aerospace Sciences, 2015, 80:18-32.
11. M M Castronuovo. Active space debris removal—A preliminary mission analysis and design[J]. Acta Astronautica, 2011, 69(9–10):848-859.
12. S I Nishida, S Kawamoto, Y Okawa, et al. Space debris removal system using a small satellite[J]. Acta Astronautica, 2009, 65(1):95-102.
13. J Jianping. Determination of the Thermo-Viscoplastic Constitutive Relations of 45~# Steel[J]. Transactions of Beijing Institute of Technology, 2008.
14. W Zhang, G Wei, X Xiao. Constitutive relation and fracture criterion of 2A12 aluminum alloy[J]. Binggong Xuebao/acta Armamentarii, 2013, 34(3):276-282.

Reliable Design: Basic Approach

Jiří Stodola[1] and Jan Furch[2]

[1]*University of Defence, Faculty of Military Technology, Kounicova 65, 662 10 Brno, Czech Republic*
[2]*University of Defence, Faculty of Military Technology, Kounicova 65, 662 10 Brno, Czech Republic*

Abstract. The knowledge and experience learnt from product designing have resulted in development of their reliability theory. The classical concept of safe – life is based on product over dimensioned design that considers safety factor or safety margin for measure. However, practical engineering has found this concept in a manner inconvenient as design fault-resistance to determine the ultimate condition and operating stress are random values. A way out is in the concept of stochastic approach to reliability design resulting from the defect-production probability distribution law. That concept allows product designing with predetermined reliability, such as in the example contained in this paper.

1 Introduction

You may question first how reliability has anything to do with design, but it's one of the most important aspects of design. There are undoubtedly objective principles about designing reflected in both the creative procedure and possible algorithmization or automation of each of the processes. Algorithmization then stands for a general procedure of multiple steps leading to the result of the questioned task. Algorithmization of designing is a systematic method possibly completed for practical effect with intuitive methods using the designer's or the creative individual's education, knowledge and experience bases for the results. The systematic and intuitive kind methods do complete rather than inhibit from each other that make us think of a new design discipline – Design Science. The first phase of a new equipment design project is dimensioning process of setting up the mechanical parts size supporting cross-section dimensions so that they stand the external load. It is practical a priori determination on design requirements of strength, rigidity and other features that include also reliability at reasonable production cost. The current practice in designing has determined the essential parameters and functions to assess the load strength of the material by modelling single-value deterministic variables [1]. The errors of calculation, if any, material defects, inaccurate knowledge of loading forces, oscillating load cycles and others are considered within the safety factor "k". The stress-restricted designs have the acceptance criterion in the following form:

$$S \geq s \cdot k, \qquad (1)$$

where $k > 1$ for the safety factor. However, safety induced in such a way to the design misses information of the construction element failure option in the form of failure probability $F(t)$. This conventional attitude supposes that sufficiently high safety factor k may completely eliminate the failure possibility. However, most design variables are actually random values that temper with the probability theory and mathematic statistics principles [2]. Therefore, the failure probability $F(t)$ may vary a lot depending on element load and strength random variables distribution at equal safety factor k. Should the machine constructions design reflect their reliability the design formulas should consider stochastic character of all the variables and use all the terms for the general design. This attitude to designing called reliability (probability) design has developed only recently. Its suitability for aircraft or automotive industries has increased as reliability of the product, such as aircraft or automobile has become an important factor besides economic operation in the competitive environment of the global market.

2 Conventional and reliability based approach to construction design

The reliability of product is strongly influenced by decisions made during the design process. Deficiencies in design affect all items produced and are progressively more expensive to correct as development proceeds. It is often not practicable or economic to change a design once production has started. It is therefore essential that design disciplines are used which minimize the possibility of failure and which allow design deficiencies to be detected and corrected as early as possible. The design process must therefore be organized to ensure that failure-free design principles are used and that any deviations from the principles are detected and corrected.

Failure-free design is the only acceptable principle for any reliability-conscious project team. Anything less will to be reflected in the acceptance of failures throughout the development and production cycle, and a low rate of improvement. The designer must produce designs which will not fail in manufactured and used as specified. In order to be able to do this test data may be needed to reduce uncertainties. Any subsequent failures can then be firmly classified as design deficiencies which escaped the review or test system, or as being due to manufacturing failures or overload. Failure-free design therefore involves prevention, check and cure. The designer must be aware of the materials, processes, components, production methods, design rules and guidelines, costs, and much else in order to create a good design.

Computer-aided engineering (CAE) methods are available to assist with a wide variety of design tasks. Their power, ease of use, and increasing availability due to reducing costs of computing equipment and software are resulting in increasing applications. CAE is also makes possible the creation of designs which would otherwise be very difficult or uneconomic, for example complex electronic circuits. CAE can also provide enormous improvements in engineering productivity. In the mechanical engineering field, software is available for stress analysis, which performs finite element analysis calculations for mechanical and thermal stress calculations, and for analysis of vibration and load responses. Drafting software is used for generating manufacturing drawings and machine tool instructions, and this can also be used to optimize the design of mechanisms. Specialist CAE software is also available for design and analysis of systems and products incorporating other technologies, such as hydraulics, magnetics, microwave electronic, etc. Multi-technology capability is now also being provided, so that mixed technology designs can be modelled and analysed. CAE provides the capability for rapid assessment different design options, and for analysing the effects of tolerances, variation, and failure modes. Therefore if used in systematic, disciplined way, with adequate documentation of the options studied and assessments performed, designs can be optimized for costs, and reliability. CAE should be considered as a powerful aid to more cost-effective and correct design, not merely a means of speeding up the design process. However, there are important limitations inherent in most CAE tools. The software models can never be totally accurate representations of all aspects of the design and of its operating environment. For example, electronic circuit simulation programs generally ignore the effects of electromagnetic interference between components, and drafting systems will ignore distortion due the stress of temperature. Therefore it is essential that engineers using CAE are aware of the limitations, and how these could affect their designs. The effective application of modern CAE places greater responsibility upon designers to be aware of the practical aspects and limitations of the relevant technologies. Otherwise they can be easily misled into placing undue faith in the accuracy and completeness of the software models, resulting in incorrect or unreliable designs.

The environments in which the product will be expected to be stored, operated and maintained must be carefully assessed, as well as the expected severity and durations. The assessment must include all aspects that could affect the product's operation, safety and reliability. Physical factors include temperature, vibration, shock, humidity, pressure, etc. Extreme values and, where relevant, rotes of change must be considered. Other environmental conditions, such as corrosive atmospheres, electrical interference, power supply variation, etc., must also be taken into account. Where appropriate, combined environmental conditions, such as temperature, corrosive atmosphere and vibration, contamination, should be assessed. An aspect of the environment often neglected is the treatment of the product by the people in storage, handling, operation and maintenance. Environmental aspects should be reviewed systematically, and the review should be properly documented. The protective measures to be taken must be identified, as appropriate to storage, transport, handling, operation and maintenance. Protective measures include packing, provision of warning labels and instructions, protective treatment of surfaces, and design features.

Protections against extreme loads is not always possible, but should be considered whenever practicable. In many cases the maximum load can be predetermined, and no special protection is necessary. However, in many other loading situations extreme external loads can occur and can be protected against. Standard products are available to provide protection against, for example, overpressure in hydraulic or pneumatic systems, impact loads or electrical overload. When overload protection is provided, the reliability analysis is performed on the basis of the maximum load which can be anticipated, bearing in mind the tolerances of the protection system. In appropriate cases, loads which can occur when the protection system fails must also be considered. However, in most practical cases it will be sufficient to design to withstand a predetermined load and to accept the fact that loads above this will cause failure. The probability of such loads occurring must be determined for a full reliability analysis to be performed. It may not always be practicable to determine the distribution of such extreme events, but data may be available either from failure records of similar items, or form test or other records. Where credible data are not available, the worst design load case must be estimated. The important point is that the worst design case is estimated and specified. A common cause of failure is the use of safety factors related to average load conditions, without adequate consideration having been given to the extreme conditions which can occur during use of the product.

Strength degradation, it is many forms, can be one of the most difficult aspect to take into account in design reliability analysis. Strength degradation due to fatigue in materials is fairly well understood and documented, and therefore reliability analysis involving metal fatigue, including the effects of stress raisers such as notches, corners, holes and surface finish, can be performed satisfactory, and parts can be designed to operate below the fatigue limit, or for a defined safe life. However, other weakening mechanisms are often more complex. Combined stresses may accelerate damage or reduce the fatigue limit. Corrosion and wear are dependent upon environments and lubrication, the effect of which is therefore often difficult to forecast. If complete protection is not possible, the designer must specify maintenance procedures for inspection, lubrication or scheduled replacement. Reliability analysis of designs with complex weakening processes is often impracticable. Test should then be designed to provide the required data by generating failures under known loading conditions.

Despites discipline, training and care, it is available that occasional oversights or errors will occur in new designs. Design analysis methods have been developed to highlight critical aspects and to focus attention on possible shortfalls. Design analyses are sometimes considered tedious expensive. In most cases the analysis will show that nearly all aspects of the design are satisfactory, and much more effect will have been expended in showing this than in highlighting a few deficiencies. However, the discovery of a very few deficiencies at an appropriately early stage can save far more than the cost that might be incurred by having to modify the design at a later stage, or by having to live with the consequences of the defect. Therefore, well-managed design analyses are extremely cost-effective. The tedium and expense can be greatly reduced by good planning and preparation and by the use of computerized methods. The main reliability design analysis techniques are:

1. Quality function deployment (QFD).
2. Load-strength analysis (LSA).
3. Reliability prediction.
4. Hazard and operability study (HAZOPS.
5. Fault tree analysis.
6. Failure modes, effects and critically analysis (FMEA, FMECA, etc.).
7. Parts materials and process review (PMP).
8. Human aspects manufacturing.
9. Aspect of maintenance.
10. Others.

For example, load-strength analysis (LSA) is a procedure to ensure that all load and strength aspects have been considered in driving the design, and if necessary in planning of tests. The load-strength analysis should include the following:

- determine the most likely worst case values and patterns of variation of load and strength,
- evaluate the safety margin for intrinsic reliability,
- determine protection methods (load limit, derating, screening, other quality control methods),
- identify and analyse strength degradation modes,
- test to failure to corroborate, analyse results,
- Correct or control (redesign, safe life, quality control, maintenance etc.).

This part of the text is taken from [3]. The book [3] contains summary information including the use of reliability theory in practice, not only for designers but also for users.

Reliability designing is based on the presumption the design variables are random values [4], [5]. Their variability rate may be used for attaining the required construction reliability. The difference of the reliability-based and conventional (deterministic) design [6] may be explained on the design example of strength S intended to transmit static service load s. The conventional approach to design introduces safety factor $k > 1$ and according to the equation (1) the design criterion takes the following form:

$$\frac{S}{S_{max}} = k \tag{2}$$

where S - characteristic material strength (such as yield point, ultimate strength, endurance limit and others), S_{max} - maximum permissible stress (such as tension etc.) [7].

The actual operating stress must comply the following condition:

$$s \leq s_{max} = \frac{S}{k} \rightarrow S - s \geq k, \tag{3}$$

to guarantee failure is avoided. Failure occurrence probability cannot be reflected in the formula or specify with a value [8]. The reliability design approach reflects reliability or design safety in the form of failure-proof probability. The design criterion is considered in the following form:

$$S > s \rightarrow (S - s) > 0\,. \tag{4}$$

Probability introduced to the in equation results in the following:

$$P(S > s) =\rightarrow (S - s) = R\,, \tag{5}$$

where $S = (\hat{S}, \sigma_S$ characteristic material strength, $s = (\hat{s}, \sigma_s)$ maximum accessible operating stress, P - probability of failure-proof operation, i.e. structural reliability.

The equation (5) may be expressed in words that due to probability R material strength is higher than the acting tension caused by the operating load and prevents the failure [9]. The difference of the conventional and reliability approaches is characterised in figures 1 and 2 [2], [10], [11], [12].

3 Reliability designing and analysis

If part strength $S = (\mu_S,\ \sigma_S^2)$ considered when the random values are distributed along normal (Gauss) rule $N(\mu,\ \sigma^2)$ where μ, σ^2 are real numbers, $\sigma^2 > 0$ thus

$$f(S) = \frac{1}{\sigma_S\sqrt{2\pi}}\exp - \frac{(S-\mu_S)}{2\sigma_S^2} \tag{6}$$

while $S \in (\pm\infty)$, or $(-\infty < S < \infty)$. This part is randomly loaded with stress $s = (\mu_s,\ \sigma_s^2)$ with normal distribution, then

$$f(s) = \frac{1}{\sigma_s\sqrt{2\pi}}\exp - \frac{(s-\mu_S)}{2\sigma_S^2} \tag{7}$$

and $(-\infty < s < \infty)$.

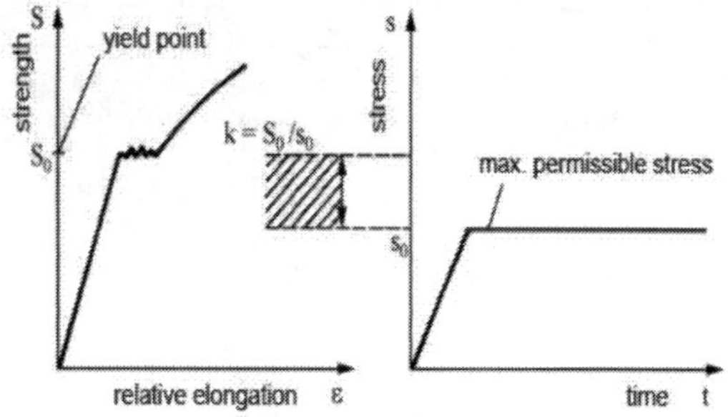

Figure 1. Conventional design

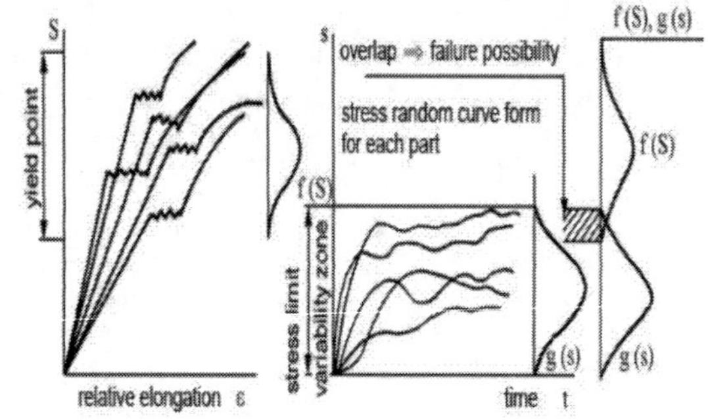

Figure 2. Reliability design

Equation (5) implies $R = P(S - s > 0)$ and the difference is $z = S - s$ with the following characteristics:

$$\mu_z = \mu_S - \mu_s \tag{8}$$

and

$$\sigma_z = \sqrt{\sigma_S^2 - \sigma_s^2} \tag{9}$$

then

$$R = \frac{1}{\sigma_z\sqrt{2\pi}}\int_0^\infty exp - \frac{(z-\mu_z)}{2\sigma_z^2}dz\,. \tag{10}$$

This equation is used in the form of standard (basic) normal distribution utilising the known transformation

$$t = \frac{z-\mu_z}{\sigma_z} \tag{11}$$

that implies the following for the variable t limits:

$$z = \infty \rightarrow t = \infty \text{ and } z = 0 \rightarrow t = \frac{\mu_z}{\sigma_z}\,.$$

$$R = \frac{1}{\sqrt{2\pi}}\int_{\mu_z/\sigma_z}^{\infty} exp\frac{t^2}{2}\mathrm{dt} = \frac{1}{\sqrt{2\pi}}\int_{-\infty}^{\mu_z/\sigma_z}\frac{\mu_z}{\sigma_z} = \Phi(\frac{\mu_z}{\sigma_z}). \tag{12}$$

The upper limit value of t is determined as follows

$$t = \frac{\mu_z}{\sigma_z} = \frac{\mu_S-\mu_s}{\sqrt{\sigma_S^2-\sigma_s^2}}. \tag{13}$$

The equation (12) is the distribution function of the standard random variable *Φ(t)* with normal distribution $N = (0, 1)$. The values of the standard random variable t distribution function variable *Φ(t)* are called Laplace's functions (probability integral) being tabled in e.g. [13], [14], [15]. There are two general approaches to the reliability calculation:

1) For the already produced and used constructions, substitution of the known values of μ_S, σ_S and μ_s, σ_s in the equation (13) determines the upper limit value of t for the analysed construction and with the standard normal distribution $N = (0, 1)$ tables [14], [16], [17] specify the questioned construction's reliability. The following analysis example shows the procedure: An automobile constructional part is loaded with static stress of normal distribution $N(\mu_s = 135 \text{ MPa and } \sigma_s = 11.6 \text{ MPa})$ and its strength is $N(\mu_S = 184 \text{ MPa and } \sigma_S = 21.5 \text{ MPa})$. The task is determination of the part defect probability. The equations (8), (9) and (11) imply: $\mu_z = \mu_S - \mu_s = 184 - 135 = 49$; $\sigma_z = \sqrt{(21.5)^2 + (11.6)^2} = 24.4$ then $t = 49/24.4. = 2$ and the standard normal distribution function tables [14] show that the automobile part reliability $R(t) = 0.9773$ and failure probability $F(t) = 1 - R(t) = 0.0227$.

2) The design of new parts issues from the prior requirement of reliability R of the structure. When the required *to* value is known, the constructional variables are set so that the required reliability is achieved. This procedure is the basic tool for designing parts of in-advance-determined reliability [18], [19], which is the principal purpose of mechanical parts reliability designing. The following is an example of the procedure.

Remark: The examples include only static stress with forces constant in time to make calculation easy and clear. However, practical calculations should consider mostly forces variable in time – deterministic (load-time history is determined by function that can be mathematically described, such as cyclic – symmetric, discontinuous, pulsating, asymmetrical or others) and stochastic (random) that cannot be described exactly by mathematics but using rather statistic assessment.

A simple automobile constructional part (rod) is only loaded with static axial load Q of normal distribution $N(\mu_Q = 12.7 \cdot 10^4 \text{N and } \sigma_Q = 1.8 \cdot 10^4 \text{ N})$. The rod is made from steel with the yield point in tension normally distributed $N(\mu_x = 350 \text{ MPa and } \sigma_x = 35 \text{ MPa})$. The design requires failure probability $F(t) = 10^{-5}$, i. e. $R(t) = 0.99999$. The rod is made in an automatic production process where the resulting rod cross-section A is known to be a random variable of the coefficient of variation $v_A = 0.05$, which implies the mean-root-square error $\sigma_A = 0.05\,\mu_A$. The designer is now to calculate the minimum nominal cross-section of the rod μ_A that would comply the reliability requirement $R(t) = 0.99999$. The solution is to use the equation (11) for unknown nominal cross-section μ_A. First, tension of the rod should be determined with the equation $N = Q/A$ [N/cm^2] if the rod cross-section is defined in cm^2. The randomly variable tension of the rod whose value is Y shall have the mean value $\mu_Y = \frac{\mu_Q}{\mu_A} = \frac{12.7 \cdot 10^8}{\mu_A}$ [Pa] and the mean-root-square error $\frac{(\mu_Q^2 . \sigma_A^2 + \mu_A^2 \sigma_Q^2)}{\mu_A^4} \rightarrow \sigma_Y^2 = \frac{1.91 \cdot 10^8}{\mu_A}$ [Pa]. The tension in the rod follows approximately normal distribution $N(\frac{12.7}{\mu_A}, \frac{1.91}{\mu_A}) \cdot 10^8$ [Pa]. The standard normal distribution function tables [3] show $t = 4.265$ at $R(t) = 0.99999$. Substitution in the equation (13) results in $4.265 = \frac{3.5 - 12.7/\mu_A}{\sqrt{(0.35)^2 + (\frac{1.91}{\mu_A})^2}}$. When this equation is solved for μ_A the required minimum cross-section $\mu_{A\,min} = 7.63$ cm^2 and its mean-root-square error $\sigma_A = 0.05\,\mu_A = 0.3815$ cm^2.

4 Conclusion

Reliability designing with variables actually represented by statistic models that us construction failure-proof operation probability R rather than the conventional safety factor offers tools to solve numerous constructional key problems. First of all, it is systematic analysis of reliability for the questioned constructional element and last but not least the reliability designs including the theoretical reliability requirements [10]. There is a very close relation between this and optimum construction as the determination of optimum dimensions may enhance with the determination of optimum production and material costs. A simple example showed the procedures of reliability designing for static load with normal distribution of random variables. It should be noted that the basic procedures remain the same also for construction design when load is dynamic in character and the random variables take more complex forms of distribution. Of course, the equations for calculation are more complex then. The world leading automobile industry companies have proved the advantage of reliability design in case of large scale production as the higher cost of reliability design pay shortly back in saved material, energy, labour and also in the increased reliability rate of the final product. Other advantages of this method include efficient planning of spare parts, maintenance and more.

Acknowledgment

Presented work has been prepared with the support of the Ministry of Defence of the Czech Republic, Partial Project for Institutional Development and Specific Research, Department of Combat and Special Vehicles, University of Defence, Brno.

References

1. F. Bohacek, *Parts and Mechanisms of Machines I. Princiles of Design*. University of Technology Brno, (1981)
2. P. Klimes, *Machine Parts and Mechanisms I. Reliability, sizing, joints and shafts*. Academic publishing house CERM Ltd., Brno, (2003)
3. P. D. T. O'Connor, D. Newton, R. BROMLEY, *Practical Reliability Engineering*. John Wiley & Sons, LTD (Fourth Edition), ISBN 978-0-470-84462-5 (2002)
4. V. K. Way, R. Prasad, A. Frank, T. Ching, L. Hwang, *Optimal Reliability Design*. Cambridge University Press. ISBN 0-52-1031-915 (2006)
5. Institute of Electrical and Electronics Engineers IEEE Standard Computer Dictionary: *A Compilation of IEEE Standard Computer Glossaries*. New York, NY ISBN 1-55937-079-3 (1990)
6. J. Sedlacek, *Theory of Reliability of Complex Mechanical Systems*. CTU Publishing Center, Prague, (1982)
7. J. T. Yang, *An Outline of Scientific Writing*. World Scientific Publishing, London, (1995)
8. RCM II, Reliability Centered Maintenance, Second edition, page 250-260, the role of Actuarial analysis in Reliability (2008)
9. Hoang Pham (editor). *Handbook of Reliability Engineering*. Springer, New Jersey. p 696, (2003)
10. J. Stodola, *The basic of Reliability Design*. Lerning text. University of Defence in Brno, (2017)
11. J. Stodola, *Operational Reliability of Automotive I: Theoretical part. Lerning text.* Military academy in Brno, (1984)
12. J. H. Saleh, K. Marais, *Highlights from the Early History of Reliability Engineering*. Reliability Engineering and System Safety, Volume 91, Issue 2, Pages 249–256 (2006)
13. D. J. Smith, *Reliability Maintainability and Risk Practical Methods for Engineers Including Reliability Centred Maintenance and Safety* (2011)
14. J. Janko, *Statistical Tables*. Publishing house of the Czechoslovak Academy of Sciences, Prague, (1958)
15. Z. Karpisek, *Matematics: Statistics and Probability*. Academic publishing house CERM Ltd. Brno, (2003)
16. JFederal Aviation Administration *System Safety Handbook* (PDF). U.S. Department of Transportation. Retrieved 2 June 2013
17. MIL-HDBK-338B Electronic Reliability Design Handbook, U.S. Department of Defense, New York, (1998)
18. S. Distefano, A. Puliafito: *Dependability Evaluation with Dynamic Reliability Block Diagrams and Dynamic Fault Trees*. IEEE Trans. Dependable Sec. Comput. 6(1): 4–17 (2009)
19. R. Denney, *Succeeding with Use Cases: Working Smart to Deliver Quality*. Addison-Wesley Professional Publishing, (2005)

An Improved Method of Geomagnetic Aided Inertial Navigation Algorithm with Gyro and Accelerometer Error Corrected Online

Song ZhongGuo[1], Gao Jiuxiang[1], Zhang Jinsheng[2], and Xi Xiaoli[1]

[1]*Faculty of Automation and Information Engineering, Xi'an University of Technology, Xi'an, China*
[2]*Xi'an research inst of hi-tech, Xi'an, China*

Abstract. In consideration of the problem that traditional geomagnetic aided navigation method cannot reduce the scaling error of indication track in inertial navigation system (INS), which will further limit the error correction precision of gyro and accelerometer, an improved geomagnetic matching algorithm based on affine transformation is proposed in this paper. A geomagnetic matching algorithm led to the optimal affine transformation solution by Procrustes analysis is presented and develops latitude and longitude reference information. Then a 13-dimensional-state extended Kalman filter which estimates the attitude misalignment angles, the position error, the velocity error, the Gyro drift, and accelerometer error is introduced to continuously update the output of INS and remove the accumulative error. The results show that geomagnetic aided navigation based on improved algorithm has better location accuracy and correction accuracy of INS than the traditional method.

1 Introduction

The orientation and velocity error of pure Inertial Navigation System (INS) will accumulate and result in divergence over time due to the existence of gyro drift and accelerometer deviation, which will further affect the positioning precision. Geomagnetic aided navigation system (GANS) utilizes the matching algorithm to compare geomagnetic field profile acquired real-time with the reference geomagnetic map stored on board for matching. The matching results will be utilized to correct gyro drift and accelerometer error of INS. GANS has characteristics of strong autonomous, good concealment and working consistently over earth surface without reliance on external radio navigation signals. Besides, the navigation errors of GANS don't cumulate over time [1]. Due to these advantages, GANS has essential application value in underwater vehicle, unmanned aerial vehicle and indoor pedestrian navigation in recent years [2].

In the 1960s, American E-System Corporation proposed contour-based magnetic matching MAGCOM system [1]. Then magnetic navigation technology has a rapidly developed in various countries from then on. The simulation platform of GANS was established and validated the iterated closest contour point (ICCP) matching algorithm better than traditional contour matching algorithm [3]. Simulation of geomagnetic/inertial integrated navigation System was presented by Zhou [4], which validated a new geomagnetic matching algorithm based on MSD algorithm and ICCP algorithm is effective. The SITAN algorithm makes use of terrain elevation information in GANS and achieves reliable results [5]. Literature [6] presents a locating method of geomagnetic/inertial integrated navigation system to improve the accuracy of initial geomagnetic matching position.

In order to solve the problem that ICCP has a high requirement no scaling on the INS indication track and enhance the accumulative error compensation accuracy and onboard performance of INS, this paper designed an information fusion filter with the combination of the improved ICCP described in [7]. The paper is organized as follows. Section 2 addresses the principle of GANS, Section 3 is devoted to key technologies of GANS, Section 4 presents the simulation and the results. Conclusions are given in Section 5.

2 Principle of GANS

The process of the GANS is shown in Figure 1. The navigation processor is mainly composed of matching algorithm and information fusion filter. The geomagnetic reference maps along the track are stored in the vehicle firstly. There are two primary inputs, one is the geomagnetic intensity processed from three-axis fluxgate magnetometers, and the other is the indication track collected from INS. During the vehicle cruise along the track, magnetometers are detecting background geomagnetic field to form the real-time magnetic sequence, which will be compared with the stored map to determine the difference and position information about the vehicle with matching algorithm. Then EKF estimates error parameters to correct INS and continuously update navigation information.

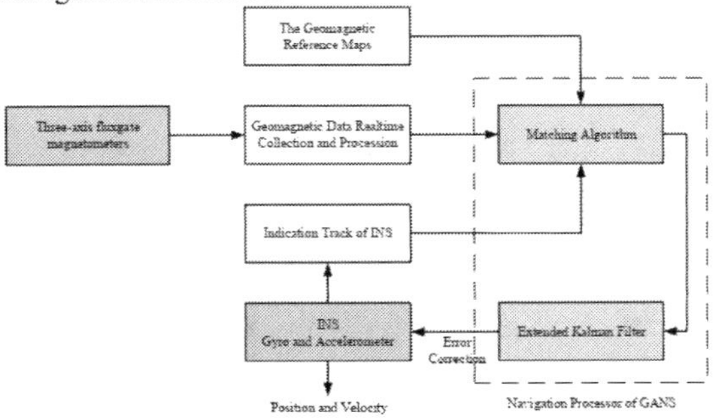

Figure 1. The process of the GANS

3 The Key Technology of GANS

3.1 Matching algorithm

The iterated closest contour point (ICCP) algorithm is a kind of matching algorithm based on contour constraint. It was derived from the iterative closest point (ICP) algorithm of image registration and often used in terrain matching and gravity matching navigation [8].

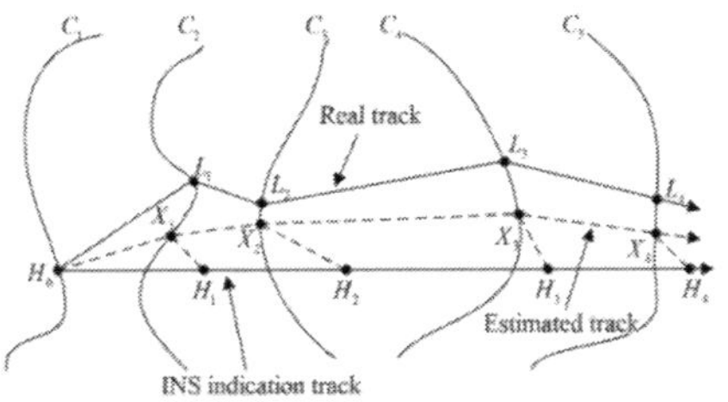

Figure 2. Illustration of INS indication track, estimated track, and real track.

The principle of ICCP algorithm is shown in Figure 2. Let us denote the INS indication track by $\{H_i\}(i=1,2,\ldots,N)$ (N is the number of points for one matching), which is deviant from the real track $\{L_i\}(i=1,2,\ldots,N)$ due to measurement errors and drifts. The magnetometer provides the corresponding measured geomagnetic intensity $\{g_i\}$ $(i=1,2,\ldots,N)$, and the corresponding geomagnetic field contour $\{C_i\}$ $(i=1,2,\ldots,N)$ can be searched out through geomagnetic values g_i from the pre-stored geomagnetic map. After which, the estimated track $\{X_i\}$ $(i=1,2,\ldots,N)$ can be obtained from C_i, which is the matching target. The navigational errors can be corrected by a rigid transformation of the indicated track into the real track. The rigid transformation T minimizes the distance between the sets $\{H_i\}$ and $\{X_i\}$.

$$Min\ E_j = \sum_{i=1}^{N} dis_j\left(X_i - X_{i-1}, H_i - H_{i-1}\right) + T\sum_{i=1}^{N} dis_j\left(X_i, C_i\right) \tag{1}$$

where $dis(X,H)=\|X-H\|^2$. The extraction of the closest contour point is discussed in [8]. The minimization process of E_j is iterated so that in the j-th iteration the set of measured points is $H^j = TH^{j-1}$ and the set of points X^j is determined from the new points H^j. The iteration process is continued until T becomes negligible.

The rigid transformation T consists of the rotation matrix R and the translation vector t as shown in (2), which can be solved with the quaternion method.

$$H_{i+1}=RH_i+t=\begin{pmatrix}\cos\theta & -\sin\theta\\ \sin\theta & \cos\theta\end{pmatrix}H_i+\begin{pmatrix}t_x\\ t_y\end{pmatrix} \tag{2}$$

where θ refers to the rotation angle. t_x and t_y are the translation.

The affine transformation T_A is used in this paper to solve the problem that ICCP with rigid transformation cannot reduce the scaling error in INS indication track. Defining scale factor S, the matching process of the indication track H_i to the real track L_i of each iterative can be performed with (3).

$$L_i = SRH_i + t \tag{3}$$

The least squares solution of this transformation can be given by solving the Procrustes problem [9]. The cost function can be written as:

$$\begin{aligned} f(S,R,t)_i &= \sum_{i=1}^{n}\left\|L_i-\left(SRH_i+t\right)\right\|^2 \\ &= n\left\|L_0-\left(SRH_0+t\right)\right\|^2+\sum_{i=1}^{n}\left\|L_i'-\left(SRH_i'+t\right)\right\|^2 \end{aligned} \tag{4}$$

where L_0 and H_0 denote the centroids of L_i and H_i, L_i' and H_i' is the relative vectors relative to their centroids. Then normalize L_i and H_i

$$L_{s,i} = L_i\Big/\sqrt{\sum L_0^2},\quad H_{s,i} = H_i\Big/\sqrt{\sum H_0^2} \tag{5}$$

Define the matrix $A = L_s H_s^T$, where L_s = [$L_{s,1}$ $L_{s,2}\ldots L_{s,n}$] and H_s = [$H_{s,1}$ $H_{s,2}\ldots H_{s,n}$]. The least squares estimate of the rotation matrix R is (6) when A is nonsingular.

$$R = A\left(A^T A\right)^{-1/2} \tag{6}$$

Regarding the singular value decomposition of A as $U\Lambda V^T$, the optimal rotation matrix R can be expressed as:

$$R = V\begin{bmatrix} 1 & 0 & 0 \\ 0 & 1 & 0 \\ 0 & 0 & \det(U)\det(V) \end{bmatrix} U^T \quad (7)$$

The scale factor S is

$$S = tr(\Lambda)\cdot\sqrt{\sum L_0^2} \Big/ \sqrt{\sum H_0^2} \quad (8)$$

Then minimize the first term contribution in Equation (4) by

$$t = L_0 - SRH_0 \quad (9)$$

3.2 Information fusion

According to positioning requirements, the system variables are defined as follows:

$$X(t)=[\phi_E,\phi_N,\phi_U,\delta V_E,\delta V_N,\delta L, \delta\lambda,\varepsilon_E,\varepsilon_N,\varepsilon_U,\nabla_E,\nabla_N,\nabla_U]^T \quad (10)$$

where ϕ_E,ϕ_N,ϕ_U is attitude misalignment angles in the direction, $\delta V_E,\delta V_N$ is velocity error in the east and north, $\delta L,\delta\lambda$ is the error of latitude and longitude, $\varepsilon_E,\varepsilon_N,\varepsilon_U$ is the drifts of the three gyroscopes, $\nabla_E,\nabla_N,\nabla_U$ is the drifts of the three accelerometers.

System state equation is followed as:

$$\dot{X}(t)=F(t)X(t)+G(t)W(t) \quad (11)$$

where $F(t)$ is System state matrix, $G(t)$ is System noise matrix and $W(t)$ is white noise in the gyros and accelerometers.

Measure equation is shown as:

$$z(t)=h(t)x(t)+v(t) \quad (12)$$

$z(t)$ is the differences of latitude and longitude between the INS and matching output. $h(t)$ is a zero matrix with dimension 2×13, except that $h(1, 6) = 1$ and $h(2, 7) = 1$. $v(t)$ is observation noise matrix, which is white noise.

4 The INS Error Propagation Model

The INS error propagation model actually used is more complex than the state equation in (11), and can be expressed as follows, with omitting the time variable t.

$$\dot{\boldsymbol{x}} = \boldsymbol{F}_{INS}\boldsymbol{x} + \boldsymbol{w} \quad (13)$$

$$\boldsymbol{x} = \begin{bmatrix} \boldsymbol{\varphi} & \boldsymbol{\delta v}^n & \boldsymbol{\delta p} & \boldsymbol{\varepsilon}^b & \nabla^b & \boldsymbol{\delta K}_g & \boldsymbol{\delta K}_a \end{bmatrix}^T \quad (14)$$

where $\boldsymbol{\varphi}=[\varphi_E \;\; \varphi_N \;\; \varphi_U]^T$ is platform misalignment angles; $\boldsymbol{\delta v}^n=[\delta v_E^n \;\; \delta v_N^n \;\; \delta v_U^n]^T$ is velocity errors; $\boldsymbol{\delta p}=[\delta L \;\; \delta\lambda \;\; \delta h]^T$ represents latitude, longitude and altitude errors respectively; $\boldsymbol{\varepsilon}^b=[\varepsilon_x^b \;\; \varepsilon_y^b \;\; \varepsilon_z^b]^T$ is the gyro drift errors expressed in body frame; $\nabla^b=[\nabla_x^b \;\; \nabla_y^b \;\; \nabla_z^b]^T$ is the accelerometer biases expressed in body frame;

$$\boldsymbol{\delta K}_g = [\delta k_{gxx} \;\; \delta k_{gyx} \;\; \delta k_{gzx} \;\; \delta k_{gxy} \;\; \delta k_{gyy} \;\; \delta k_{gzy} \;\; \delta k_{gxz} \;\; \delta k_{gyz} \;\; \delta k_{gzz}]^T$$

$$\boldsymbol{\delta K}_a = [\delta k_{axx} \;\; \delta k_{ayx} \;\; \delta k_{azx} \;\; \delta k_{ayy} \;\; \delta k_{azy} \;\; \delta k_{azz}]^T$$

and δk_{gii}, $i=x,y,z$ are gyro scale factor errors; $\delta k_{gij},(i,j=x,y,z,\ i\neq j)$ are gyro actual axis misalignment angles with respect to ideal body frame axis; δk_{aii}, $(i=x,y,z)$ are accelerometer scale factor errors; $\delta k_{aij},(i,j=x,y,z,\ i\neq j)$ are accelerometer actual axis misalignment angles with respect to ideal body frame axis. The components $\boldsymbol{\delta K}_g$, $\boldsymbol{\delta K}_a$ are all assumed to be constant vectors. $\boldsymbol{w}$ are white Gaussian noise sources including the first-order Gauss-Markov bias process for the accelerometer and gyro.

The $\boldsymbol{F}_{INS}$ in (13) is expressed as:

$$\boldsymbol{F}_{INS} = \begin{bmatrix} M_{aa} & M_{av} & M_{ap} & -C_b^n & 0_{3\times3} & M_{ag} & 0_{3\times6} \\ M_{va} & M_{vv} & M_{vp} & 0_{3\times3} & C_b^n & 0_{3\times9} & M_{vf} \\ 0_{3\times3} & M_{pv} & M_{pp} & 0_{3\times3} & 0_{3\times3} & 0_{3\times9} & 0_{3\times6} \\ & & & 0_{21\times30} & & & \end{bmatrix} \quad (15)$$

Assume the gyro sensed angular rate and accelerometer sensed specific force are $\boldsymbol{\omega}_{ib}^b=[\omega_{ibx}^b \;\; \omega_{iby}^b \;\; \omega_{ibz}^b]$ and $\boldsymbol{f}_{sf}^b=[f_{sfx}^b \;\; f_{sfy}^b \;\; f_{sfz}^b]$, then the components in (15) are given as below:

$$\boldsymbol{M}_{aa} = (-\boldsymbol{\omega}_{in}^n\times) \quad (16)$$

$$\boldsymbol{M}_{av} = \begin{bmatrix} 0 & -1/R_{Mh} & 0 \\ 1/R_{Nh} & 0 & 0 \\ \tan L/R_{Nh} & 0 & 0 \end{bmatrix} \quad (17)$$

$$\boldsymbol{M}_{ap} = \begin{bmatrix} 0 & 0 & v_N^n/R_{Mh}^2 \\ -\omega_{ie}\sin L & 0 & -v_E^n/R_{Nh}^2 \\ v_E^n\sec^2 L/R_{Nh} + \omega_{ie}\cos L & 0 & -v_E^n\tan L/R_{Nh}^2 \end{bmatrix} \quad (18)$$

$$\boldsymbol{M}_{ag} = -\begin{bmatrix} \omega_{ibx}^b\boldsymbol{C}_b^n & \omega_{iby}^b\boldsymbol{C}_b^n & \omega_{ibz}^b\boldsymbol{C}_b^n \end{bmatrix} \quad (19)$$

$$\boldsymbol{M}_{va} = (\boldsymbol{f}_{sf}^n\times) \quad (20)$$

$$\boldsymbol{M}_{vv} = (\boldsymbol{v}^n\times)\boldsymbol{M}_{av} - ((2\boldsymbol{\omega}_{ie}^n + \boldsymbol{\omega}_{en}^n)\times) \quad (21)$$

$$M_{vp} = \left(v^n \times\right) \begin{bmatrix} 0 & 0 & v_N^n / R_{Mh}^2 \\ -2\omega_{ie} \sin L & 0 & -v_E^n / R_{Nh}^2 \\ v_E^n \sec^2 L / R_{Nh} + 2\omega_{ie} \cos L & 0 & -v_E^n \tan L / R_{Nh}^2 \end{bmatrix} + \begin{bmatrix} 0 & 0 & 0 \\ 0 & 0 & 0 \\ -g_0 \beta_1 \sin L \cos L & 0 & \beta_3 \end{bmatrix} \tag{22}$$

$$M_{vf} = \left[f_{sfx}^b C_b^n \quad f_{sfy}^b C_b^n(:,2) \quad f_{sfy}^b C_b^n(:,3) \quad f_{sfz}^b C_b^n(:,3) \right] \text{ with } C_b^n(:,i) = i^{\text{th}} \text{column of } C_b^n \tag{23}$$

$$M_{pv} = \begin{bmatrix} 0 & 1/R_{Mh} & 0 \\ \sec L / R_{Nh} & 0 & 0 \\ 0 & 0 & 1 \end{bmatrix} \tag{24}$$

$$M_{pp} = \begin{bmatrix} 0 & 0 & -v_N^n / R_{Mh}^2 \\ v_E^n \sec L \tan L / R_{Nh} & 0 & -v_E^n \sec L / R_{Nh}^2 \\ 0 & 0 & 0 \end{bmatrix} \tag{25}$$

The " $\times$ " in $\omega_{in}^n \times$, $f_{sf}^n \times$, $v^n \times$, etc. is the skew-symmetric operator of the vector, and where $f_{sf}^n = C_b^n f_{sf}^b$, $\omega_{nb}^b = \omega_{ib}^b - {C_b^n}^T \omega_{in}^n$, and $\omega_{in}^n = \omega_{ie}^n + \omega_{en}^n$,with

$$\begin{cases} \omega_{ie}^n = \begin{bmatrix} 0 & \omega_{ie} \cos L & \omega_{ie} \sin L \end{bmatrix}^T \\ \omega_{en}^n = \begin{bmatrix} -\dfrac{v_N^n}{R_{Mh}} & \dfrac{v_E^n}{R_{Nh}} & \dfrac{v_E^n}{R_{Nh}} \tan L \end{bmatrix}^T \end{cases} \tag{26}$$

$$\begin{cases} R_{Mh} = R_M + h, \; R_{Nh} = R_N + h \\ R_M = \dfrac{R_N (1-e^2)}{(1-e^2 \sin^2 L)}, \; R_N = \dfrac{R_e}{(1-e^2 \sin^2 L)^{1/2}} \\ e = \sqrt{2f - f^2} \end{cases} \tag{27}$$

$$\begin{cases} g^n = \begin{bmatrix} 0 & 0 & -g \end{bmatrix}^T \\ g = g_0 (1 + \beta_1 \sin^2 L + \beta_2 \sin^4 L) - \beta_3 h, \\ \text{with} \begin{cases} \beta_1 = 5.27094 \times 10^{-3} \\ \beta_2 = 2.32718 \times 10^{-5} \\ \beta_3 = 2 g_0 / R_e = 3.086 \times 10^{-6} (1/s^2) \end{cases} \end{cases} \tag{28}$$

C_b^n is transformation direct cosine matrix from body frame to navigation frame; $v^n = \begin{bmatrix} v_E^n & v_N^n & v_U^n \end{bmatrix}^T$ is velocity along east, north and up-vertical direction; $p = \begin{bmatrix} L & \lambda & h \end{bmatrix}^T$, where L, λ, h is latitude, longitude and altitude; $R_e = 6371.2km$ is the Earth's semi-major axis; $f = 1/298.257$ is the Earth's flattening; $\omega_{ie} = 7.2921151467? 10^5$ rad/s is the Earth's angular rate; $g_0 = 9.7803267714 m/s^2$ is gravity magnitude at the equatorial sea-surface.

5 Simulation

The effectiveness of the proposed matching algorithm and data fusion algorithm can be examined by this numerical simulation. The reference matching maps are generated with Enhanced Magnetic Model of order 720 and disturbed by Gaussian noise. The simulated magnetic measurement data are acquired based on the geomagnetic model above and disturbed by certain level external interferences, as shown in Table 1.

According to the simulation results in figure 3, the traditional ICCP algorithm cannot effectively rectify the indication track to the real track, and also cannot get the correct matching results when there is a scaling error between the INS indication track and the real track. However, the improved algorithm achieves accurate matching results. At the same time, the improved ICCP algorithm can also effectively estimated gyro and accelerometer error as shown in Figure 4. The improved algorithm not only solves the scaling problem in the indication track during the geomagnetic matching process, but also accurately correct gyro drift and accelerometer error of INS.

Table 1. The parameters of simulation.

Parameter	Parameter Values
Matching points	50
Reference map noise	10 nT
Matching region size	10 km × 8 km
Center location of matching region	(108.4°E, 34.35°N)
Average altitude	380m
Gird size of matching region	100 × 80
Measurement Gaussian noise	10 nT
Matching Tolerance	10m
Platform misalignment angles	1'/axis
Initial velocity error	10m/s,30m/s
Initial position error	-800m,1000m
Gyro constant bias	0.1deg/h
Gyro random walk	0.1deg/sqrt(h)
Accelerometer constant bias	50μg
Accelerometer random walk	5μg/sqrt(Hz)
Scale factor error	10ppm
Askew installation error	10′

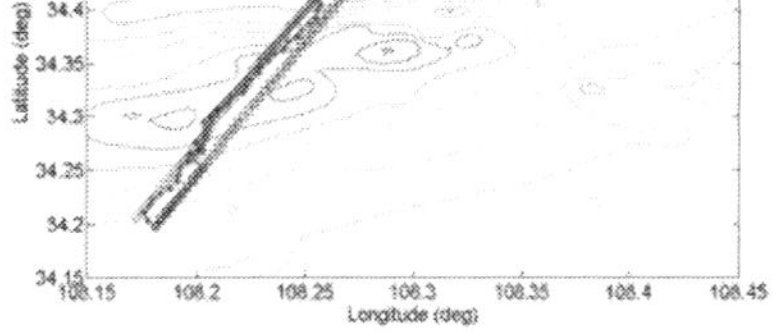

(a)

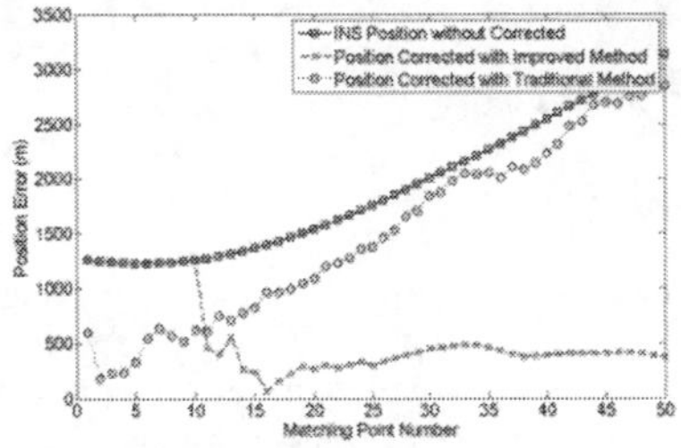

(b)

Figure 3. The comparison of the track and position error with proposed method and traditional method. The first ten points are not matched for collecting geomagnetic data.

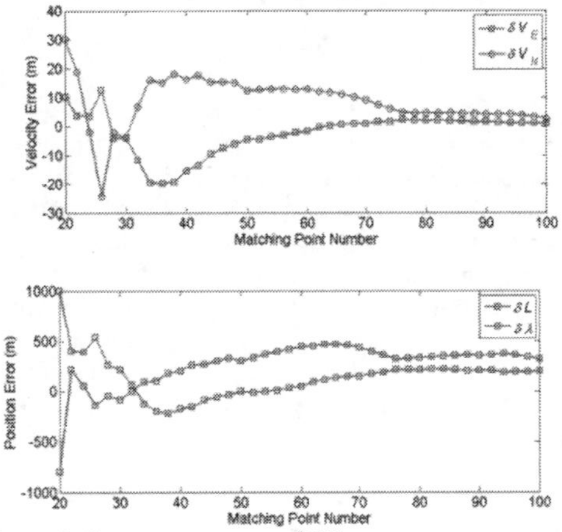

Figure 4. Convergence curve of INS

5 Conclusion

This paper proposed an improved method of geomagnetic aided inertial navigation algorithm with gyro and accelerometer error corrected, the affine transformation was adopted to ICCP to increase adaptability with indication track error characteristics of INS, and then the gyro drift and accelerometer deviation of INS were accurate estimated with EKF. The best achievable positioning accuracy was less than 300 m and velocity was less than 2 m/s in case of existing scale error.

Acknowledgment

This work was supported in part by the National Natural Science Foundation of China (51707152), in part by the China Postdoctoral Science Foundation funded project (2014M562522XB), in part by the Shaanxi Province Natural Science Foundation of China (2016JQ4023), in part by the Scientific Research Program Funded by Shaanxi Provincial Education Department (15JK1523), and in part by the Scientific Research Project Foundation of Xi'an University of Technology (2016TS024).

References

1. F. Goldenberg, 2006 IEEE/ION Position, Location, And Navigation Symposium 14, 7 (2016)
2. H. Xin, X. Zhi, X. Jianxin and X. Limin, 2017 29th Chinese Control And Decision Conference (CCDC), (2016)
3. F. Haonan Feng, Y. Zhaohua, F. Jiancheng, 2008 2nd International Symposium on Systems and Control in Aerospace and Astronautics (2008)
4. Z. Yingying, Z. Yanshun and G. Lei, 2011 Third International Conference on Measuring Technology and Mechatronics Automation (2011)
5. Z. Dai and C. Kang, The 2014 2nd International Conference on Systems and Informatics (2014)
6. Y. Zhang, C. Li, S. Liu, M. Jiang and L. Guo, Proceedings of 2014 IEEE Chinese Guidance, Navigation and Control Conference (2014)
7. S. Zhongguo, Z. Jinsheng, Z. Wenqi, X. Xiaoli, Sensors, 7 (2016)
8. L. Byung-Uk, K. Chul-Min, P. Rae-Hong, IEEE Transactions on Pattern Analysis and Machine Intelligence, 22 (2000)
9. K. P.Behzad, K. P. Behrooz, Proceedings of SPIE-The International Society for Optical Engineering (1999)
10. L. Dorst, IEEE Trans. Pattern Anal. Mach. Intell, 27 (2005)

Simulation of flow field in steam ejector and analysis of shock wave characteristics

Yi Cui

Shanghai Aircraft Design and Research Institute, Commercial Aircraft Corporation of China Ltd, 5188 Jinke Road, Shanghai, China

Abstract: Fluent software is used to simulate the two-dimensional flow field in the steam ejector. The generation, development and end of shock in the flow field are analyzed. The influence of the shock wave characteristics on the ejector performance is discussed. The results show that the working back pressure in the steam ejector and the throat diameter of the nozzle all affect the position and intensity of the shock wave. When the shock wave ends prematurely or belatedly, it all affect the performance of the injector.

1 Introduction

Although the structure of the steam ejector is simple, the supersonic mixing process of the gas inside the equipment is very complicated, and there are extremely complicated flow phenomena such as turbulence, entrainment, mixing, and shock waves. Wang et al. analyzed the influence of suction pressure on the pumping performance of the steam jet pump and found that the performance of the jet pump is closely related to the pressure of the pumped gas. The simulation results show that the ejector coefficient increases with the pressure of the pumped gas [1]. Xinzhuang Wu etc. analyzed the influence of working back pressure on the ejector performance and found that critical back pressure exists at the ejector outlet. Simulation results show that when the working back pressure changes, the internal shock position of the ejector also changes. It has a certain influence on the performance of the injector, but it does not do a deep research on the impact of the shock wave position on the ejection coefficient [2]. Shah etc. analyzed the effect of nozzle outlet location and mixing section structure on ejector performance. The results show that different angles of contraction can result in a larger difference in ejector ejection coefficient [3].Wu etc. analyzed the length of the mixing section and the influence of the shrinkage angle. The simulation results show that the length and the contraction angle of the mixing section have optimal values and the entrainment ratio is the largest [4]. However, the impact of the position of the shock wave in the injector on the performance of the injector has not been analyzed. The steam ejector working fluid generates a series of shock waves by generating supersonic fluid through the Laval nozzle, and position of the shock wave greatly influence the performance of the ejector. Research on the shock wave characteristics of steam ejectors has important theoretical significance and engineering value for improving the entrainment ratio of steam ejectors.

In this paper, a two-dimensional axisymmetric model of the steam ejector was established, and we simulated the complex flow inside the injector. On this basis, the effect of the back pressure of the injector and the throat diameter of the nozzle on the ejection performance was analyzed and the effects of complex flow conditions such as shock waves inside the steam ejector on the performance of the jet pump were discussed in detail.

2 The working principle of the steam ejector

2.1 The working principle of the steam ejector

The steam ejector consists mainly of the nozzle, the suction chamber, the mixing chamber, the ejector throat and diffuser, as shown in Figure 1. Its working principle is that the high pressure working steam adiabatically expands through the nozzle, converts the pressure energy into kinetic energy, and ejects it at a high speed. The pressure is reduced from p to p_1, and the velocity is increased from u to u_1 (supersonic). A local low pressure appears in the nozzle and the ejected steam is sucked. Driven by the pressure difference, the ejected steam is drawn into the suction chamber. Under the action of the viscous force, the expanding core of the working steam introduces the injected steam into the mixing chamber of the jet pump and accelerates it, and the two airflows perform energy exchange in the mixing. The working steam at the rear of the chamber is completely mixed with the ejected steam to reach the same speed u_3, and the pressure also rises from p_2 to p_3. Then, the mixed fluid generates a positive shock at a section of the throat and produces a compression effect on the mixed steam. The speed of the mixed fluid is suddenly reduced from the supersonic speed to u_3 (subsonic speed), and the pressure value also has a step increase from p_3 to p_4. Finally, the mixed gas exits from the injector. A multi-stage injection pump is made up of a series multiple injectors. The pumped gas is compressed step by step at the injector. At the last level (i.e. atmospheric level), the pressure is higher than the atmospheric pressure and is discharged from the jet pump [5].

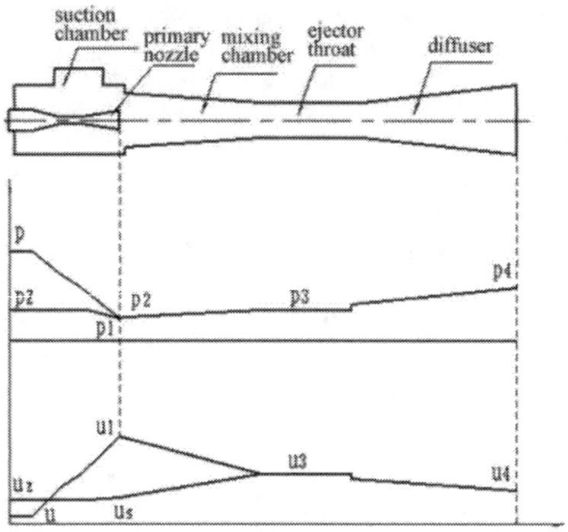

Figure 1. The working principle of the steam ejector

2.2 Shock effect in steam ejectors

In the steam ejector, the shock effect has a significant effect on the working performance of the ejector. Steam does the compressional expansion in the nozzle, forming a lower pressure than the ejected fluid at the outlet of the nozzle. Therefore, after the supersonic air flows out of the nozzle, it will be subjected to high back pressure compression to form an oblique shock wave. Because the oblique shock wave at the outlet of the nozzle can normally intersect at the center, and the compression wave is formed again, the air flow is compressed twice, so the Mach number is reduced, and the pressure rises to a level higher than the surrounding medium. As a result, the gas flow generates an expansion wave at the periphery of the boundary between the compression wave and the boundary layer. and then the expansion wave intersects again to form the expansion wave. After the air flow expands twice, the velocity increases and the pressure decreases, usually lower than the ambient pressure [6]. As a result, in the course of its continued flow, it must be oppressed by the surrounding medium, and the compression wave is formed again on the boundary layer. The wave is reflected by the boundary layer and multiple reflections form a very regular diamond grid pattern [5], as shown in Figure 2. Between the flow area of the working steam and the flow area of the periphery of the ejected steam, there is a layer with a large velocity gradient, namely a shock-mixing layer. The high-speed working steam transmits its kinetic energy to the ejected steam through the viscous-carrying effect of the shock-mixing layer to increase its speed, and the ejected steam passes through the shock mixing layer to trap the working steam, making the working steam decline gradually. At the end of the pump chamber, the velocity of the two- strand fluid is close or even common.

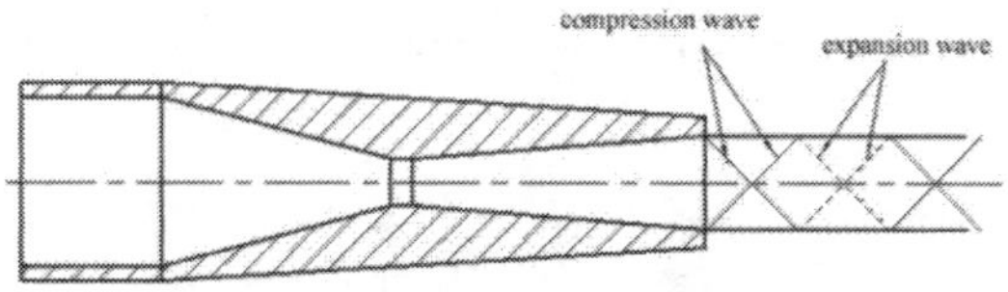

Figure 2. Supersonic ejector outlet shock effect

The flow inside the steam injection pump is very complicated, including the subsonic, sonic, and supersonic flow conditions. It is generally difficult to observe through experiments, but the flow parameters can be obtained through numerical simulation of the flow field.

3 CFD Model and Internal Flow Simulation

3.1 Computational model and boundary conditions

Injector are two-dimensional axisymmetric structures, except that the suction inlet is often non-axisymmetrically arranged so that the flow of suction fluid into the injector is three-dimensional. Considering that the gas flow velocity in the inlet region of the ejection fluid is very small compared with the flow rate of the working gas at the outlet of the nozzle, so the lateral inlet of the ejection fluid can be simplified into an axial annular inlet. Thus, the three-dimensional structure of the injector is simplified into a two-dimensional axisymmetric model. The mesh is properly encrypted at the nozzle, and grids are refined for areas with large velocity gradients and pressure gradients[7,8].

In this paper, the inlet steam and ejector steam are both considered saturated and are assumed as ideal gas. The calculation uses pressure inlet and outlet boundary conditions. Turbulence model uses Realizable k-ε model. The wall surface of the injector is set to a non-slip condition, and the near wall surface treatment mode is set to a standard wall surface function. In the calculation, all residual items are below 10^{-6}, and the mass flow is stable[8,9].

3.2 Internal flow field simulation

In the simulation, the working steam pressure is p_p=0.2Mpa, the ejected steam pressure is p_h=0.02Mpa, and the working back pressure is p_c=0.035Mpa.

The Mach cloud diagram of the flow field inside the ejector is shown in Figure 4. It can be seen from Figure 4 that the working fluid forms a supersonic high-speed jet after it is ejected through the nozzle. There is a velocity gradient between the high-speed jet and the low-velocity jet and momentum exchange occurs. The result of the momentum transformation accelerates the flow of the low-speed ejector fluid, decelerates the high-speed jet, and fully develops the fluid mixture in the mixing chamber. The two fluids become more and more uniform in speed, completing the mixing process. The velocity of the mixed fluid in the diffuser chamber is further reduced, the dynamic pressure is converted into static pressure, and the pressure is increased to the back pressure requirement.

Figures 4 and 5 are the Mach numbers and pressure distribution along the axis of the fluid inside the ejector. It can be seen from the figure that after the high-pressure working fluid passes through the nozzle, the speed rises sharply along the axis. At the outlet of the nozzle, the Maher number is over 2.5, the supersonic speed is reached, the pressure is rapidly reduced, the local low pressure area is formed, sucks the injection fluid to flow forward. The intermediate high-speed fluid collide with the two sides of the low-speed fluid, mix and transmit energy, and the two form the speed change rule of the change. Finally, the speed tends to be consistent. After the working fluid passes through the nozzle, the pressure decreases and the speed changes, resulting in an oblique shock wave. Thus, it was verified that the supersonic jet of the nozzle undergoes the compression wave, and the expansion wave repeatedly compresses and expands the shock wave characteristics. In the diffusion section of the ejector, the mixed fluid is disturbed to produce a positive shock, resulting in a sharp drop in the velocity and the pressure rising to the back pressure.

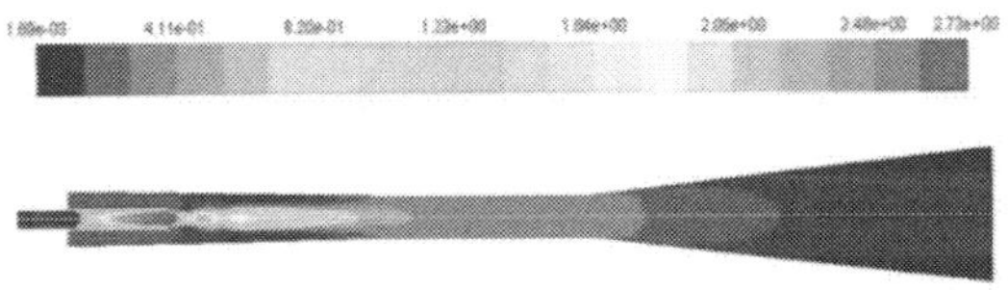

Figure 3. Mach number internal cloud of ejector

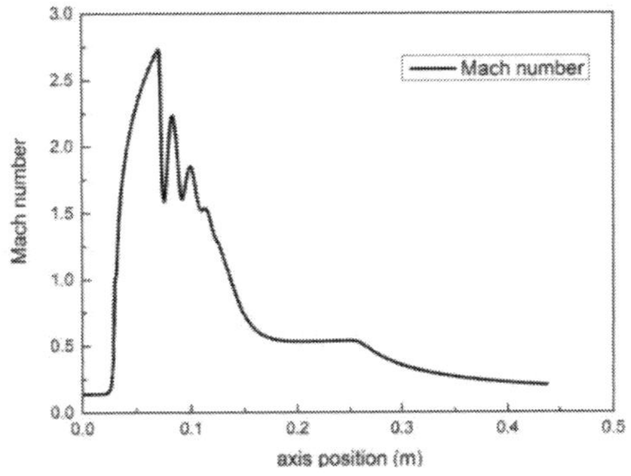

Figure 4. Mach number distribution along the axis

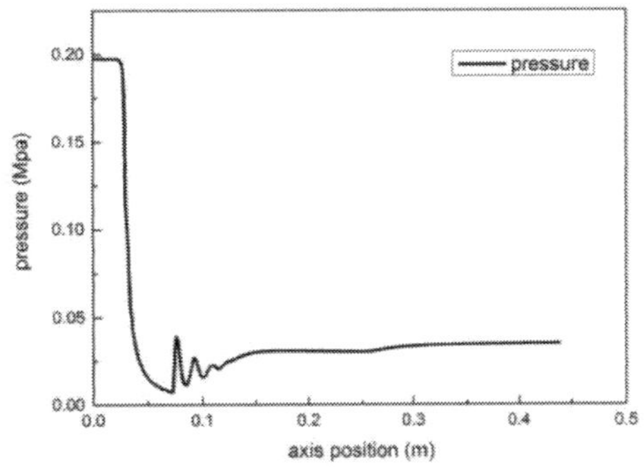

Figure 5. Pressure distribution along the axis

4 Analysis of shock wave characteristics in steam ejectors

4.1 Influence of back pressure on shock wave characteristics

In the simulation process, the working steam is ensured, and the inlet pressure of the stimulating steam does not change and the working back pressure is changed. Work back pressure p_c varies between 0.02Mpa and 0.04Mpa.

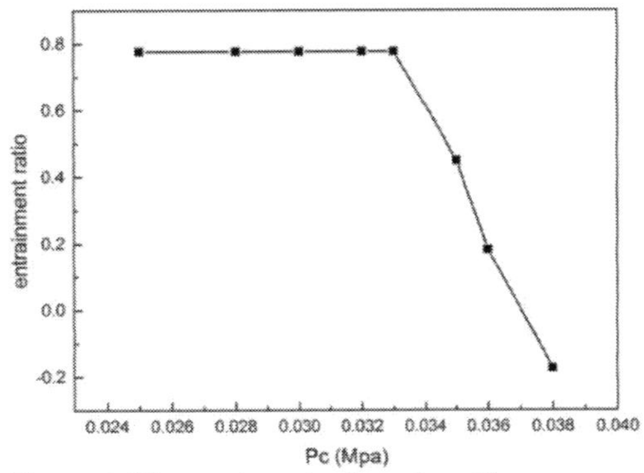

Figure 6. The entrainment ratio varies with p_c

Figure 6 shows the variation of the entrainment ratio u with the back pressure p_c. Obviously seen from the figure. entrainment ratio basically remains unchanged in the initial stage, when the back pressure is greater than a certain value, entrainment ratio drops abruptly, so the critical back pressure p_{cr} exists in the steam ejector. When $p_c< p_{cr}$, entrainment ratio basically remains unchanged, at this time the steam ejector is in a stable working state; when $p_c> p_{cr}$, entrainment ratio decreases rapidly with p_c, and even reflux occurs. At this point, the working performance of the injector gradually deteriorated.

Figure 7 presents the Mach cloud under different mixed vapor pressures p_c. It can be clearly seen from the cloud image that a double shock wave phenomenon has appeared in the steam ejector. The first shock wave occurs near the nozzle exit section. The second shock wave appears in the throat section of the steam ejector or in the inlet section of the diffuser chamber. This position is closely related to the total energy of the mixed fluid. When the fluid energy is high, the second shock wave moves toward the entrance of the diffusion chamber, and the injector can work at a higher critical back pressure.

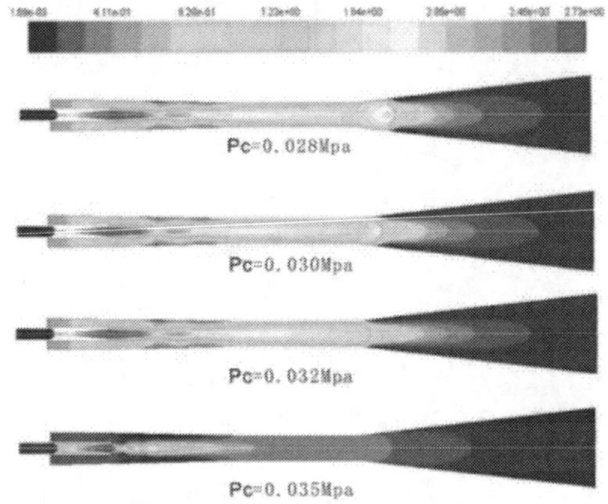

Figure 7. Maher number cloud under different p_c

Figure 8 presents the distribution of Mach numbers along the axis under different p_c. It can be clearly seen from the figure that when $p_c<p_{cr}$, the change of p_c has less influence on the first shock wave and can be ignored, but it has a greater impact on the second shock wave. When the back pressure is vary from 0.028Mpa to 0.032Mpa, the shock wave characteristics appear at the inlet of the diffuser chamber, which is reflected in the compression shock characteristic of pressure step increase and speed step decrease. Steam velocity decreases from supersonic to subsonic. The working fluid and the ejecting fluid are choked at the injector's main nozzle and the diffuser inlet so that the flow of working fluid and ejector fluid does not change with the outlet pressure. Therefore, entrainment ratio remains unchanged, and the position of the second shock wave increases as p_c moves toward the upstream of the diffuser. When $p_c > p_{cr}$, the mixing action of the fluid causes the position of the second shock wave to move toward the mixing chamber. The shock wave at this time interferes with the weakening or even elimination of the jamming phenomenon of the effective mixing of the two fluids at the inlet of the diffuser chamber. As a result, a small amount of ejected fluid is sucked into the ejector and the entrainment ratio decreases. When the p_c continues to rise, and reaches or exceeds the back pressure, the second shock wave moves toward the nozzle. At this point, the working fluid is disturbed by it, this no longer produces swell wave even produces reflux phenomenon.

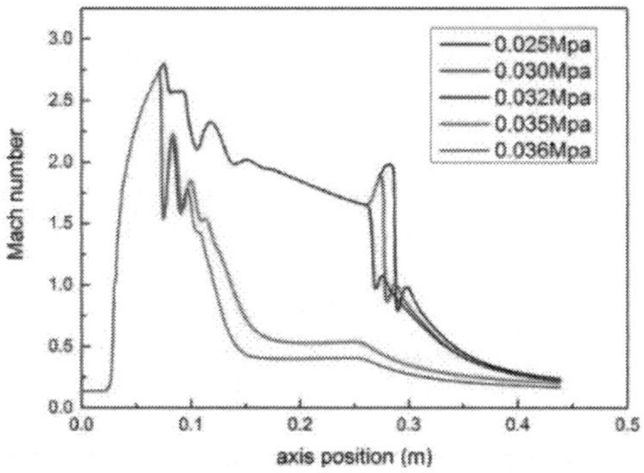

Figure 8. Distribution of Mach numbers along the axis under different p_c

4.2 Effect of throat diameter on shock wave characteristics

The working fluid reaches the supersonic velocity at the outlet of the nozzle, and the Mach number at the nozzle outlet cross section is determined by the ratio of the nozzle outlet sectional area to the throat sectional area. The Mach number at the outlet of the nozzle is also a very important parameter reflecting the working capability of the injector model. Therefore, changing the diameter of the nozzle throat, examining the influence of the diameter of the throat on the shock characteristics and the influence on the ejection coefficient is particularly important.

In the simulation, the working steam pressure is p_p =0.2Mpa, the ejected steam pressure is p_h =0.02Mpa, and the working back pressure is p_c =0.035Mpa. Keep the diameter of the nozzle outlet constant, adjust the diameter of the nozzle throat d, and observe the effect of changes in the structure of the injector on the nozzle outlet flow field. Figure 9 shows the relationship between the entrainment ratio and the diameter of the throat of the nozzle. The diameter of the throat of the nozzle has a great influence on the entrainment ratio. For a specific design condition, there is an optimal nozzle throat diameter.

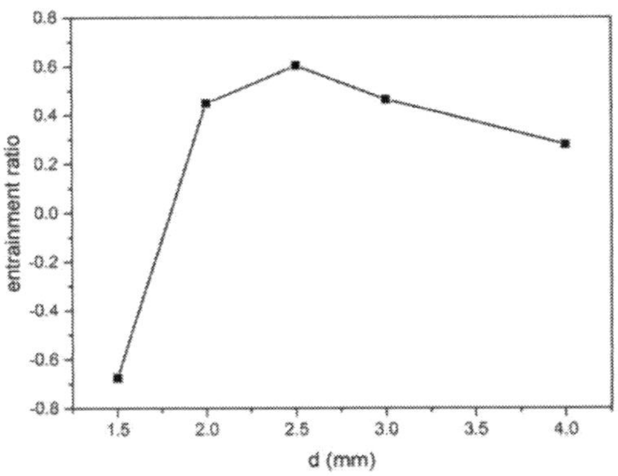

Figure 9. Distribution of Mach numbers along the axis under different p_c

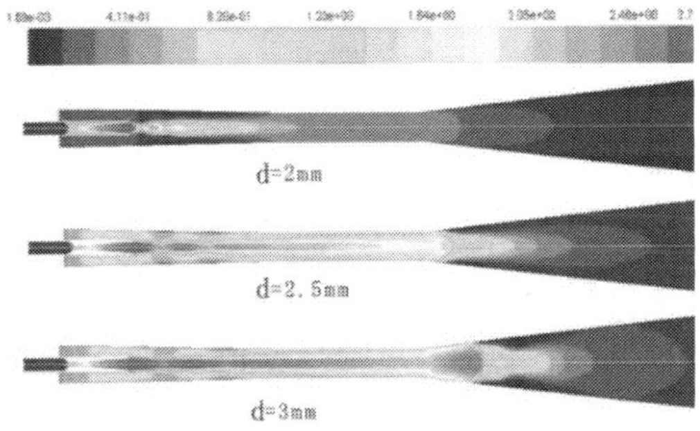

Figure 10. Mach Clouds with different diameters

Figure 10 shows the Mach number in the throat diameter of different nozzles. It can be clearly seen from the figure, the difference in the internal flow field of the different throat diameter injectors. The shock waves generated mainly by the supersonic fluid at the outlet of the nozzle are different and the distribution of the flow field in the diffusion chamber is different. When the nozzle throat diameter is small, two kinds of fluid in the diffuser entrance choking phenomenon is very weak or there is no second shock wave, resulting in a lower entrainment ratio and even backflow. When the nozzle throat diameter is large, the two fluids are not completely mixed in the mixing chamber, the speed and pressure have strong fluctuations, resulting in greater loss of mechanical energy. By observing the shape and length of the shock wave, it can be seen that when the shock wave is generated at the outlet of the nozzle and extends to the vicinity of the outlet of the mixing section, the entrainment ratio reaches the highest, which is the best state of the ejector performance. When the shock wave ends prematurely or belatedly , it affects the performance of the injector.

5 Conclusion

In this paper, numerically simulates the complex flow of the flow field inside the injector. By changing the working back pressure of the steam ejector and the diameter of the throat of the nozzle, the internal shock characteristics of the steam ejector are analyzed and the influence of shock variation on the overall content of the ejector is studied. Get the following conclusions:

Verify that the steam ejector working fluid generates supersonic fluid through the laval nozzle, and generate a series of shock trains inside the injector. The location of the shock wave generation and termination has a great influence on the ejector performance.

There is a critical back pressure inside the steam ejector. When the working back pressure is lower than the critical back pressure, the entrainment ratio remains unchanged, and the working fluid and the ejecting fluid generate shock waves at the main nozzle of the injector and the entrance of the diffuser chamber, causing a jamming phenomenon. With the increase of the back pressure, the position of the second shock moves toward the upstream of the diffuser. When the working back pressure is higher than the critical back pressure, the position of the second shock wave moves to the mixing chamber as the back pressure increases. The jamming phenomenon at the inlet of the diffuser is weakened or even relieved, the entrainment ratio is drastically reduced, and the ejector even returns.

The Mach number at the nozzle outlet cross section of the steam ejector has a large relationship with the nozzle throat diameter, and the nozzle throat diameter has an optimal value. When the diameter of the throat of the nozzle is small, the congestion at the inlet of the diffuser chamber is very weak and there is no second shock wave, and the entrainment ratio is low. When the throat diameter of the nozzle is larger, the energy loss is higher and the entrainment ratio is low. When the nozzle diameter is optimal, the shock wave is generated by the nozzle outlet and ends when it reaches near the outlet of the mixing section. The ejection entrainment ratio the maximum.

References

1. X. D. Wang, J. L. Dong, J. Y. Tu. Numerical analysis of the effect of suction pressure on pumping performance of steam injectionpump [J]. *Journal of Vacuum Science and Technology*, **30**, 695-698 (2010).
2. X. Z. Wu, S. Xia, J. Qiu. CFD Analysis of Internal flow Field and Variable Conditions of Supersonic Jet[J]. *Machinery Design & Manufacture*, **09**, 49-51 (2014).
3. Shah A, Chughtai I R, Inayat M H. Experimental and numerical investigation of the effect of mixing section length on direct-contact condensation in steam jet pump[J]. *International Journal of H*eat & Mass Transfer, **72**, 430-439 (2014).
4. H. Wu, Z. Liu, B. Han, et al. Numerical investigation of the influences of mixing chamber geometries on steam ejector performance[J]. *Desalination*, **353**, 15-20 (2014).
5. Oishi T, Watanabe T, Undagawa Y. Mixing Characteristics for a Supersonic Ejector[J]. *ASME-FED*, High Speed Jet Flow, **214**, 157-162 (1995).

6. Y. Yang, Y. Q. Li, S. Q. Shen, P. Li, et al. Shock wave effects in steam ejectors[J]. *Journal of Engineering Thermophysics*, **35**, 1419-1423 (2014).
7. J. Yao. CFD numerical simulation of steam injectors and process state analysis [D]. North China Electric Power University (Beijing), 2009.
8. H. J. Li, S. Q. Shen, B. Zhang. Numerical simulation of two-dimensional flow field in steam ejector[J]. *Thermal Science and Technology*, **04**, 353-357 (2004).
9. D. A. Da. Vacuum Design Manual [M]. National Defense Industry Press, 2004.

Unsteady aerodynamics modeling using Volterra series expansed by basis function

Senlin Chen[1], Zhenghong Gao [1]

[1]*School of Aeronautics, Northwestern Polytechnical University, Xi'an 710072, China*

Abstract. Unsteady aerodynamics modeling must accurately describe nonlinear aerodynamic characteristics in addition to unsteady aerodynamic characteristics. The Volterra series has attracted increasing attention as a powerful tool for nonlinear system modeling. It is essential to incorporate the influence of the second-order Volterra kernel or higher-order kernels to build a nonlinear unsteady aerodynamics model. The main difficulty in the identification of higher-order kernels is that the number of parameters to be identified increases exponentially with the order of a kernel. This paper expands the Volterra kernels with the four-order B-spline wavelet on the interval as the basis function, converts the problem into the solution of low-dimensional equations, and obtains a stable solution. A nonlinear unsteady aerodynamics model is built by identifying the second-order and third-order kernels of the lift, drag, and pitching moment coefficients of the NACA0012 airfoil. Then the model is verified at different reduced frequencies using CFD.

1 Introduction

Aerodynamics exhibit evident nonlinear unsteady hysteresis characteristics when a complex flow of shock waves, flow separation, vortices, etc. occurs in a flow field [1]. At this point, aerodynamics depend not only on the current exercise state but also on their time history [2], and the conventional aerodynamic derivative modeling method becomes difficult to apply [3]. Thus, the study of unsteady aerodynamic modeling methods is required.

Several methods have been proposed for unsteady aerodynamic modeling [4-6]. The Volterra series [7] method is a mathematical model that describes the nonlinear time-invariant system proposed by Italian mathematician Volterra. There have been a few studies on aerodynamic modeling in recent years [8-10]. The Volterra series method is a series model whose ability to describe the nonlinear characteristics of a system increases with the number of kernels involved. Most existing studies only use the first-order Volterra kernel or the truncated second-order Volterra kernel [10]. Hence, such methods are only suitable for aerodynamic modeling with linear or weak nonlinearities and are difficult to apply for strongly nonlinear aerodynamic modeling. Second-order or even higher-order kernels are rarely used because the identification of higher-order kernels is difficult. For example, second-order kernel identification based on an impulse response is extremely complicated and time consuming [9]. As the number of parameters to be identified increases exponentially with kernel order, the introduction of higher-order kernels leads to the so-called dimensionality disaster, which results from solving high-dimensional ill-posed equations.

Regarding the difficulties in Volterra kernel identification, Silva [11] and Balajewicz [12] proposed a truncation method that preserves the main diagonal of a high-order kernel or its nearby elements. As this approach neglects a large number of coupling factors in the Volterra kernel, it reduces the ability of the Volterra series to express nonlinearities. The other method is the basis function method [13, 14], which uses a set of basis functions to expand the Volterra kernel, thereby transforming the problem into the estimation of limited expansion coefficients.

In this paper, the Volterra kernel is expanded using the B-spline wavelet on the interval (BSWI) as a basis function. The identification of Volterra kernels is converted into a small number of wavelet coefficients by utilizing its good approximation characteristics. This considerably reduces the number of parameters to be identified. Finally, the ability of the Volterra series to describe nonlinear unsteady aerodynamics is verified by predicting the lift coefficient, drag coefficient, and pitching moment coefficient of the NACA0012 airfoil in plunging motion at transonic speed.

2 Volterra series

The Volterra series theory [15] shows that the input output relation of any continuous-time, causal, time-invariant, fading memory, and nonlinear system can be formulated as an infinite sum of multidimensional convolution integrals.

$$y(t)=y_0+\sum_{n=1}^{\infty}\int_0^t\int_0^t L\int_0^t h_n(\tau_1 L\ ,\tau_n)\bullet \prod_{k=1}^{n}u(t-\tau_k)\mathrm{d}\tau_1 L\ \mathrm{d}\tau_n \tag{1}$$

where $y(t)$ is the response of a nonlinear system to input $u(t)$, y_0 is the steady-state response, $h_1(\tau)$ is the first-order Volterra kernel and the linear unit impulse response, and $h_n(\tau_1,\tau_2 L\ ,\tau_n)$ is the nth-order Volterra kernel. Without loss of generality, higher-order kernels $h_n(n\geq 2)$ are supposed to be symmetric, namely, $h_n(\tau_1,\tau_2 L\ ,\tau_n)=\ h_n(\tau_{\pi(1)},\tau_{\pi(2)} L\ ,\tau_{\pi(n)})$, where $\pi(1)$, $\pi(2) L\ ,\pi(n)$ are all the permutations of integers 1 through n [15].

Silva [11] showed that for a linear system, higher-order kernels are zero and the system is completely characterized by the first-order kernel. For a nonlinear system, the first-order kernel reflects the linear part of the system and higher-order kernels represent the nonlinear part of the system. Even though the Volterra series is the sum of infinite terms, the nonlinear influence for an actual physical system decays rapidly with the increase in kernel order. Thus, the first few terms can typically represent the system quite accurately [16]. This study is limited to the first three order kernels; hence, Eq. (1) reduces to

$$y(t)=y_0+y_1(t)+y_2(t)+y_3(t) \tag{2}$$

where

$$\begin{aligned} y_1(t)&=\int_0^t h_1(\tau)u(t-\tau)\mathrm{d}\tau \\ y_2(t)&=\int_0^t\int_0^t h_2(\tau_1,\tau_2)u(t-\tau_1)u(t-\tau_2)\mathrm{d}\tau_1\mathrm{d}\tau_2 \\ y_3(t)&=\int_0^t\int_0^t\int_0^t h_3(\tau_1,\tau_2,\tau_3)u(t-\tau_1)u(t-\tau_2)\times \\ &\quad u(t-\tau_3)\mathrm{d}\tau_1\mathrm{d}\tau_2\mathrm{d}\tau_3 \end{aligned} \tag{3}$$

where $y_1(t),y_2(t),y_3(t)$ are the responses of the first-order kernel, second-order kernel, and third-order kernel, respectively.

It should be noted that this identification problem is ill-posed in that the objective is to determine the structure of the system from input and output measurements [17]. Therefore, to obtain stable kernel estimation, a regularization method must be used to solve the least square problem. In this study, we use truncated singular value decomposition.

3 Volterra kernel identification based on BSWI

3.1 B-spline wavelet on the interval (BSWI)

On a bounded interval, the BSWI basis function has the good characteristics of semi-orthogonality, symmetry, and smoothness [18]. The BSWI function is $\phi_{m,k}^{j}(t)\ m,k,j\in ¢$, where $¢$ is the set of all integers, m is the order of the B-spline, and j is the dilation index that determines the scale or resolution of the function. $k\in[-m+1,2^j-1]$ is the translation index that provides the position of the function. The four-order BSWI $\phi_{4,k}^{3}(t)$, where $j=3$, is shown in Figure 1. Concrete expressions are provided in [18].

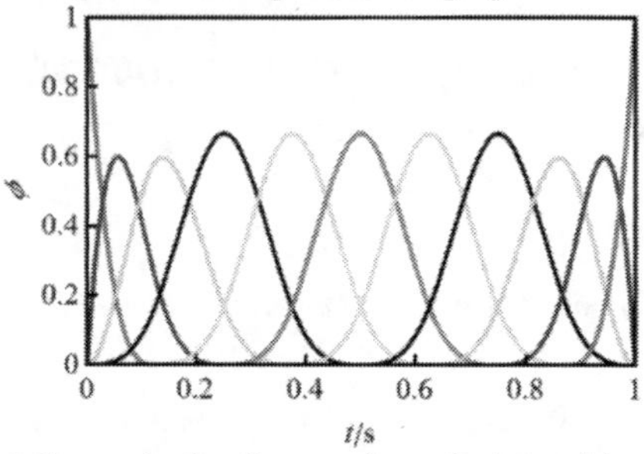

Figure 1. Four-order B-spline wavelet on the interval ($j=3$)

The basis function in the two-dimensional case is the tensor product of two one-dimensional basis functions

$$\Phi_{m,(k_1,k_2)}^{j}(x,y)=\phi_{m,k_1}^{j}(x)\phi_{m,k_2}^{j}(y) \tag{4}$$

where $\Phi_{j,(k,m)}^{(s,v)}(x,y)$ is the two-dimensional basis function. Similarly, the basis function in the three-dimensional case is the tensor product of three one-dimensional basis functions, the complete expression for which is provided in [13].

3.2 Volterra kernel identification

The first-order kernel part of Eq. (3) is discretized for the identification of the first-order Volterra kernel. The response at $t_n=2^{-j}n(n=1,2 L\ ,N)$ is

$$y_{1,j}(t_n)=\int_0^{t_n}h_1(\tau)u(t_n-\tau)\mathrm{d}\tau \tag{5}$$

where N is the number of sample data.

The first-order kernel can be represented as a linear combination of basis functions

$$h_{1,j_1}(\tau)=\sum_{k=1-m}^{2^{j_1}-1}\alpha_{m,k}^{j_1}\phi_{m,k}^{j_1}(\tau) \tag{6}$$

Substituting Eq. (6) into Eq. (5), the following simplified matrix form can be obtained:

$$\boldsymbol{y}_{1,j}=\boldsymbol{M}_{\alpha_{j_1}}\boldsymbol{\alpha}_{j_1} \tag{7}$$

where $\boldsymbol{\alpha}_{j_1}$ is a vector of $\alpha_{m,k}^{j_1}$ in Eq. (6), namely, the coefficient of the basis function that represents the first-order kernel. $\boldsymbol{M}_{\alpha_{j_1}}$ is a coefficient matrix of $\boldsymbol{\alpha}_{j_1}$; it is composed of $\phi_{m,k}^{j_1}(\tau)$ and $u(t_n-\tau)$ in Eq. (5). Eq. (7) is solved for $\boldsymbol{\alpha}_{j_1}$ to obtain h_1 from Eq. (6).

The second-order kernel part of Eq. (3) is discretized for the identification of the second-order Volterra kernel. The response at t_n is

$$y_{2,j}(t_n)=\int_0^{t_n}\int_0^{t_n}h_2(\tau_1,\tau_2)u(t_n-\tau_1)\bullet u(t_n-\tau_2)\mathrm{d}\tau_1\mathrm{d}\tau_2 \tag{8}$$

The second-order kernel can be represented as a linear combination of two-dimensional basis functions.

$$h_{2,j_2}(\tau_1,\tau_2)=\sum_{k_1,k_2=1-m}^{2^{j_2}-1}\alpha^{j_2}_{m,(k_1,k_2)}\phi^{j_2}_{m,k_1}(\tau_1)\phi^{j_2}_{m,k_2}(\tau_2) \quad (9)$$

Substituting Eq. (9) into Eq. (8), the following simplified matrix form can be obtained:

$$\boldsymbol{y}_{2,j}=\boldsymbol{M}_{\alpha_{j_2}}\boldsymbol{\alpha}_{j_2} \quad (10)$$

where $\boldsymbol{\alpha}_{j_2}$ is a vector of $\alpha^{j_2}_{m,(k_1,k_2)}$ in Eq. (9), namely, the coefficient of the basis function that represents the second-order kernel. $\boldsymbol{M}_{\alpha_{j_2}}$ is a coefficient matrix of $\boldsymbol{\alpha}_{j_2}$; it is composed of $\phi^{j_2}_{m,k_1}(\tau_1)\phi^{j_2}_{m,k_2}(\tau_2)$ and $u(t_n-\tau_1)$ $u(t_n-\tau_2)$ in Eq. (8). Eq. (10) is solved for $\boldsymbol{\alpha}_{j_2}$ to obtain h_2 from Eq. (9).

The identification of the third-order kernel is similar to that of the second-order kernel. Hence, it is not repeated here.

4 Aerodynamics modeling with Volterra series

Considering the plunge motion of the NACA0012 airfoil at transonic speed as an example, this study verified the ability of the Volterra series to predict nonlinear unsteady aerodynamics and the effectiveness of BSWI basis functions for Volterra kernel identification.

4.1 Input

Input selection and design is extremely important in system identification, and it directly determines the accuracy of parameter identification. In Volterra kernel identification, first-order kernel identification is a linear system identification problem and there are several input signals such as impulse [11] and sweep frequency [19]. Higher-order kernel identification is a nonlinear system identification problem, and currently, there is no universal input signal. This study used an orthogonal phase-optimized multisine input designed by Morelli[20].

$$u=\sum_{n=1}^{M}A_n\sin\left(\frac{2\pi nt}{T}+\phi_n\right) \quad (11)$$

where u is the input, A_n is the amplitude, T is the time length of input excitation, ϕ_n is the phase shift, and M is the total number of available harmonic frequencies. The first step is to select the time period, T, for the input, which determines the smallest harmonic frequency resolution, $\Delta f=1/T$ Hz, and the limit on the minimum frequency, $f_{\min}\geq 2/T$. The second step is select the frequency band, $[f_{\min},f_{\max}]$ Hz, for the input. The total number of frequencies is $M=\text{fix}\left[(f_{\max}-f_{\min})/\Delta f\right]+1$, where "fix" indicates rounding to the nearest integer toward zero. The third step is select the input power spectrum by specifying and the amplitude of the input to determine the amplitude, A_n, of each component. The last step is to assign the starting values for phase angles, $\phi_n\in[-\pi,\pi]$, and then apply an optimization algorithm to adjust ϕ_n to achieve the minimum relative peak factor (RPF) for the input.

$$\text{RPF}(u)=\frac{\left[\max(u)-\min(u)\right]/2}{\sqrt{2(u^Tu)/N}} \quad (12)$$

where N is the number of sample data. The orthogonal phase-optimized multisine input is a sum of selected harmonic sinusoids, and it is orthogonal. This input can reflect the coupling of different frequencies, and it is suitable for the identification of nonlinear systems [14].

When the airfoil is in plunge motion, only the angle of attack changes. Considering the angle of attack as an independent variable, the relationship between the angle of attack and displacement is

$$s(t)=-V\int_0^t\tan\left[u(\tau)\right]\mathrm{d}\tau \quad (13)$$

where V is the freestream velocity, u is the angle of attack, and s is the displacement of the airfoil. The output used for identification is calculated through CFD. The conditions are, $M_\infty=0.8$ and $Re=7.2\times10^6$, and a rigid dynamic grid is used.

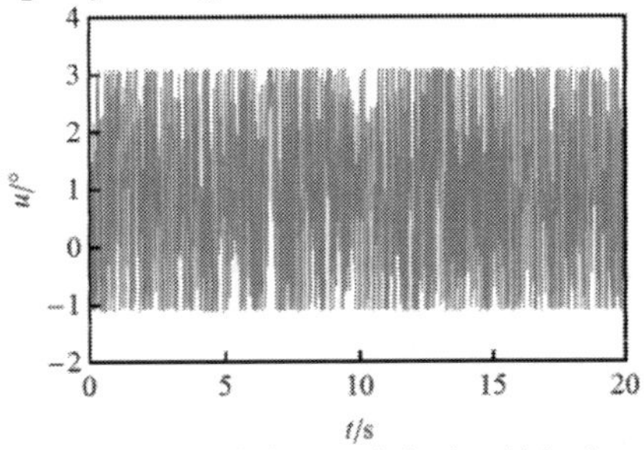

Figure 2. Orthogonal phase-optimized multisine input

4.2 Volterra kernel identification of aerodynamics

The orthogonal phase-optimized multisine input is considered, and then the outputs with different amplitude inputs are used to separate the responses generated by various order kernels to identify them separately. The initial angle of attack, $\alpha_0=1°$, is selected as the reference state, the input duration is $T=20$ s, the minimum and maximum frequencies are $f_{\min}=3/20$ Hz and $f_{\max}=30$ Hz respectively, and the amplitude of each component of the input is $A_n=2/\sqrt{M}$. The input can be seen in Figure 2. The first three order kernels identification results of the lift coefficient, drag coefficient, and pitching moment coefficient are shown in Figure 3-Figure 5, where the third-order kernels, $h_3(\tau_1,\tau_2,\tau_3)$, are plotted at $\tau_3=0.01$ s.

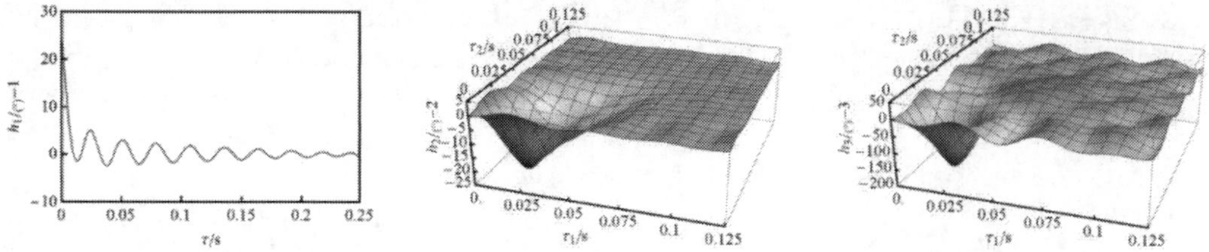

(a) $h_1(\tau)$ (b) $h_2(\tau_1,\tau_2)$ (c) $h_3(\tau_1,\tau_2,0.01)$

Figure 3. Volterra kernel of lift coefficient when angle of attack $\alpha_0 = 1°$

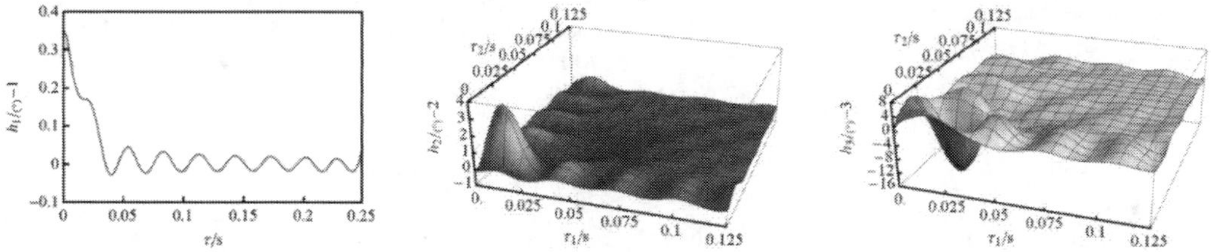

(a) $h_1(\tau)$ (b) $h_2(\tau_1,\tau_2)$ (c) $h_3(\tau_1,\tau_2,0.01)$

Figure 4. Volterra kernel of drag coefficient when angle of attack $\alpha_0 = 1°$

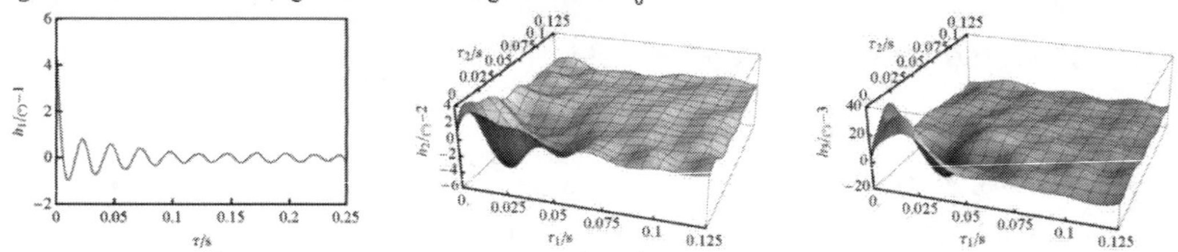

(a) $h_1(\tau)$ (b) $h_2(\tau_1,\tau_2)$ (c) $h_3(\tau_1,\tau_2,0.01)$

Figure 5. Volterra kernel of pitching moment coefficient when angle of attack $\alpha_0 = 1°$

4.3 Transonic aerodynamics validation

The aerodynamics model has been established as we have obtained the Volterra kernels. The next step is to validate the ability of the Volterra series to predict nonlinear unsteady aerodynamics. The displacement of the airfoil in plunge motion at the initial angle of attack, $\alpha_0 = 1°$, is

$$s(t) = 0.5236(\cos(23.612t)-1)$$
$$s(t) = 0.1309(\cos(94.448t)-1) \quad (14)$$

The corresponding reduced frequencies are 0.05 and 0.1.

The lift coefficient, drag coefficient, and pitching moment coefficient calculated by the Volterra series and CFD are shown in Figure 6 and Figure 7. When the reduced frequency is 0.05, the aerodynamics calculated by the first-order kernel reflect the basic features. However, there are clear errors. The system is nonlinear at this time. Prediction accuracy improves considerably with the introduction of the second-order kernel. Prediction results become extremely close to CFD results after the third-order kernel is introduced. When the reduced frequency is 0.1, the errors in aerodynamics calculated by the first-order kernel are reduced, and prediction accuracy is gradually improved after the introduction of higher-order kernels.

In summary, higher-order kernels characterize the nonlinearity of aerodynamics. To accurately predict aerodynamics, it is necessary to incorporate the influence of higher-order kernels based on the first-order kernel.

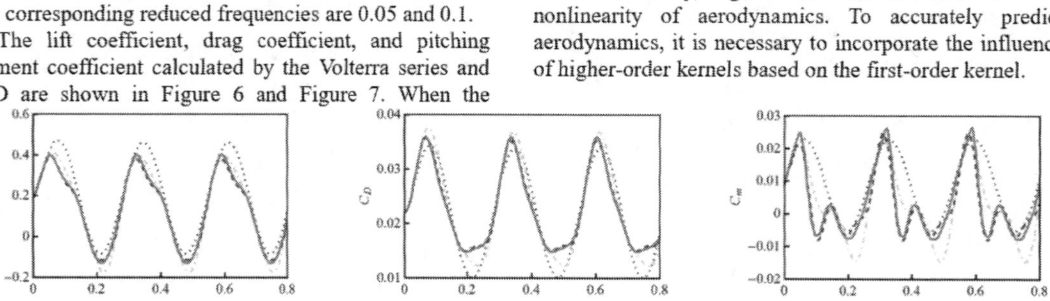

(a) C_L (b) C_D (c) C_m

Figure 6. Aerodynamics in plunge motion when angle of attack $\alpha_0 = 1°$ and reduced frequency $k = 0.05$

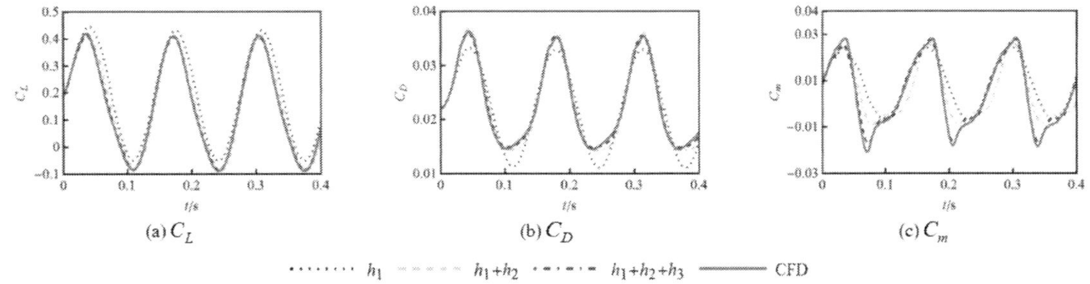

Figure 7. Aerodynamics in plunge motion when angle of attack $\alpha_0 = 1°$ and reduced frequency $k = 0.1$

5 Conclusion

1) In this study, the Volterra series is developed for nonlinear unsteady aerodynamics modeling. The first-order Volterra kernel can only describe the linear characteristics of the system. It is critical to incorporate the influence of higher-order kernels to accurately predict nonlinear aerodynamics.

2) Regarding the difficulties in Volterra kernel identification, the Volterra kernels are expanded with the BSWI as the basis function. The problem is transformed into the solution of low-dimensional equations, and a stable solution is obtained.

3) A nonlinear unsteady aerodynamics model is built by identifying the first three order kernels of the lift, drag, and pitching moment coefficients of the NACA0012 airfoil in plunging motion at transonic speed. The aerodynamics at different reduced frequencies computed by the Volterra series and CFD verify the ability of the Volterra series to predict nonlinear and unsteady aerodynamics.

The aerodynamics modeling in this study focused on the plunge motion. At this time, only the angle of attack changes, and this is a single variable situation. Actually, aerodynamics have multiple influencing factors such as pitch rate. The next step is to investigate the application of multivariate Volterra series methods in unsteady aerodynamics modeling.

References

1. M. Ghoreyshi, K. J. Badcock, A. D. Ronch, et al., J Aircr, *Framework for establishing limits of tabular aerodynamic models for flight dynamics analysis* **48**, 42-55 (2011)
2. M. Ghoreyshi, A. Jirasek, R. M. Cummings, AIAA J, *Computational investigation into the use of response functions for aerodynamic-load modeling* **50**, 1314-1327 (2012)
3. A. J. Mccracken, D. J. Kennett, K. J. Badcock, et al., AIAA Paper 2013-4978, *Assessment of tabular models using CFD* (2013)
4. M. Ghoreyshi, A. Jirasek, R. M. Cummings, Prog Aerosp Sci, *Reduced order unsteady aerodynamic modeling for stability and control analysis using computational fluid dynamics* **71**, 167-217 (2014)
5. A. D. Ronch, K. J. Badcock, A. Khrabrov, et al., AIAA Paper 2011-6524, *Modeling of unsteady aerodynamic loads* (2011)
6. D. J. Lucia, P. S. Beran, W. A. Silva, Prog Aerosp Sci, *Reduced-order modeling: New approaches for computational physics* **40**, 51-117 (2004)
7. S. Boyd, L. O. Chua, C. A. Desoer, Ima Journal of Mathematical Control and Information, *Analytical foundations of Volterra series* **1**, 243-282 (1984)
8. T. Skujins, C. Cesnik, AIAA Paper 2010-8127, *Reduced-order modeling of hypersonic vehicle unsteady aerodynamics* (2010)
9. W. A. Silva, AIAA Paper 1999-1262, *Reduced-order models based on linear and nonlinear aerodynamic impulse responses* (1999)
10. M. Balajewicz, E. Dowell, J Aircr, *Reduced-order modeling of flutter and limit-cycle oscillations using the sparse Volterra series* **49**, 1803-1812 (2012)
11. W. A. Silva, J Aircr, *Application of nonlinear systems theory to transonic unsteady aerodynamic responses* **30**, 660-668 (1993)
12. M. Balajewicz, F. Nitzsche, D. Feszty, AIAA Paper 2009-2319, *Reduced order modeling of nonlinear transonic aerodynamics using a pruned Volterra series* (2009)
13. C. M. Cheng, Z. K. Peng, W. M. Zhang, et al., Nonlinear Dyn, *Wavelet basis expansion-based Volterra kernel function identification through multilevel excitations* **76**, 985-999 (2014)
14. R. J. Prazenica, A. J. Kurdila, Nonlinear Dyn, *Multiwavelet constructions and Volterra kernel identification* **43**, 277-310 (2006)
15. M. Schetzen, *The Volterra and Wiener theories of nonlinear systems* (2006)
16. S. Boyd, L. O. Chua, IEEE Trans Circ Syst, *Fading memory and the problem of approximating nonlinear operators with Volterra series* **32**, 1150-1161 (1985)
17. M. Brenner, Y. Jiang, Y. S. Xu, Adv Comput Math, *Multiparameter regularization for Volterra kernel identification via multiscale collocation methods* **31**, 421-455 (2009)
18. J. W. Xiang, X. F. Chen, Z. J. He, et al., Comput Mech, *A new wavelet-based thin plate element using B-spline wavelet on the interval* **41**, 243-255 (2008)
19. L. Tronchin, J Audio Eng Soc, *The emulation of nonlinear time-invariant audio systems with memory by means of Volterra series* **60**, 984-996 (2012)
20. E. Morelli, S. Derry, M. Smith, AIAA Paper 2005-5921, *Aerodynamic parameter estimation for the X-43a (Hyper-X) from flight data* (2005)

Static analysis on spherical magnetically suspending rotor based on finite element

Wang Weijie , Ren Yuan , and Yin Zengyuan

Space Engineering University, Beijing 101416, China

Abstract. In this paper, the static mechanical properties of a spherical magnetically suspending rotor are studied based on the finite element analysis method. The structure of the spherical magnetically suspending rotor is designed. The characteristics of the spherical magnetically suspending rotor are analysed and the necessity of static analysis is pointed out. The main content and general flow of statics analysis are discussed. The two working modes (GUI mode and command flow mode based on APDL) of the finite element software ANSYS are compared and analysed. Then the finite element model of the spherical magnetically suspending rotor is established by the command flow method, and the detailed modelling steps are given. On this basis, the static characteristics of the spherical magnetically suspending rotor are simulated and analysed. The simulation results show that the maximum equivalent stress and rigid body displacement of the rotor are within the safety threshold at a design speed of 10000rpm.

1 Introduction

Compared to flywheels supported by mechanical bearings, magnetically suspending flywheel (MSFW) relies on electromagnetic force to support them. They have the advantages of no friction, low power consumption, high reliability, and long lifespan. They are ideal actuators for spacecraft attitude control [1][2][3]. The MSFW is a complex system integrating mechanical, electronic, electromagnetic, and automatic control. Its performance directly affects the overall performance of the attitude control system, and thus the quality, volume and power consumption of the flywheel, output torque magnitude, accuracy, and bandwidth are all strict requirements[4][5]. As the core component of the MSFW, the rotor is a key research object. The usual research method is to use finite element software ANSYS to analyze the mechanical properties of the rotor.

The literature [6] discusses the basic method and process of using ANSYS software to analyze the dynamic characteristics of AMB rotor system. The literature [7] performs the finite element modal analysis under the conditions of multiple groups of parameters for a magnetically suspending flexible rotor system, verifying the correctness of the finite element analysis. The literature [8] uses the Subspace method to calculate the first four natural frequencies and mode shapes, and compares them with experimental modal analysis results. The literature [9] uses the Block Lanczos method to extract the first 20 modes of vibration, displacement and natural frequency of the flywheel rotor. The literature [10] uses SolidWorks to establish a three-dimensional model of a magnetic bearing rotor and introduces it into ANSYS workbench for finite element analysis. The magnetically suspending rotors studied in the above documents are all conventional disc or column structures, and the finite element modeling methods are all GUI (Graphical User Interface) or directly imported from third-party software. The literature [11] optimizes the design of the magnetically suspending rotor to reduce its quality.

At present, the research on magnetic bearings [12][13][14] focuses more on how to optimize the configuration relationship between the permanent magnet magnetic circuit and the electromagnetic magnetic circuit to reduce the power consumption and the volume of the magnetic bearing. The rotors that are targeted are usually traditional disc or spoke structure. The literature [15] proposes a spherical MSFW rotor, which realizes full five-degree-of-freedom control of the MSFW rotor and full-channel decoupling among the magnetic bearings. The universal deflection of the rotor can be precisely controlled, but it has not been studied in depth to the static characteristics of a spherical magnetically suspended rotor (SMSR).

In view of the above mentioned research insufficiency, this paper studies the static characteristics of a spherical magnetically suspending rotor based on the finite element analysis method. The structure of the spherical magnetically suspending rotor is designed. The characteristics of the spherical magnetically suspending rotor are analyzed and the necessity of static analysis is pointed out. The main content and general flow of statics analysis are discussed. The two working modes (GUI mode and command flow mode based on APDL) of the

Foundation item: National Natural Science Foundation of China (No. 51605489)

finite element software ANSYS are compared and analyzed. Then the finite element model of the spherical magnetically suspending rotor is established by the command flow method, and the detailed modeling steps are given. On this basis, the static characteristics of the spherical magnetically suspending rotor are simulated and analyzed. The simulation results show that the maximum equivalent stress and rigid body displacement of the rotor are within the safety threshold at a design speed of 10000rpm.

2 SMSR Structure Design

The spherical magnetically suspending rotor is an inner rotor structure. The rotor system is mainly composed of a rotary table, a Lorentz force magnetic bearing rotor, a motor rotor, and an inductance displacement sensor detection disk. The main structure is shown in Figure 1.

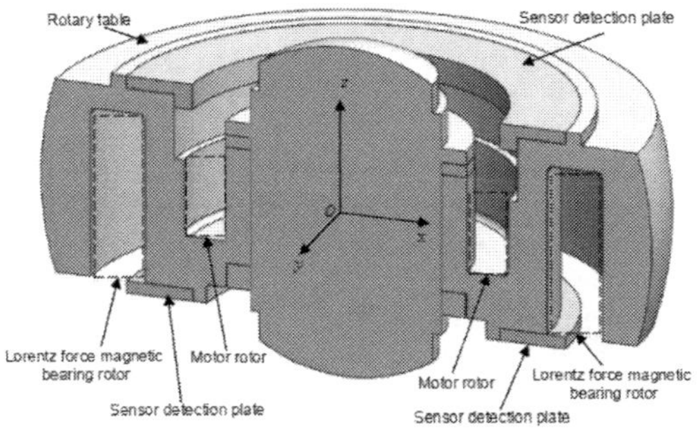

Figure 1. Rotor structure

The motor-driven rotor rotates to output the moment of inertia. The sensor detects the working surface of the disk symmetrically. The inductance displacement sensor is located on the axial outer side of the sensor detection disk and is used for the displacement of the sensitive rotor system along the axial direction. The radial spherical surface is used as the radial displacement of the rotor. The test reference is that four eddy current displacement sensors are arranged on the radial outer side. Compared with the ordinary MSFW rotor, the characteristics of the spherical magnetically suspending rotor are:

(1) From the configuration point of view, the rotor rotary table consists of two spheres with a common sphere centered in the direction of the rotor's polar axis and the radial direction of the equator plane. The two spheres form the envelope of the rotor system. The sphere constitutes the rotor portion of the axial reluctance magnetic bearing and the radial sphere constitutes the rotor portion of the radial reluctance magnetic bearing.

(2) From the material point of view, the entire rotor rotary as a component of the magnetic reluctance magnetic bearing rotor, Lorentz magnetic bearing rotor and motor rotor, the need to provide magnetic medium, so should be selected with high permeability, high saturation Magnetically dense soft magnetic material. At present, commonly used soft magnetic materials mainly include electrical pure iron, iron cobalt vanadium (IJ22) and iron-nickel soft magnetic alloys. Among them, electrical pure iron has advantages such as high saturation magnetic density, low coercive force, and high magnetic permeability, but it is easy to oxidize and rust. The rotor rotary serves as the main body of the rotor. Once rusted, the reliability of the flywheel cannot be guaranteed; In soft magnetic materials, 1J22 has the highest saturated magnetic density [16], which is up to 2.1T. However, the alloy is expensive, easy to oxidize, and has poor processing performance. It is a soft state hardness up to HRB90, and the rotor rotary spherical surface is used as a magnetic bearing. On the other hand, manual grinding is required to achieve the required accuracy. If 1J22 is selected, the difficulty and cost of the processing are greatly increased. In the iron-nickel soft magnetic alloy, 1J50 has high magnetic permeability and high saturation magnetization (1.56T). Anti-rust performance and processing performance are also good, so select 1J50 as the rotary table material.

(3) From a functional point of view, the rotary table as the main body of the rotor, not only to provide most of the rotor moment of inertia, but also as a structural component bearing Lorentz force magnetic bearing rotor, motor rotor and sensor detection disk and other components. Considering that the preliminary design of the motor, Lorentz force magnetic bearing, sensor, axial magnetic bearing, and radial magnetic bearing has been determined, the mechanical performance of the magnetically suspended rotor is more determined by the rotary table. Taking into account the influencing factors of the design speed, moment of inertia, and geometry of the flywheel rotor, the basic structure of the rotary table is finally determined as shown in Figure 2.

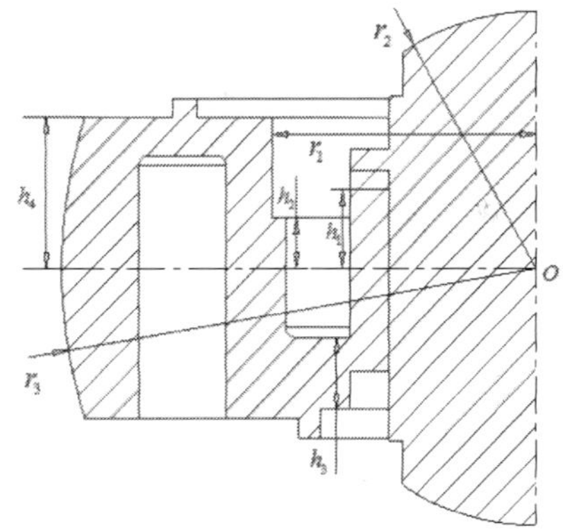

Figure 2. Parameters of rotary table

As the key part of the rotor, the rotary table structural and mechanical properties have the following effects on the flywheel system.

(1) In the high-speed rotation state, the moment of inertia of the rotary table determines the inertia moment of the MSFW. Under the rated speed 10000rpm, the angular momentum of the flywheel output depends on the polar moment of inertia.

(2) Rigid body displacement of the rotary table at high speed directly affects the reliability of the radial magnetic bearing. Spherical MSFW adopts inner rotor structure, especially for radial magnetic bearing, the gyration radius of the rotor is large, and deformation is more likely to occur when rotating at a high speed. Since the air gap

between the stator and the rotor is only 0.35mm, the displacement of the rigid body may cause the air gap to change, thereby increasing the difficulty of system control and threatening the safety of the flywheel. This requires the structural design to ensure that the maximum rigid body displacement of the wheel is small enough compared to the air gap within the working speed range of the flywheel. .

(3) As the magnetic medium for each magnetic bearing and motor rotor, the rotary table must ensure sufficient static redundancy for its static performance.

In summary, the core of the spherical magnetically suspending rotor is the rotary table, which is based on the special role of the rotary table in the spherical MSFW rotor system, it is necessary to carry out static analysis.

3 SMSR Finite Element Modeling

3.1. Static model

The so-called static force is the structure by the static load (fixed load) role, inertia and damping effects can be ignored, under the static load, the structure is in static equilibrium. Static analysis is to calculate the response of a structure under static loading, such as displacement, stress, strain, and reaction force. Static load is an ideal assumption, but in many cases, if the load cycle is much larger than the period of the structure's natural vibration, the inertia effect of the structure can be ignored. This situation can be simplified to static analysis. The static analysis of a SMSR mainly includes displacement (field) analysis, stress (field) analysis. The loads imposed on it mainly include externally applied forces and pressures, steady-state inertial forces, and displacement loads.

Static analysis includes linear analysis and nonlinear analysis [17]. Linear static analysis is the most basic but also the most widely used type of analysis for linear elastic materials, static loading conditions. For linear analysis, the material is linear, the stress-strain relationship is linear, the deformation is recoverable, and the small displacement, small strain, and small rotation of the structure occur, and the structural stiffness does not change due to deformation.

According to classical mechanics theory, the general equation of the dynamics of an object is[18]

$$[\boldsymbol{M}]\{x''\}+[\boldsymbol{C}]\{x'\}+[\boldsymbol{K}]\{x\}=\{F(t)\} \tag{1}$$

Where, $[\boldsymbol{M}]$ is a mass matrix; $[\boldsymbol{C}]$ is a damping matrix; $[\boldsymbol{K}]$ is a stiffness matrix; is a displacement vector; $\{x'\}$ is the velocity vector; $\{x''\}$ is the acceleration vector; $\{F(t)\}$ is the force vector.

In the static analysis, the time-related quantities will be ignored, so the above equation can be simplified as

$$[\boldsymbol{K}]\{x\}=\{F(t)\} \tag{2}$$

The general steps of static analysis are usually as follows [19].

(1) Modeling. First, the user should specify the working directory, file name, and analysis title, and then define the model geometry elements, cell types, real constants, material parameters, etc. in the preprocessing.

(2) Set up the solution control. Setting solution control includes defining analysis types, setting general analysis options, specifying load step options, and more. Users can set solution control options through the Solution Control dialog box.

(3) Apply a load.

(4) Solve.

(5) Check the analysis results. The static analysis results are saved in the structural analysis result file and include the following contents: basic solutions (node displacement, temperature, voltage, etc.); derived solutions (node and element stress, node and element strain, element force, node reaction force, etc.).

3.1. Finite Element Modelling

ANSYS finite element software provides two working modes: human-computer interaction mode (GUI mode) and command flow input mode [20]. The GUI mode does not need to memorize the usage rules of the programming language and the use format of the commands. As long as the mouse picks up the operation on the graphic, it is relatively easy for the beginner to grasp some of them. However, when faced with a complex finite element model, the disadvantages of using the GUI approach will be revealed. For example, there will be a large number of simple and repetitive operations, which will greatly increase the computation time, affect the quality of the model analysis, and generate a large number of data files.

The command flow method (Batch method), ANSYS Parameter Design Language (APDL), is a scripting language that is used to complete finite element routine analysis operations or to establish analysis models through parameterized variables. It uses intelligent analysis. The method provides users with the function of automatically completing the finite element analysis process. That is, the input of the program can be determined according to the specified function, variable, and the selected analysis type. It is the most important basis for completing the optimization design and adaptive grid. APDL allows complex data entry, allowing the user to actually have control of any design or analysis attribute. APDL extends the capabilities beyond the scope of traditional finite element analysis and extends more advanced operations, including sensitivity studies, parametric modelling of parts libraries, design modifications, and design optimization.

Compared with the GUI mode, the command flow is an ASCII text file. Its advantages are: it can reduce a lot of repetitive work, and it is especially suitable for occasions that require repeated calculations after minor modifications; it is not limited by the ANSYS software version and operating system. Easy to save, carry and communicate; users can easily establish a parameterized part library to facilitate the rapid generation of finite element analysis model; use APDL to write some commonly used macro commands to engage in secondary development. In addition, the APDL file system must be used for optimization design and adaptive mesh analysis.

Based on the above analysis, a finite element model of a magnetically suspending spherical rotor was established using the command flow input method based on APDL. Specific modelling steps are as follows.

Step1, set the unit system, and use the centre of the rotor sphere as the origin of the coordinates, and define the variables as shown in Figure 2.

Step2: The plane model of the rotor rotary table is defined by a combination of points, lines, and planes. The forward profile of the rotary table must be composed of a series of quadrilaterals (edges may be arcs) to ensure the consistency of the topological structure of the wheel, as shown in the Figure. 3, the commands involved include: a point definition command K, a straight line definition command LSTR, an arc definition command LARC, and an area definition command AL.

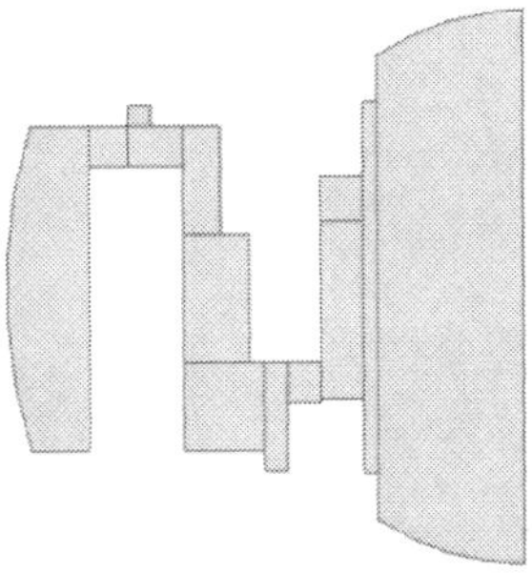

Figure 3. Roulette half-section finite element model

Step3: Use the VROT command to rotate the generator sphere. As shown in Figure 4, it should be noted that the rotor rotary table generated by the rotation is a combination of multiple bodies. Therefore, it is necessary to use the VGLUE command to bond the assembly into one new whole.

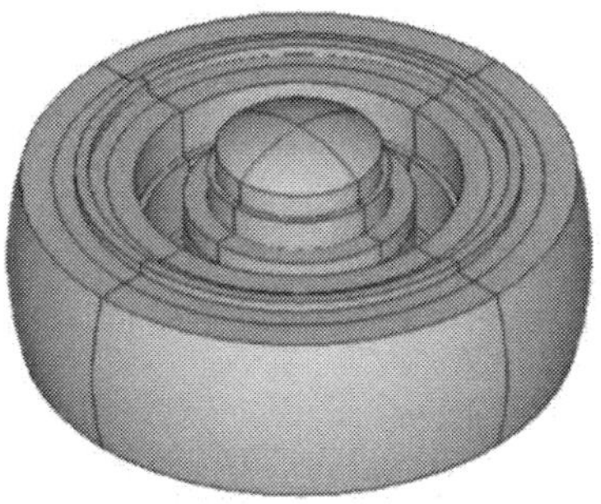

Figure 4. Roulette body finite element model

Step4, set the material properties of 1J50. The unit type is set to Solid185 using the command ET, which is used to construct a three-dimensional, eight-node solid structure, each node having three degrees of freedom in translation along the xyz direction. Use the MP to set the elastic modulus, Poisson's ratio, and density, and associate the material properties with the 3D model.

Step5, divide the grid. In order to balance the calculation accuracy and time, the grid size is set to 3mm, and the VSWEEP command is used to divide the grid. The cells formed in this way are almost all regular hexahedron elements, which can effectively guarantee the accuracy of finite element calculations. As shown in Figure 5.

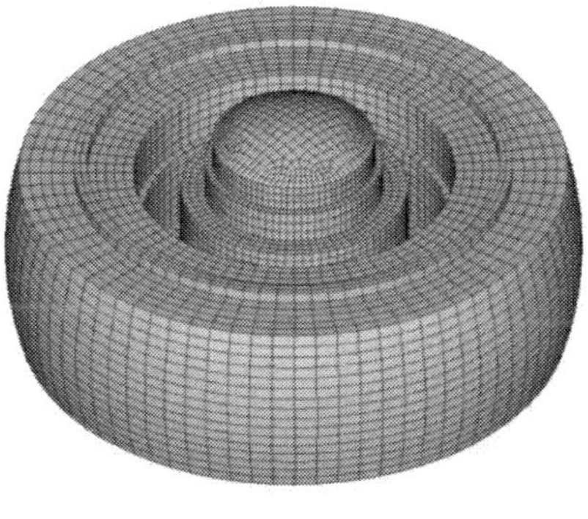

Figure 5. Finite element model of rotary table

Step6, calculate the stress. The inertia release calculation control is set to output only the pre-calculated total mass regardless of inertial release. The type of analysis is set to static analysis. Select the appropriate node, set the constraint type to 3DOF, and use the command OMEGA to set the rotor speed to 10000rpm.

Step7, post-processing. Use the NSORT command to arrange the equivalent stress and displacement vectors in ascending order, and then use the *GET command to extract the maximum equivalent stress and maximum displacement vector, respectively.

4 Simulation Analysis

At the rated speed of 10000rpm, the maximum stress distribution of the disk is shown in Figure 6. The maximum equivalent stress obtained from the static analysis is 52.2MPa, which occurs at the radially inner corner of the mounting surface of the upper plate. The yield limit of soft magnetic material 1J50 is 147MPa, so the safety factor is 2.8, which is greater than the design threshold (usually is 2), indicating that the rotary table has sufficient strength and safety redundancy at the design speed.

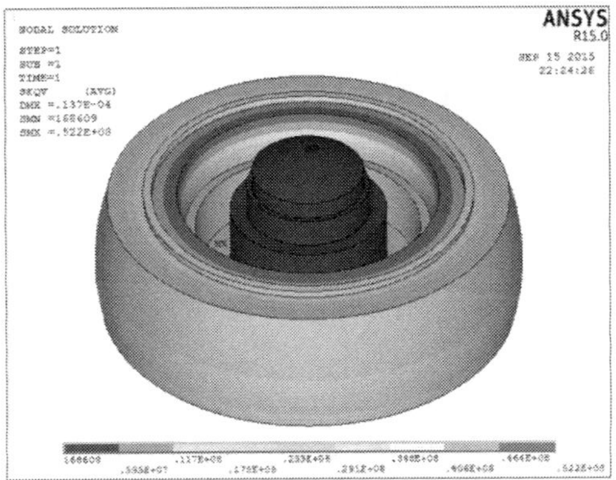

Figure 6. Stress nephogram of rotary table

The rigid disk displacement distribution profile is shown in Figure 7, the maximum rigid body displacement occurs at the lower side of the radial sphere, and the value

https://doi.org/10.1051/matecconf/201819804001

thereof is 0.014 mm. That is, when the rotor is operated at the design speed, the air gap at the lower side of the radial magnetic bearing is reduced by 0.014 mm. The initial air gap of 0.35mm, air gap reduction rate of 4%, this result is in line with the requirements of engineering practice, indicating that the rotor can maintain the stability of the structure at high speed rotation.

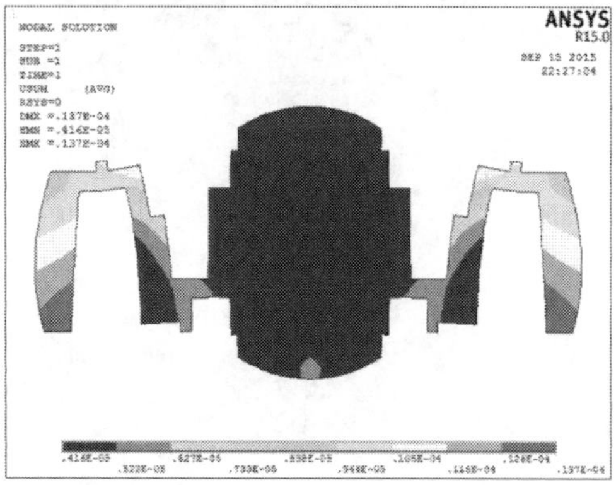

Figure 7. Rigid body displacement of rotary table

5 Conclusion

The paper studies the static characteristics of a spherical magnetically suspending rotor based on the finite element analysis method. The design of the spherical magnetically suspending rotor is designed. The characteristics of the spherical magnetically suspending rotor are analyzed and the necessity of static analysis is pointed out. The main content and general flow of statics analysis are discussed. The GUI mode and command flow mode based on APDL of the finite element software ANSYS are compared and analyzed. Then the finite element model of the spherical magnetically suspending rotor is established by the command flow method, and the detailed modeling steps are given. On this basis, the static characteristics of the spherical magnetically suspending rotor are simulated and analyzed. The simulation results show that the maximum equivalent stress and rigid body displacement of the rotor are within the safety threshold at a design speed of 10000rpm.

References

1. HOU E Y, LIU K, SHAN X Q. Analysis of dynamics characteristics of hybrid magnetically suspending flywheel with oblate external rotor [J]. Journal of Astronautics, 2011, 32(5): 998-1004.
2. WANG X, SHI Y, FAN Y H, et al. A New type of heteropolar radial magnetic bearing biased with permanent magnet for magnetically suspended flywheel [J]. Journal of Astronautics, 2014, 35(12): 1457-1464.
3. WANG C E, FANG J C, TANG J Q, et al. Thermal design method and experimental research of magnetically suspended reaction flywheel [J]. Acta Aeronautica et Astronautica Sinica, 2011, 32(4): 598-607.
4. LIU Q, FANG J C, HAN B C. Novel electromagnetic repeated launch locking/unlocking device (RLLUD) based on self-locking for magnetic bearing flywheel [J]. Sensors and Actuators A, 2012(175): 116-126.
5. LIU Q, FANG J C, WU D Y. Optimization and experiment of magnetic bearing wheel rotor [J]. Machinery Design & Manufacture, 2015(2): 111-114.
6. WAN J G, WANG X P, JIANG P, et al. Application of ANSYS in analyzing dynamic characteristics of active magnetic bearing-rotor System [J]. Machinery Design & Manufacture, 2009(4): 87-89.
7. WAN J G, WANG X P, GAO Q, et al. Research on Dynamic Characteristics of Rotor System Supported by AMB Based on ANSYS Modal Analysis [J]. Bearing, 2010(6): 1-5.
8. LI K L, XIE Z Y. Model analysis of the magnetically suspending rotor based on ANSYS [J]. Mechanical & Electrical Engineering Magazine, 2008, 25(1): 1-3.
9. TANG S Q, HUANG P, KE Y W, et al. Modal Analysis of Magnetic bearing Flywheel Rotor Based on ANSYS [J]. Journal of China Three Gorges Univ. (Natural Sciences), 2017, 39(6): 85-89.
10. SONG J C, OUYANG H M, ZHANG G M. Rotor System Of Magnetic Suspending Flywheel Of Dynamic Research Based On Finite Element Analysis [J]. Journal of Mechanical Strength, 2015, 37(3): 403-407.
11. HAN B C, HU G, FANG J C. Optimization design of magnetic suspended gyroscope rotor [J]. Optics and Precision Engineering, 2006, 14(4): 663-666.
12. YOHJI O, HIROAKI K, KOUICHI K. New concept of miracle magnetic bearings[C]. Proceeding of the 9th International Symposium on Magnetic Bearings, USA, August, 3-6, 2004: 89-95.
13. CHRISTIAN E, TILO S, RAINER N. Comparison of active magnetic bearings with and without permanent magnet bias[C]. Proceeding of the 9th International Symposium on Magnetic Bearings, USA, August, 3-6, 2004: 420-425.
14. SUN J J, FANG J C, WANG Xi, et al. A new permanent magnet biased radial magnetic bearing [J]. Transactions of China Electrotechnical Society, 2009, 24(11): 53-60.
15. WANG W J, REN Y, LIU Q, et al. Optimal design of rotary table for spherical rotor of magnetically suspending gimbaling flywheel [J]. Acta Aeronautica et Astronautica Sinica, 2016, 37(9): 2874-2883.
16. China aeronautical materials handbook editorial board. China aeronautical materials handbook (second edition vol.5) [M]. Beijing: China Standard Press, 2002: 176, 187.
17. LIU G Q, YANG Q D. ANSYS Engineering Application Tutorial ANSYS——Mechanical articles [M].Beijing: China Railway Press, 2003: 185.
18. HUANG Z X, LIU C Z. ANSYS WORKBENCH14.0 Super study manual [M]. Beijing: Post & Telecom Press, 2013: 167.
19. ZHANG Y Q, ZHANG H C. ANSYS Finite Element Analysis from Getting Started to Proficient [M]. Beijing: Post & Telecom Press, 2010: 194-195.
20. GONG S G, XIE G L, HUANG Y Q. ANSYS Parametric Programming and Command Manual [M]. Beijing: Machinery Industry Press, 2009: 2-3.

Analysis and Experiment of Eddy Current Loss in Radial Magnetic Bearings with Solid Rotor

Zengyuan Yin [1], Yuanwen Cai [1], Weijie Wang [1] and Yuan Ren [1]

[1] Space Engineering University, Beijing 101416, China

Abstract. In order to reduce power losses for the aerospace applications, this paper analyzes the eddy current losses produced by high-speed rotating solid rotor core magnetic bearings of magnetically suspended control & sensitive gyroscope (MSCSG). An analytical model of the eddy current loss of a solid rotor radial magnetic bearing (RMB) is presented. Considering the difference between NNNN and NSNS magnetic circuits of RMB, the magnetic field expressions of stator magnetic poles are listed. The magnetic field of stator poles is replaced by Fourier series expansion. According to the magnetic field distribution around the magnetic pole surface and the boundary conditions of the rotor surface, the mathematical expression of eddy current loss is obtained. The measurement method of rotational power loss in radial magnetic bearing is proposed, and the results of the theoretical analysis are verified by experiments in the prototype MSCSG. The experimental results show the correctness of calculation results. Eddy current loss models and test methods provide theoretical support for analyzing eddy current losses in solid rotors and reducing power consumption.

1 Introduction

Active magnetic bearings can eliminate the friction of traditional mechanical bearings, no lubrication, high torque accuracy, micro-vibration, active vibration suppression and other advantages [1-2]. Not only the output torque accuracy improve by 1~2 orders of magnitude compared to the mechanical bearing flywheel, but also has a long life. Therefore, the inertial actuator supported by the magnetic bearing is an ideal inertial actuator for the high-precision earth observation satellite. When the rotor used for the magnetic bearing support rotates at a high speed, in addition to the friction caused by the air friction, the rotor will also generate eddy current loss and hysteresis loss. Generally, the eddy current loss is far greater than the hysteresis loss, which seriously affects the inertial actuator performance. In order to reduce the eddy current loss, the laminated structure is usually adopted, but in some cases, a solid rotor is still needed. The focus of the research in this paper is to analyze and experimentally verify the eddy current loss of the magnetic bearing with the solid rotor.

Eddy current loss model and analysis has been very mature in the field of motor research [3-5]. There are mainly finite element methods, parameter predictions or experimental tests. However, research in the field of magnetic bearings needs to be improved. Some literature has proposed that the eddy current loss generated by the rotor is proportional to the frequency of the applied current [6-7]. However, the current applied by the RMB is a DC power. In order to reduce the eddy current loss, it is necessary to establish a precise eddy current loss model. David C. Meeker analyzes the rotation loss of a heteropolar radial magnetic bearing [8]. The loss of the rotor during rotation is analyzed by numerical analytical methods. Finally, the correctness of the numerical analysis results is verified by experimental methods. Mizuno et al. tested the magnetic bearing magnetic poles in the NSNS and NNSS configuration with solid rotor iron loss [9]. The test results show that the iron loss in the NSNS magnetic pole distribution is slightly less than that in the NNSS configuration. Sun Jinji analyzed and experimented the eddy current loss of the rotor [10]. In the article, the loss of the rotor was divided into four categories: wind resistance, eddy current loss produced by a radial magnetic bearing, loss caused by axial magnetic bearing, and loss generated by the motor. Hu Xiaofei [11] analyzed and optimized the eddy current loss of a Homopolar radial magnetic bearing rotor. The influence of the number of magnetic poles on eddy current loss is analyzed. It is concluded that increasing the number of magnetic poles of the magnetic bearing will cause the eddy current loss of the rotor to increase first and then decrease. In addition, the pole width of the magnetic bearing and the increase of the air gap length can effectively reduce the rotor eddy current loss. Based on the analysis results, he proposed a method to reduce the eddy current loss of the magnetic bearing rotor.

This paper presents an analytical model of magnetic pole eddy current loss of a solid rotor of RMB. Considering the difference between the NNNN and NSNS magnetic circuits, the magnetic field expression of the stator magnetic pole is discussed. The Fourier series expansion formula is used to substitute the magnetic field of the stator magnetic pole. According to the magnetic field distribution around the magnetic pole surface and the boundary conditions of the rotor surface, the mathematical expression of eddy current loss is obtained. By solving the eddy current loss expression, the eddy current loss value under different arrangement of magnetic poles is obtained. The eddy current loss value of the RMB can be obtained through the free Run-down experiments of the rotor under the vacuum condition. The results of the eddy current loss obtained by the experiment agree well with the numerical solution results. Eddy current loss models and test methods provide theoretical support for analyzing eddy current losses in solid rotors and reducing power consumption.

2 Analysis of eddy-current loss

RMB structure shown in Figure 1 (a), mainly including the stator (shown in Figure 1 (b)), stator winding and rotor system. The stator pole adopts the spherical structure, and the rotor pole adopts the spherical structure. This structure can avoid the interference torque caused by the rotor assembly error. The radial magnetic bearing bias flux is generated by the bias current and the control flux is generated by the control current. The rotor is differentially controlled by a bias current and a control current, and is stably suspended in an equilibrium position Magnetic bearing structure parameters are shown in Table 1.

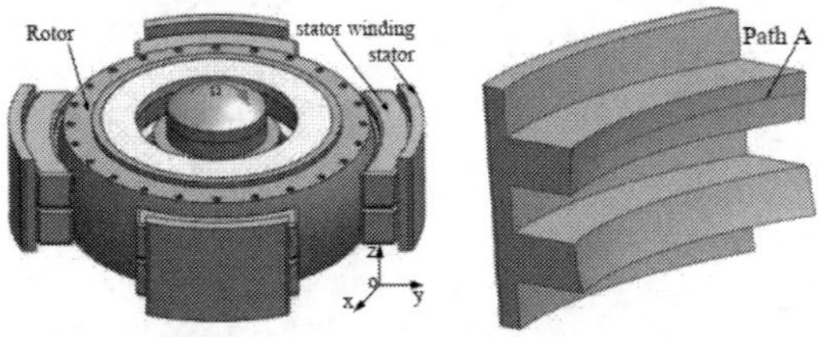

(a) (b)

Figure 1. RMB structure

Table 1. Magnetic bearing structure parameters.

Design Parameters	Parameter
Stator inner diameter (mm)	158
Stator core circumferential length (mm)	70
Very wide stator (mm)	10
Rotor outside diameter (mm)	157.3
Inner diameter of rotor (mm)	100
Distance between stator poles on both sides (mm)	10
Number of turns per coil	200
Coil window area (mm*mm)(Radial* circumferential)	15.5*4.5
Unilateral air gap (mm)	0.35
Protection air gap (mm)	0.2
Rotor weight (kg)	6
Bias current (A)	0.5

When the rotor rotates at high speed, the magnetic field on the surface of the rotor changes periodically and eddy current loss occurs around the surface of the rotor. This eddy current loss will increase the temperature of the rotor and cause deformation of the rotor structure, which will seriously affect the performance of the rotor. radial magnetic bearing of MSCSG adopts upper and lower double-layer magnetic pole structure to form a closed loop, so the magnetic field distribution of four magnetic poles in each layer generally has two forms, NNNN and NSNS. The eddy current loss produced by these two distributions is different. The eddy current loss analysis of these two structures follows.

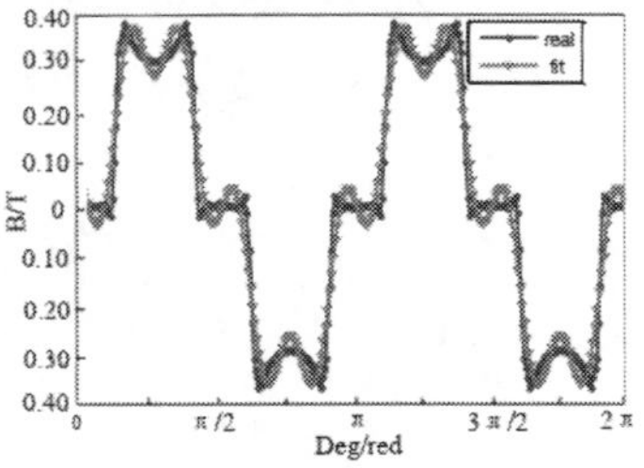

(a) NSNS arrangement

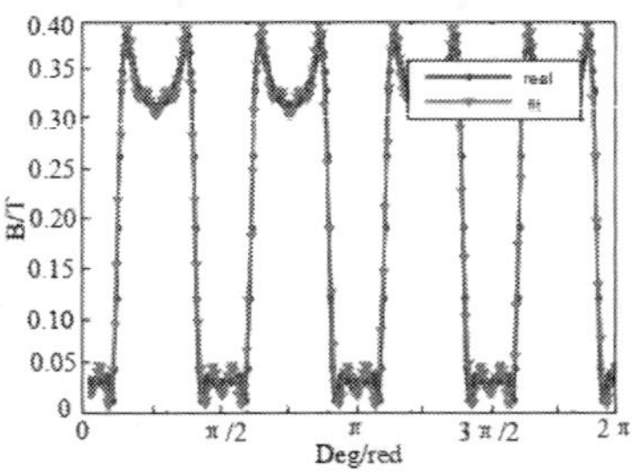

(b) NNNN arrangement

Figure 2. Path A magnetic field distribution

Fig 2 shows the distribution of the magnetic field when the magnetic fields NNNN and NSNS are arranged in the stator magnetic path A (Fig. 1b). The blue lines in the figure represent the real value. Due to the periodic rotation of the rotor, it can be assumed that the rotor surface magnetic field periodically change. From the above figure, we can see that there is a change of magnetic density in the air-gap of the magnetic bearing, and the magnetic flux density of the rotor is unevenly distributed in the circumferential direction. Therefore, when the rotor rotates, the magnetic field on the rotor surface changes and core loss occurs.

Magnetic field of path A expand using Fourier series. The red line represents the result of the Fourier series expansion. The green line According to the fig 2 we can see that the result of Fourier expansion can represent the magnetic field in the path A. The expression of the magnetic field represented by the Fourier series is shown in the following formula (1).

$$B_0 = \sum_{n=1}^{\infty} A_n \cos(\omega n x) + B_n \sin(\omega n x) \quad (1)$$

According to euler equation, formula(1) can be simplified as the following formula(2).

$$B_0 = \sum_{n=1}^{\infty} C_n e^{i\omega Tx} + C_n^{'} e^{-i\omega Tx} \quad (2)$$

For the following calculation, it is assumed that all magnetic poles of RMB are same. In order to solve Maxwell equations, boundary conditions must be defined and boundary conditions are as follows:

1.It is assumed that the magnetic flux density is constant (B_0) at the boundary between the stator and air gap and the applied flux density is periodic.

2. The conductivity and permeability are constant, and the rotor magnetization curve is linear.

3. Magnetic fields and induced currents tend to concentrate on the surface of the rotor. The penetration depth(δ) of the magnetic field is much smaller than the diameter of the rotor.

The steady state MAXWELL equation of a moving conductor can be represented by the second-order partial differential equation as follows:

$$\frac{1}{\mu\delta}\nabla^2 B + \nabla\times(\upsilon\times B) = 0 \quad (3)$$

When the rotor rotates at high speed, the polarities of the magnetic poles will change and an alternating magnetic field will be generated in the rotor which will generate eddy current losses.

Compared with the rotor radius, the skin effect depth of the magnetic field is much smaller than the rotor radius. Therefore, the outer surface of the rotor is expanded in the circumferential direction,therefore using the spherical coordinate system instead of Cartesian Cartesian coordinate system produces little error. As shown in Figure 3 is Circumferential expansion diagram of RMB.

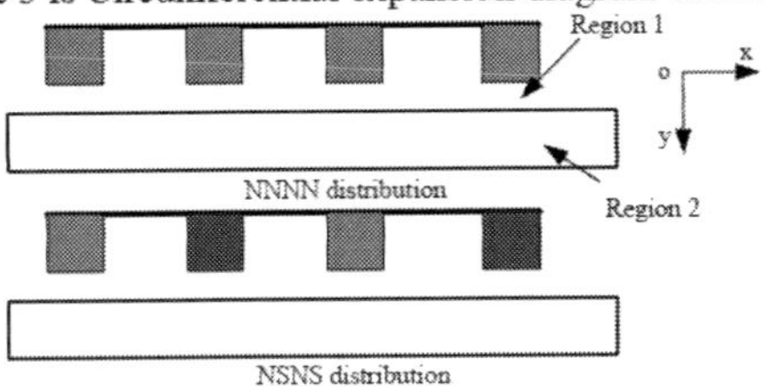

Figure 3. Circumferential expansion diagram

The magnetic field B and the rotor speed υ have the following expression in the Cartesian coordinate system:

$$B = B_x\mathbf{i} + B_y\mathbf{j} + B_z\mathbf{k} \quad (4)$$

$$\upsilon = \upsilon_x\mathbf{i} \quad (5)$$

In this type of radial magnetic bearing, the magnetic field produces components in the X and Y directions and the Z direction component is 0. So, $B_z = 0$,the above equation can be simplified as shown in the following equation:

$$\begin{cases} \frac{\partial^2 B_x}{\partial x^2} + \frac{\partial^2 B_x}{\partial y^2} - \mu\sigma\upsilon_x\frac{\partial B_x}{\partial x} = 0 \\ \frac{\partial^2 B_y}{\partial x^2} + \frac{\partial^2 B_y}{\partial y^2} - \mu\sigma\upsilon_x\frac{\partial B_y}{\partial x} = 0 \end{cases} \quad (6)$$

According to the stator magnetic field periodic boundary conditions, the magnetic field produced by the stator magnetic pole can be expressed by the following formula:

$$B = \begin{pmatrix} B_x \\ B_x \end{pmatrix} = \begin{pmatrix} \tilde{B}_x(y,z) \\ \tilde{B}_x(y,z) \end{pmatrix} e^{ikx} \quad (7)$$

Because of $\nabla \bullet B = 0$,combining (4)~(7) lead to

$$\frac{\partial^2 \tilde{B}_x}{\partial y^2} - q^2\tilde{B}_x = 0$$
$$\frac{\partial^2 \tilde{B}_y}{\partial y^2} - q^2\tilde{B}_y = 0 \quad (8)$$

In the above formula, $q = \sqrt{k^2 + k\sigma\mu\upsilon_x}$, The above differential equation can be obtained:

$$\tilde{B}_x = -\frac{q}{ik}(ae^{ky} - be^{-ky})$$
$$\tilde{B}_y = ae^{ky} + be^{-ky} \quad (9)$$

The magnetic bearing model in the Cartesian coordinate system is divided into two areas. The region 1 is the air gap. At this time, the conductivity $\sigma=0$ .so $q = k$ .Region 2 is the rotor of the movement. For the result of the above formula 9 in region 1, the result is:

$$B_{x1} = ie^{ikx}(a_1e^{ky} - b_1e^{-ky})$$
$$B_{y1} = ie^{ikx}(a_1e^{ky} + b_1e^{-ky}) \quad (10)$$

The results for region 1 can be simplified as:

$$B_{x2} = b_2\frac{q}{ik}e^{ikx}e^{-ky}$$
$$B_{y2} = b_2e^{ikx}e^{-ky} \quad (11)$$

According to the boundary conditions, $y = 0$, magnetic field is generated by magnetic poles. $y = \Delta$, Here, the magnetic field in Region 1 and Region 2 must be equal. So get the following formula:

$$B_{y1}(y=0) = ce^{ikx}$$
$$B_{y1}(y=\Delta) = B_{y2}(y=\Delta) \quad (12)$$
$$B_{x1}(y=\Delta) = \frac{1}{\mu_r}B_{x2}(y=\Delta)$$

From the formula (12) , a_1, b_1 and b_2 can be obtained. As shown in the formula (13):

$$a_1 = c(1 - \frac{q}{\mu_r k})\frac{e^{-k\Delta}}{2\gamma}$$
$$a_2 = c(1 + \frac{q}{\mu_r k})\frac{e^{k\Delta}}{2\gamma} \quad (13)$$
$$a_1 = c\frac{e^{q\Delta}}{\gamma}$$

Where $\gamma = co\cosh(k\Delta) + \frac{q}{\mu_r k}\sinh(k\Delta)$

According to the MAXWELL stress tensor, the force acting on the moving conductive rotor is:

$$F_x = -\frac{1}{\mu_0}\int_s B_x B_y da$$
$$F_x = -\frac{1}{2\mu_0}\int_s (B_x^2 - B_y^2) da \quad (14)$$

The formula (14), S is a closed surface that arbitrarily surrounds a conductor. The face selected in this article is $y=\Delta$, $0\le x\le 2\pi$, $0\le z\le b$, Where z is the axial length of the radial magnetic bearing.According to Formula 14, the force applied to the rotor is:

$$F_x=\frac{ibU}{\mu_0}\sum_{n=1}^{\infty}\frac{c_n\bar{c}_n}{\gamma_n\bar{\gamma}_n}\frac{q_n-\bar{q}_n}{\mu_r k_n}$$
$$F_y=\frac{bU}{\mu_0}\sum_{n=1}^{\infty}\frac{c_n\bar{c}_n}{\gamma_n\bar{\gamma}_n}\frac{q_n\bar{q}_n-\mu_r^2k_n^2}{\mu_r^2k_n^2} \quad (15)$$

In the above formula $\bar{c}_n$、$\bar{q}_n$ and $\bar{\gamma}_n$ are the conjugate complex numbers of c_n、q_n and γ_n. The subscript n indicates the n-th harmonic of the Fourier transform of the applied magnetic field. The rotor eddy current losses can be obtained by combining Equations 5 and 15. As shown in the following formula:

$$P=F_x\upsilon_x \quad (16)$$

3 Experiment setup and measurement

The cross-section of the MSCSG prototype is shown in Fig. 4. It mainly consists of RMB ,axial magnetic bearing, LFMB, motor system, rotor and gyro room, as shown in the figure 4.

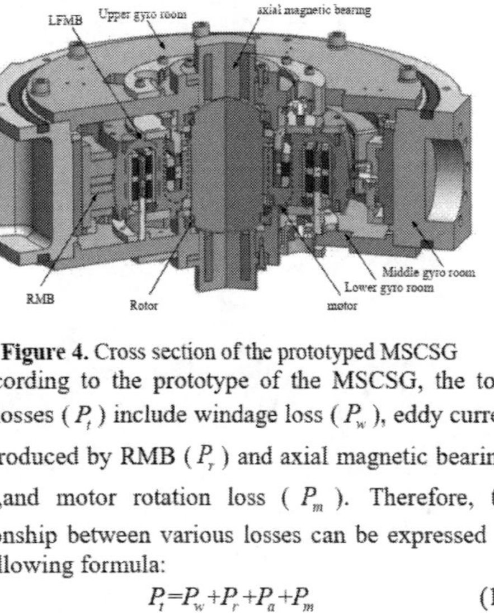

Figure 4. Cross section of the prototyped MSCSG

According to the prototype of the MSCSG, the total rotor losses (P_t) include windage loss (P_w), eddy current loss produced by RMB (P_r) and axial magnetic bearings (P_a),and motor rotation loss (P_m). Therefore, the relationship between various losses can be expressed by the following formula:

$$P_t=P_w+P_r+P_a+P_m \quad (17)$$

The loss of each part of the rotor can be verified by the motor's run-down test. The calculation steps are as follows:

1.windage loss P_w

When the MSCSG is evacuated internally, the pressure $P=0$, and the influence of the wind resistance can be ignored at this time.

2.motor loss P_m

The motor used by MSCSG is a brushless DC motor. The magnet is divided into two layers of inner and outer magnets, and the middle of the magnet is a stator of polyimide material. Compared with the traditional structure motor, when the MSCSG rotor rotates at high speed, the magnet and the rotor rotate together, so the magnetic field generated by the motor is invariable in space, so the motor loss is far less than the loss produced by the radial magnetic bearing. So, let's assume the motor loss is equal to 0, $P_m\approx 0$. Under vacuum conditions, when the rotor is accelerated by the motor to the specified speed, the motor suddenly loses power, and its kinetic energy is mainly converted into magnetic bearing eddy current loss to increase the temperature of the rotor.

3.Axial magnetic bearing loss P_a

Stator magnetic pole of axial magnetic bearing of MSCSG is spherical structure, axial magnetic bearing rotor is spherical structure, when the rotor rotates, the stator of axial magnetic bearing produces a stable magnetic field, So assume the loss caused by the axial magnetic bearing is $P_a\approx 0$.

4.Radial magnetic bearing loss

According to the above analysis, the reason for the speed reduction of the rotor only comes from the eddy current loss produced by the RMB. By setting the rotor rundown test, the eddy current loss generated by the RMB can be measured.

When vacuuming the MSCSG, the rotor speed is ω. The kinetic energy of the rotor is:

$$E_r=\frac{1}{2}I\omega^2 \quad (18)$$

Where I is the axial moment of inertia of the rotor. According to the above analysis, when the rotor decelerates to 0, its kinetic energy is mainly converted into heat generated by eddy current losses.so $E_r=P_rt$. The eddy current loss of the radial magnetic bearing can be calculated by controlling the rotor speed and calculating the time of rotor deceleration. The eddy current loss p of the RMB is as follows:

$$P_r=\frac{I\omega^2}{2t} \quad (19)$$

The RMB eddy current loss experiment platform is shown in Figure 2. It Mainly include magnetic MSCSG prototype, vacuum pumping system, magnetic bearing control circuit, and motor control system.

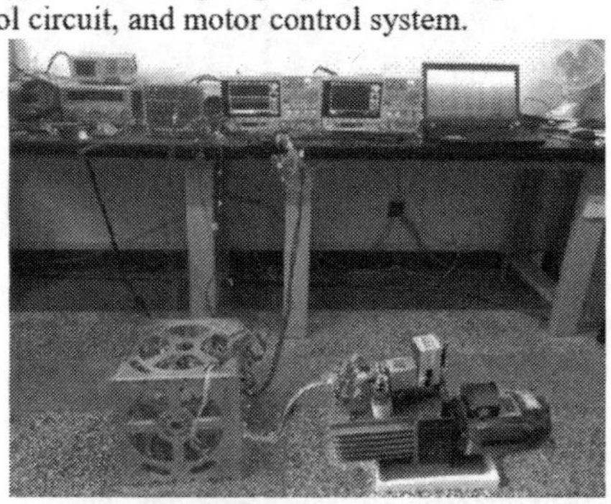

Figure 5. MSCSG experiment platform

The measurement result is shown in Fig 6. According to Figure 6(a), The arrangement of magnetic pole NSNS is more obviously than that of NNNN arrangement. Figure 6(b) shows the eddy current loss of the experimental results and the calculated results. The experimental results are basically consistent with the calculation results.

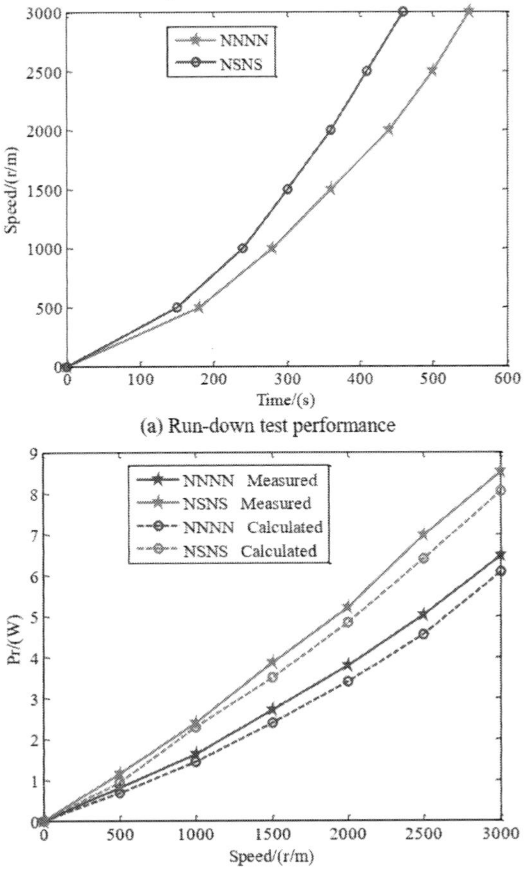

(a) Run-down test performance

(b) power loss of RMB

Figure 6. Calculated and measured eddy current losses

4 Conclusion

This paper presents an eddy current loss analysis model of a solid rotor core magnetic bearing. Considering the difference in the arrangement order of the magnetic poles in the same layer, the magnetic field expression of the rotor core is discussed. According to the magnetic field distribution around the pole surface and the boundary conditions of the rotor core surface, the analytical solutions by solving the electro- magnetic field approach is processed. In the MSCSG prototype, the eddy current loss of the RMB is separated by experimental methods. By comparison, the calculated results is good agreement with the experimental results. The eddy current loss model and test method proposed in this paper provide theoretical support for analyzing the eddy current loss of solid rotors.

References

1. Le Y, Sun J, Han B. Modeling and Design of 3-DOF Magnetic Bearing for High-Speed Motor Including Eddy-Current Effects and Leakage Effects[J]. IEEE Transactions on Industrial Electronics, 2016, 63(6):3656-3665. Holl, B.G. Orr, Eur. Phys. J. E **14**, 7 (2004)
2. J. Fang, S. Zheng, and B. Han, "AMB Vibration Control for Structural Resonance of Double-Gimbal Control Moment Gyro with High-Speed Magnetically Suspended Rotor," *IEEE/ASME Trans. Mechatronics*, vol.18, no. 1, pp. 32–43, Sep. 2013.Luigi T.De Luca, *Propulsion physics* (EDP Sciences, Les Ulis, 2009)
3. Dlala E, Belahcen A, Arkkio A. Efficient magnetodynamic lamination model for two-dimensional field simulation of rotating electrical machines[J]. Journal of Magnetism & Magnetic Materials, 2008, 320(20):e1006-e1010.
4. Alatawneh N, Pillay P. Rotational Core Loss and Permeability Measurements in Machine Laminations with Reference to Permeability Asymmetry[J]. IEEE Transactions on Magnetics, 2012, 48(4):1445-1448.
5. Komeza K, Dems M. Finite-Element and Analytical Calculations of No-Load Core Losses in Energy-Saving Induction Motors[J]. IEEE Transactions on Industrial Electronics, 2012, 59(7):2934-2946.
6. Bishop J E L. The analysis of eddy-current-limited magnetic domain wall motion, including severe bowing and merging[J]. Journal of Physics D Applied Physics, 1973, 6(6):97.
7. Kendall D, Piercy A R. The frequency dependence of eddy current losses in Terfenol - D[J]. Journal of Applied Physics, 1993, 73(10):6174-6176.
8. Meeker D C, Maslen E H. Prediction of Rotating Losses in Heteropolar Radial Magnetic Bearings[J]. Journal of Tribology, 1998, 120(3):629-635
9. Meeker D C, Filatov A V, Maslen E H. Effect of magnetic hysteresis on rotational losses in heteropolar magnetic bearings[J]. IEEE Transactions on Magnetics, 2004, 40(5):3302-3307.
10. Sun J, Dong C. Analysis and experiment of eddy current loss in Homopolar magnetic bearings with laminated rotor cores[J]. Acta Astronautica, 2013, 89(8):229-235.
11. Hu X F, Liu G, Sun J J, et al. Analysis and optimization of rotor eddy current loss in homopolar radial magnetic bearings[J]. Journal of System Simulation, 2013.

Dynamic Simulation Studies of Control Design for Engine Bleed Air Test Facility

Zheng Dai[1], Yi Cui[1]

[1]*Shanghai Aircraft Design and Research Institute, Commercial Aircraft Corporation of China Ltd, Shanghai, China*

Abstract. This paper describes a model-based method for the control design of engine bleed air test facility for the bleed air system. In order to perform the control design, a dynamic simulation model for the facility is developed to simulate the dynamic behaviour. The main equipment dynamic modelling approaches are described. This simplified dynamic model may be used to perform trade studies and to optimize component characteristics for performance and cost comparison purposes. With the performance shown, this engine bleed air test facility should be quite capable to emulate the engine for bleed air system testing and evaluation.

1 Introduction

The environmental control system (ECS) of commercial aircraft contains several main subsystems as shown in figure 1. The atmospheric environment within the flight envelope of the commercial aircraft varies drastically. For instance, the ambient temperature varies from -55°C to 40°C, the pressure varies from 20 kPa to 101 kPa. The environmental control system is designed to control and regulate the interior environment of the aircraft cabin. The bleed air system is a highly coupled control system which regulates temperature and pressure of compressed engine bleed air and serves as the pneumatic air source for the downstream ECS subsystems.

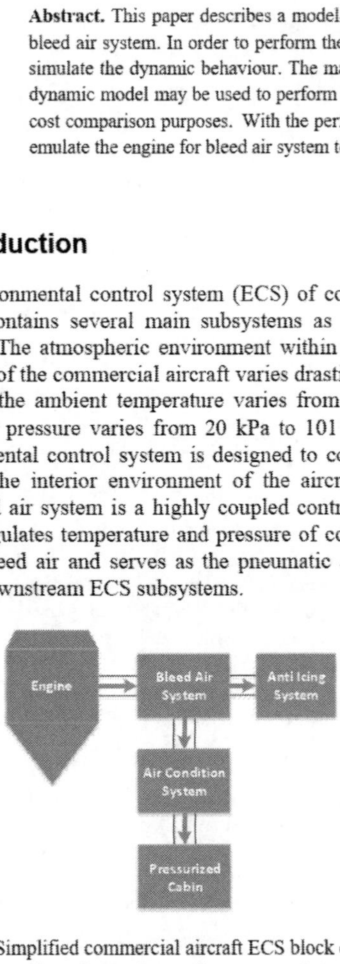

Figure 1. Simplified commercial aircraft ECS block diagram

The excellent dynamic performance of the bleed air system is the guarantee for the normal operation of the environmental control system. Poor dynamic characteristics may cause uncontrollability of the system. The parameters may oscillate with serious overshoot, causing a system alarm[1].

The Engine Bleed Air Test Facility (EBATF) is an essential infrastructure for the performance and verification test of the bleed air system. The objective of EBATF is to provide airflow to simulate engine bleed air for bleed air system test bench[2-3]. The engine bleed air pressure and temperature can change rapidly, the pressure and temperature controlled by EBATF are therefore necessarily fast responding.

As shown in Figure 2, the general system architecture contains the main equipment:

- Compressed air source: to provide high pressure air;
- Heater: to heat the air for desired temperature;
- Heat exchanger: to cool the air;
- Control valve 1, 2: to perform temperature control;
- Outflow valve 1, 2: to perform pressure control.

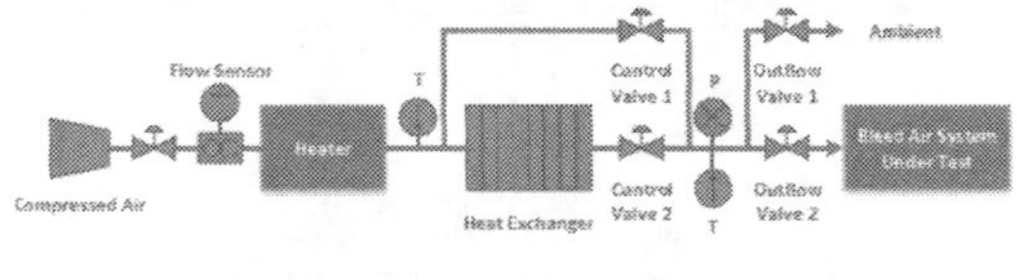

Figure 2. Simplified EBATF architecture diagram

In this paper a dynamic model of the EBATF is developed for the control design based on the fundamental of mechanics and heat transfer. And numerical simulation study is performed using the model. Several different control design methods are used and studied to realize the facility and meet the control requirements.

2 Dynamic Requirements of The Facility

The dynamic performance test of the bleed air system requires that the air temperature and pressure at the inlet of the pipeline can be rapidly changed. Therefore, the control system must overcome the inherent thermal inertia, volume inertia, and coupling characteristics between the control parameters. For ECS applications,

https://doi.org/10.1051/matecconf/201819804003

engine bleed air pressure and temperature from the engine high pressure port typically range from 40 psia to 500 psia and 230 °C to 650 °C. During takeoff, the pressure and temperature could change from 55 psia to 550 psia and 260 °C to 650 °C rapidly. During cruise at 40000 ft, the pressure and temperature are around 160 psia and 550 °C. At the same altitude for descend, they are about 50 psia and 360 °C. For the low pressure port, engine bleed air pressure and temperature typically range from 10 to 150 psia and 95 °C to 375 °C. During takeoff, the pressure and temperature could change from 10 psia to 150 psia and 45 °C to 355 °C rapidly. During cruise at 40000 ft, the pressure and temperature are around 35 psia and 280 °C. At the same altitude for descend, they are about 10 psia and 130 °C.

However, during rapid engine pressure transient like take off operation, both pressure and temperature can change rapidly and form one operation state to another operation state in about 5 to 10 seconds. Hence, it would be necessary for the EBATF to emulate this rapid pressure and transient performance on ground.

Despite the difficulty, the ideal dynamic performance of the EBATF is described in detail as follows:

- Maximum temperature change :
  - Rise from 260 °C to 600 °C in 10 sec
  - Drop from 650 °C to 300 °C in 10 sec
- Maximum pressure change :
  - Rises from 100 psia to 500 psia within 5 sec
  - Drop from 450 psia to 50 psia within 5 sec

In order to meet these requirements, a preliminary dynamic simulation model for EBATF is developped according to the general system architecture. Common equipment performance is chosen to reflect the real system state to the greatest extent. The temperature and pressure control design will be studied based on this model to make a preliminary assessment about the dynamic control performance.

3 Modelling Approach

3.1 Heater model

The heater is represented by a simple nonlinear model based on energy balance approach with a time constant at maximum flow rates to simulate warm up time at full airflow rate.

The heater is modeled by equation (1).

$$M_hC_h\,(dT_o/dt) = WC_p\,(T_o - T_i) + Q \qquad (1)$$

Where M_hC_h is the thermal mass of the heater, W is the airflow rate through the heater, T_i is the inlet air temperature, C_p is the specific heat of air and Q is the heating power. The model assumes perfect heat transfer between the electrical heating element and airflow with no energy losses.

3.2 Valve Model

All the valves are butterfly type with specific leakage area. The flow of the valve is calculated based on the inlet and outlet pressure, temperature, and the opening of the valve. The calculation expression is as shown in equation (2)[4-6] below.

$$W = KA\frac{p_i}{\sqrt{T_i}}\phi(\varsigma) \qquad (2)$$

Where $\zeta = p_o/p_i$, $\phi(\varsigma)$ can be calculated by equation (3) and (4). K is flow coefficient, A is flow area, p_i, T_1 are valve inlet air pressure and temperature, p_o is valve outlet air pressure.

$$\phi(\varsigma) = (\varsigma^{\frac{2}{\kappa}} - \varsigma^{\frac{\kappa+1}{\kappa}})^{0.5}\left[\frac{2\kappa}{R(\kappa-1)}\right]^{0.5}, \varsigma > 0.528 \qquad (3)$$

$$\phi(\varsigma) = (\frac{2}{\kappa+1})^{\frac{\kappa}{\kappa-1}}\left[\frac{2\kappa}{R(\kappa-1)}\right]^{0.5}, \varsigma \le 0.528 \qquad (4)$$

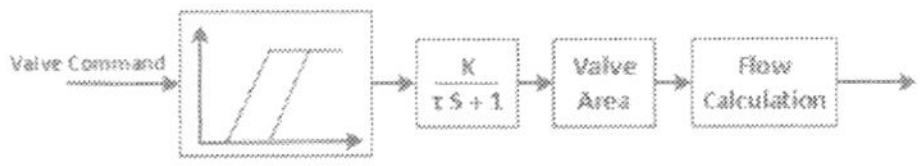

Figure 3. Valve model block diagram

The valve actuation dynamics will be represented by using a first order transfer function with a time constant. And a simple backlash function block is also used as shown in figure 3.

3.3 Air Duct Model

The air ducts are assumed to be made of stainless steel and modeled by equation (5), (6), (7), (8). The pressure loss of air flow in the duct is calculated according to equation (9)

$$M_mC_m\,(dT_m/dt) = Q_f + Q_a \qquad (5)$$

$$Q_f = W_fC_f\,(1\text{-}exp(\text{-}H_f/W_fC_f))(T_i - T_m) \qquad (6)$$

$$Q_a = H_a\,(T_a - T_m) \qquad (7)$$

$$T_o = T_m + (T_i - T_m)\,exp(\text{-}H_f/W_fC_f) \qquad (8)$$

$$\sigma\Delta P = kW_f^{\alpha},\ \alpha = \rho/\rho_o \qquad (9)$$

Where M_mC_m , T_m is the thermal mass and temperature of the duct, W_f is the airflow rate through the duct, C_f is the specific heat of air, T_i, T_o are the inlet and outlet air temperature. H_f , Q_f are the heat transfer coefficient and the heat transfered between air and the duct. H_a , Q_a are the heat transfer coefficient and the heat transfered between ambient and the duct. ΔP is the pressure loss of the duct, and ρ, ρ_o are air density in specific state and standard atmospheric density.

3.4 Heat Exchanger Model

Heat exchanger is cross flow type and assumed to be made of stainless steel. The hot side and cold side pressure drop are modeled according to equation (9). Subcell lumped parameter model is used for the modeling of temperature dynamic using equation (10), (11), (12).

More detailed information can be found from reference [7-9]

$$r_h\rho_h C_h\,(\partial T_h/\partial t) + W_h C_h(\partial T_h/\partial x) = H_h(T_m - T_h) \quad (10)$$

$$r_c\rho_c C_c\,(\partial T_c/\partial t) + W_c C_c(\partial T_c/\partial y) = H_c(T_m - T_c) \quad (11)$$

$$M_m C_m(\partial T_m/\partial t) - H_h(T_h - T_m) - H_c(T_c - T_m) = 0 \quad (12)$$

Where W_h , W_c are the hot side and cold side flow rate. H_h, H_c are hot side and cold side heat transfer rate. H_h, H_c are the specific heat of hot side and cold side air. T_m , T_h , T_c are temperature of the heat exchanger core, hot side and cold side. x, y are independent spatial variables and t is the time variable. ρ_h , ρ_c are air density of hot side and cold side.

4 Simulation and Analysis

4.1 Control Operating Mode

As shown in Figure 2, the compressed air is heated by the heater, and the outlet temperature can be controlled to reach a set value. The heated hot air is then divided into two air paths. One of the path is cooled by the heat exchanger and mixed with the other path, the outlet temperature can be controlled through the cooperation adjustment of the control valve 1 and control valve 2.

Considering the heat capacity and volume effect of the pipeline/heat exchanger and high-pressurized air, the mixing temperature control is preferred to use a cascade control channel as shown in Figure 4[10]. As for the heater temperature, bang-bang control techniques are applied to perform the heater outlet temperature control.

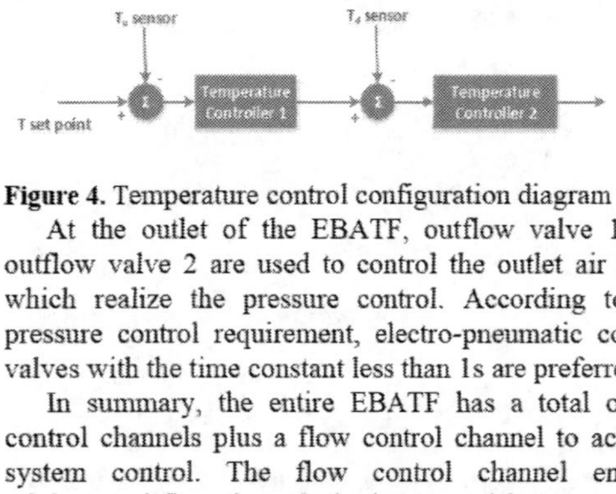

Figure 4. Temperature control configuration diagram

At the outlet of the EBATF, outflow valve 1 and outflow valve 2 are used to control the outlet air flow, which realize the pressure control. According to the pressure control requirement, electro-pneumatic control valves with the time constant less than 1s are preferred.

In summary, the entire EBATF has a total of six control channels plus a flow control channel to achieve system control. The flow control channel ensures minimum airflow through the heater and is to set a low limit in the case of warming up the system is necessary.

For this analysis, the control performance has not been optimized and the prime concern is to achieve stable operation. The control structure should be quite representative of what may actually be required for controlling the facility. It is set up so as to illustrate the dynamic performance of the system closed loop behavior only.

4.2 Analysis of The Heater Control

The controller uses bang-bang control techniques since the heating electrical power can only be full on or full off. Figure 5 shows the heater outlet temperature response using bang-bang control. The heater outlet temperature drops in about 3 sec during T_d (mixing outlet temperature) rises, causing a surge of airflow through the heater. And it takes longer to warm up the heater to restore the maximun temperature due to the long heater time constant. Therefore it might be necessary to keep the heater temperature at its maximum for all operating conditions to meet all airflow surge demands.

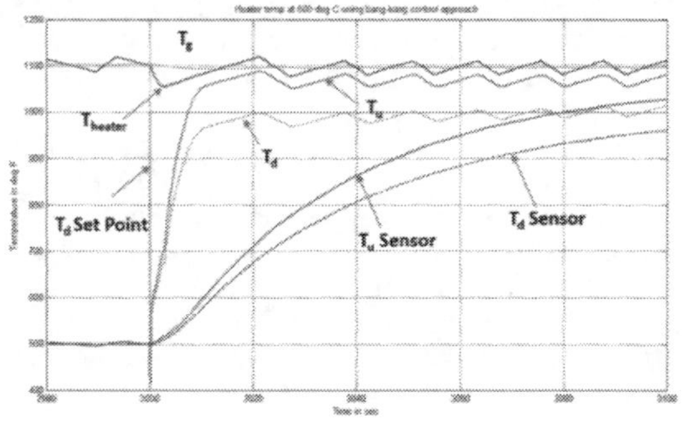

Figure 5. The heater outlet temperature response in °F using bang-bang control

4.3 Analysis of Temperature and Pressure Control

The cascade control techniques are used during the mixing outlet temperature control to overcome the heat capacity and volume effect of the pipeline and the heat exchanger, as well as the high-pressurized air. Figure 6 shows that the mixing outlet temperature (T_u) changes rapidly as the set point changes between different states. Comparing Figure 6 with Figure 7, the temperature control is stable and has no severe interactions with the pressure control, showing the facility structure mechanically decouples the temperature and pressure controls.

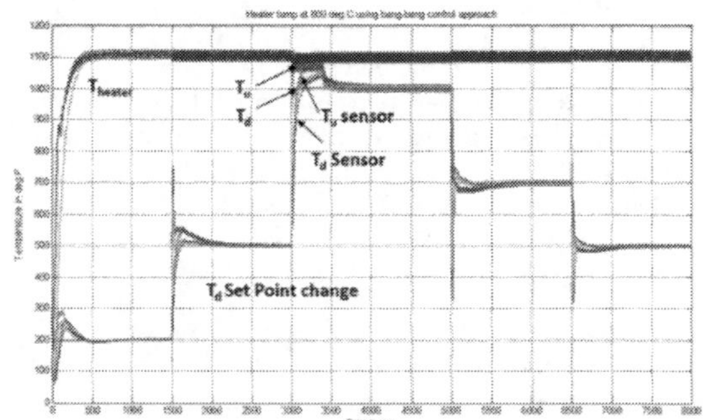

Figure 6. The mixing outlet temperature response in °F using cascade control.

As shown in Figure 7, the pressure control is stable and the response is rapid and accurate.

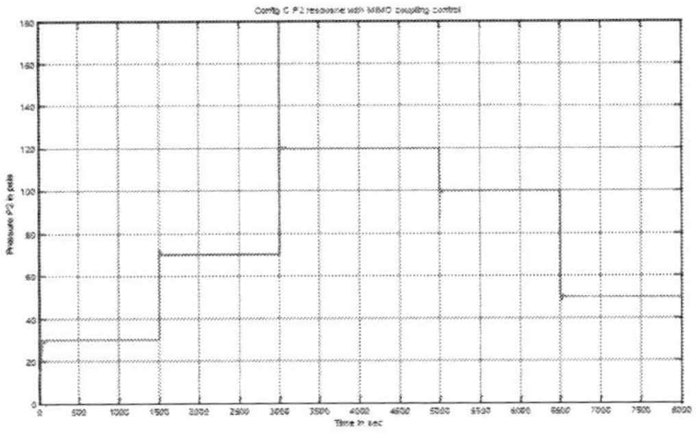

Figure 7. The pressure response in psia.

According to Figure 5, the mixing outlet temperature settling time is approximately 18 seconds, which cannot meet the dynamic requirement. Raise the temperature set point of the heater to 700 °C and get the temperature response as shown in Figure 8. The temperature settling time is approximately 8 seconds, meets the requirement. Therefore, a heater with higher temperature output should be used whenever physically possible.

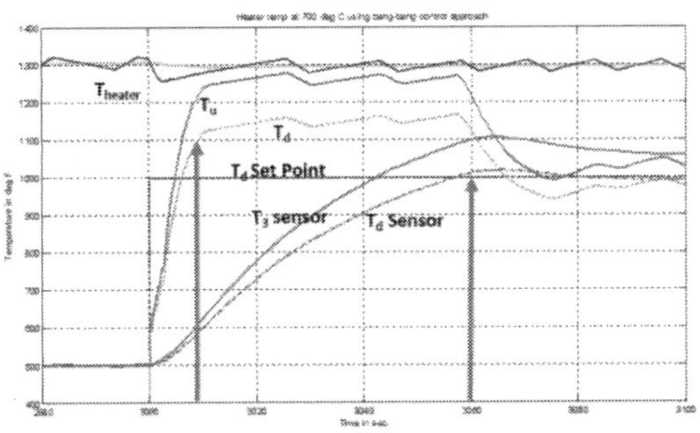

Figure 8. The temperature response after raising the heater outlet temperature.

5 Conclusion

For the purpose to perform the control design of Engine Bleed Air Test Facility, a dynamic simulation model for the facility was developed to simulate the dynamic behaviour. Common equipment performance is chosen to reflect the real system state to the greatest extent. Detailed modeling approaches have been introduced within several main equipments of EBATF. This simplified dynamic model may be used to perform trade studies and to optimize component characteristics for performance and cost comparison purposes.

Different control techniques are used to evaluate the dynamic performance and to find the best control design method.

The pressure control performance is close to ideal and likely could emulate the engine dyanmic performance closely. The mixing outlet temperature control is also close to ideal when the heater outlet temperature raises up to 700 °C. Bang-bang control produces limit cycling and has little effects upon the overall control performance and stability of the system.

With the performance shown, this Engine Bleed Air Test Facility should be quite capable to emulate the engine for bleed air system testing and evaluation.

References

1. SAE ARP 217D: Testing of Airplane Installed Environmental Control Systems. *Warrendale, PA : Society of Automotive Engineers*, (1999) .
2. J. HE, J. ZHAO. Dynamic Simulation of the Aircraft Environ-mental Control System. *Chinese Journal of Aeronautics*, Vol.**14**, No. 3, 129-133 (2001)
3. G. Fang, Z. Yao, Y. Hu. Modeling and Simulation for Temperature Control System of Environmental Control System Function Test. *Machine Design and Research*, **4**, 82-85 (2012).
4. D. SCHOLZ, C. MÜLLER, T. GIESE, C. ERDMANN. FLECS: Functional Library of the Environmental Control System-A Simulation Tool for the Support of Industrial Processes. *Proceedings of the 1st International Workshop on Aircraft System Technologies*. 143-157 (2007)
5. B. Andersen. The Analysis and Design of Pneumatic Systems. *R.E.Krieger Publishing Co.* (1967).
6. L. Zalmanson. Components for Pneumatics Control Instruments. *Pergamon Press*. (1965).
7. P. Hodal and G. Liu, Bleed Air Temperature Regulation System: Modeling, Control, and Simulation, *Proceedings of the 2005 IEEE International Conference on Control Applications*, Toronto, Canada, 1003-1008 (2005)
8. J. Mason. Heat Transfer in Cross Flow. *II Proceedings 2nd US National Congress of Applied Mechanics*. New York: ASME, 801-803(1995)
9. W. Roetzel and Y. Xuan. The Dynamic Behaviour of Heat Exchangers, *WIT Press*, (1998).
10. T. Marlin. Process Control. *McGraw Hill*. (1995).

Sustainable Power Generation by Tidal Current Turbine in Straits of Malacca

P. T. Ghazvinei[1], H. H. Darvishi [2], A. Bhatia[3]

[1]*Department of Civil Engineering, Technical and Engineering College, Ale Taha Institute of Higher Education, Tehran, Iran, Email: p.taherei@gmail.com*
[2]*Department of Civil Engineering, Engineering and Management of Water Resources, Shahr-e-Qods Brach, Islamic Azad University, Tehran, Iran, E-mail: hhassanpour87@gmail.com*
[3]*Department of Civil Engineering, Faculty of Engineering, University of Malaya, 50603 Kuala Lumpur, Malaysia, Email: aalisha9990@gmail.com*

Abstract. Marine current power is a significant energy resource which is yet to be exploited for efficient energy production. Malaysia, being a tropical country is rich in renewable sources and tidal power is one of them. In Malaysia, Straits of Malacca is a potential site to establish a tidal current turbine. In the current study, the potential sites of the Straits of Malacca are discussed. A detailed review about the generator suitable for the Straits of Malacca with the associated challenges has also been discussed. Furthermore, the suitable solution for such challenges is proposed. The role of simulation in choosing an appropriate site and generator has also been reviewed. The focus of the study is to propose a generator suitable for the flow characteristics of the Straits of Malacca.

1 Introduction

Marine energy has become an issue of significant research achieving a spectacular increase in the last years [1]. Studies reveal that the theoretical potential of oceans using different techniques can produce 20,000 TWh to 92000 TWh, however, the energy demand of our planet is 16000 TWh [2]. Hence, oceans alone are capable of meeting the electricity demand of the whole planet. It is currently the focus of much industrial and academic research around the world [3]. Indeed, the astronomic nature of this resource makes it predictable, to within 98% of accuracy for decades which makes it more reliable than wind and solar energy [4]. This predictability is critical for the successful integration of renewable energy in the electrical grid. There are various ways to extract energy from oceans: Saline gradient temperature gradient tides, waves and tidal currents [5] are the prominent ones.

Malaysia is a tropical country in South-east Asia. It is rich in both renewable and non-renewable resources. The potential and the amount of electricity produced in Malaysia by various resources have already been discussed elsewhere. Malaysian government has done various efforts in the area of renewable and sustainable resources. In the 8th Malaysian plan, it has introduced renewable energy as the fifth fuel. Presently, Malaysia is highly dependent upon non-renewable sources for electricity production. However, owing to the negative impacts and the limited reserves of fossil fuels it has also started exploring the renewable sources. Malaysia is rich in biomass (palm oil), solar energy, wind, and marine energy. The resource map of Malaysia is discussed in [6].

Researchers have worked on the various resources available in Malaysia and marine energy is one of them Studies on the potential sites in Malaysia to install tidal current turbine had been conducted. Three out of four potential sites are located in the Straits of Malacca (SoM). Hence, it is important to study about the crucial aspects to establish the tidal current turbine at this site. SoM is a water stretch between peninsular Malaysia and the Indonesian island of Sumatra. Figure 1 shows the location of SoM. SoM is famous for the heritage values and business development. The potential of SoM is found in literature. The site selection for the turbine installation has been studied in. Even after a lot of research on marine energy assessment in Malaysia only a few systems are practically tested. This is because of the lack of the study of technical aspects related to the field [7]. Therefore, the study of the generator for this site is crucial.

Generator is one of the important parts of the tidal current turbine. The selection of the generator is done on the basis of tidal current velocity, power produced and number of turbines. There are various methods to arrange the generator in tidal current turbine. That is, either one generator can be used for various turbines or a single generator can be used for each turbine. In this study, these factors and the associated challenges are discussed. The remedies for these problems are also proposed. Tides change direction in every 6-12 hrs depending on the types of tide. The rise and fall in the height of the tides and the change in direction cause instability in the produced power. This instability causes power fluctuations which is one of the major challenges for the generator of the tidal current turbine. Hence, a thorough study about the marine current generator, associated problems and solutions is required.

Selection of generator can be done by practical testing, simulation in laboratory and by using simulation software. Among all these three methods software testing is most convenient and cost-effective. Water velocity, wake effect, blade structure are the main factors which can be easily studied using simulations. The study of the wake effect helps in investigating the performance of the turbine farm. The electromagnetic field, response of the generated power, power fluctuations and other electrical factors can also be studied by using various software. In this study a brief discussion is made to review the role of simulations. However for the detailed discussion about the role of simulation techniques is made [8].

Figure 1. Geographical location of the Straits of Malacca.

The Straits of Malacca (SoM) is the longest international navigation route from the straits. It connects the Indian Ocean to the north with the South China Sea on the south via Andaman Sea and Straits of Singapore respectively. The straits is named after the Malacca Sultanate that ruled over the archipelago between 1400 and 1511. Fig 1 shows the geographical position of the Straits of Malacca. It is about 500 miles long, the width and depth varies from 11 to 200 miles and 10 to 70 meters respectively. The current flows in almost constant direction with the average current speed of 1-2m/s [9]. According to Peter Frankel the requirement of the site for tidal turbine is specific and the straits between islands & main lands are suitable locations to install the tidal turbine [10]. Thus, SoM can be an appropriate location for the installation of tidal power plant for generating electricity.

A number of studies have been made to analyze the nature of SoM and to examine the impact of installing a tidal current turbine in it. A study also shows that the nature of SoM is tropical. However, the studies on the flow characteristics of the SoM are limited. The termination of various rivers in the SoM makes its environment similar to an estuary. This means SoM has low salinity, high spatial and temporal variation in many physical and chemical elements. In addition to this the SoM also have rich primary productivity and bio-diversity. The hydro-graphic conditions of SoM are mainly because of the diurnal tides as well as by the northeast and southwest monsoons. The chemical & biological conditions of the straits and the effect of human activities on the hydro-graphics and marine environment have been discussed. The investigation of the aforementioned factors indicates that bio-fouling will have a significant impact in the SoM. Research has been carried out to produce electricity from the SoM. The tidal current turbine is the most efficient way of extracting power from the SoM. It was found that the area around the Pangkor Island has steady current that moves northward for most part of the year with the current speed of 1-2 m/s [11]. The speed may be variable as many rivers bring sediments which makes the straits shallower and reduces the speed. Another study was done using TPXO software to identify high stream current velocity sites in Malaysia. The study was carried out to test the main spots in Malaysia with highest tidal speed. There were 4 main sites in Malaysia: Sarawak, Pankgor Island, Malacca and Klang. Three out of these four sites are part of the SoM.

Pangkor Island, Malacca and Klang are the ports and islands of SoM. Among these three sites only Pangkor and Malacca seems to be appropriate options for establishing the turbines. The maximum velocity at these two sites is 2.4 m/s and 1.8 m/s respectively. The power output from the islands and ports of the SoM using the tidal stream approach were also studied in 2007 [7]. According to Lee and co-workers tidal turbine at Pangkor Island and Malacca can deliver power for 19 hours per day. As per the calculation done by Lee and co-workers the potential of SoM is to deliver 7.86 MW of power per year. However, the extractable power will be lower than this. Hence, a further detailed study on these sites will be helpful in determining an appropriate site for the installation of tidal turbine.

2 Tidal Current Turbine Generator

The potential of marine energy has attracted the attention of various industrial and academic researchers. The review about the tidal current turbines has been made in. Studies are being made on different turbine designs, to reduce the problem of sea scouring, to ease the procedure of installation and transportation and to improve the over-all efficiency. Along with all the aforementioned factors, generators also play a vital role in the working and efficiency of the turbine. The selection of generator depends upon the available water velocity, nature of tides and the energy output [12]. Along with the limited studies on the flow characteristics of the SoM, no study up to date has been done to select a suitable generator for this straits.

One research group from Upsalla University in Sweden is working on the low speed tidal current generator and suggested that the direct drive permanent magnet synchronous generator (PMSG) is an optimum solution for the sites with low water velocity. Most of the research from this group is for vertical axis tidal current turbine. According to a study by this group of researchers the iron losses for generator are more than the copper losses for low velocity sites. Ben Elghali and co-workers from France have been actively working on the horizontal axis tidal current turbine generator. Ben Elghali and co-workers suggest that the PMSG has better power control option for horizontal axis marine current turbines compared to the doubly fed induction generator (DFIG). Elghali proved this by introducing the second-order sliding mode control that takes into consideration the turbulence and swell effects of marine current flow. However, it is remarkable to note that most of the installed turbines used the induction generator [13].

In this study generators are classified into two main categories: alternating current (AC) generators (synchronous and asynchronous) and direct current (DC) generators.

2.1 DC Generator

In later half of the eighties, it was proposed to have a power system with 3-phase, 50 Hz A.C generation, transmissions and distribution networks. Once A.C system was adopted the transmission of large power (MW) at higher transmission voltage became feasible by using transformers. Level of voltage could be changed virtually to any other desired level with a transformer which was hitherto impossible with D.C system. This is the reason that A.C generator is used in power generation in place of D.C generators [13]. The absence of commutators in A.C generators makes the construction simpler.

2.2 AC Generator

The generator which produces alternating power is known as a.c. generator. The usage of generator for tidal current turbine is not specific and varies as per the requirement of the site. SeaGen, SeaFlow and Kinetic hydropower system tidal current turbines used induction generator while Encurrent turbine and Cycloidal turbine used synchronous generator. Kobold and Gorlov turbines used both induction and synchronous generator. For better understanding A.C generators are classified into two categories: Induction (Asynchronous) generators and Synchronous generators [13].

2.3 Induction generators

These generators use the principle of electromagnetic induction for power generation. In such generators the rotor rotates faster than the synchronous speed and thus has a negative slip. These generators mostly require initial external excitation. This can be done by connecting it to the electrical grid. However, such generators can be made self-excited by using phase correcting capacitors. Induction generator has various advantages: no voltage regulation, voltage and frequency are controlled by the utility, excitation is provided by the utility, synchronization of the generator with the system is not hampered by large power swings and simple construction with no brushes [13].

These are of two types singly fed and doubly fed induction generators. The main difference between these two generators is the single fed induction generator is self-excited while the DFIG requires external excitation. Even if DFIG is not self-excited its use is common in the renewable industry especially for power generation by wind turbine. The reason can be attributed to the constant amplitude and frequency of the output power. It is remarkable to note that the working of both the wind and tidal current turbine is similar but the density of air is 1000 times lower than that for water. This is the reason that the wind velocity in the range of 5-14 m/s is required for power generation while it is 1-3 m/s for tidal current turbine [14].

2.4 Synchronous generator

The other class of generator is synchronous generator. A synchronous generator consists of a stator containing a three-phase winding with each of individual phases positioned 120° apart, and a rotor containing a field winding which is magnetized by direct current. This current can either be drawn from a brush exciter, brushless exciter or from the grid. The synchronous generator rotates at a constant speed at steady state. Controlled reactive power and fixed speed regulation are the major advantages of synchronous generator. Synchronous generator supplies reactive power rather than absorbing it which is the key significance of such generators. The supply of reactive power increases the apparent power and thus the power factor of the system. A small change in the value of power factor has a major impact on the system stability and cost [15].

Synchronous generator is of four types: Permanent magnet, wound rotor, salient pole and cylindrical rotor synchronous generator. From these four types, cylindrical rotor synchronous generator are beneficial for high speed applications like steam generators. Thus these are not applicable for marine technology where the speed is low or medium. The usage of wound rotor is more inclined towards the traction systems. Salient pole and permanent magnets are widely used for the wind and tidal turbines. However, the cost of PMSG is higher than the salient pole synchronous generator (SPSG). The reason can be attributed to the usage of rare earth magnets, which are expensive. The difference in the cost and the performance of permanent magnet and salient pole generators is explained [16].

In the current study the focus is on the low speed generators. Both PMSG and SPSG can be used for low current speed [16] and use classic three phase windings. However, PMSG have high flux density, high torque, high acceleration, easy launch and high efficiency [17]. PMSG are well known with the advantages of no field winding, slip-rings and exciter generator, avoiding heat dissipation in rotor and providing over-all efficiency and the removal of the conductor losses of the rotor conductor [16]. Earlier PMSG were used for low or medium power generation but because of above advantages now these are frequently used for high power generation [17] also.

Efficiency and output of PMSG is higher than the SPSG. A comparison of the efficiency and cost of both the generators have been made [16]. After the analysis, it is understood that while the efficiency of the SPSG is lower than the PMSG's, the SPSG is more economical in terms of cost and the electronic drive circuits used to control the machines. Even though the use of PMSG may be advantageous in lower power applications, SPSG may be more advantageous in high power applications with a better design factor.

Additionally, the processing and disposal of rare earth materials is not environmental friendly [17]. It also has the disadvantages of limited flux density regardless of high coercivity, core saturation and poor electronic control reliability. The use of PMSG is inclined more towards high efficiency, low current velocity and light weight turbine whilst the use of SPSG is more favourable for medium water current sites and is more environmental friendly during disposal. However, the water velocity plays an important role in the selection of generator. Thus PMSG with the direct drive system seems to be a suitable choice for the SoM. Nevertheless, a more in-depth study and experimentation on this will led to a more evident solution for this.

3 Challenges for the tidal current turbine generator

In Section 3, the selection of generator based on the requirements is explained. However, there are various challenges for the tidal turbine generator. The impact of these varies for ocean water and tropical straits water. These challenges are discussed in detail in the below sections and the effect on different type of water is compared in table 1. The impact of power fluctuation will be lower for SoM as the speed of water is constant in the same direction, however, it is variable in case of oceans. This is the reason that it will be easier to control the output power from the generator installed at the SoM. The chemical composition of water effect the turbines parts exposed to water by causing corrosion and bio-fouling. Salinity of SoM is lower than ocean water; however, due to the presence of various organic components bio-fouling will be higher for SoM. The effect of heating of the windings depends upon the load and hence it is site dependent.

Table 1. Response of tropical and cold water towards various challenges.

Challenges	Tropical Water	Cold Water
Power fluctuations	Lower	Higher
Corrosion	Higher	Lower
Control System	Easy to control	Difficult to control
Heating of windings	Site dependent	Site dependent

3.1 Power Fluctuations

Power fluctuation is one of the major problems for power generated by renewable sources (tidal current turbines) and makes the grid integration difficult. These fluctuations are due to the swells and direction change of the tides [18]. There are two types of power fluctuations: short-time scale and long-time scale power fluctuations. Large time scale power fluctuations occur after six or twelve hours and are mainly because of the astronomical phenomenon. These fluctuations are predictable and depend upon flood and ebb tide, tidal speed, tidal acceleration, water depth, turbulence strength and intensity.

Small scale power fluctuations occur for a few seconds and are produced due to the swell effect. Swells are the waves which are travelled over a long distance after its area of generation. The marine current speed fluctuation by underwater swells is considerable. In contrast to this, long term power fluctuations influence the generation reserves and energy dispatch of the power system while short term power fluctuations influence the power system frequency control [18]. The short term power fluctuations have high frequency while the long term power fluctuations persist for a long time. The solution for this problem has been discussed in the later sections.

3.2 Corrosion

Sea water is a corrosive medium. The effect of corrosion on the parts of tidal turbines has been studied by many researchers as an important consideration for tidal turbine [14]. Generator of the tidal turbine is either completely or partly immersed in water. Hence, a protective coating should be applied on the winding of the generator. This prevents the untimely outages. There are various materials available for this purpose [19]. However, the effect of insulations on the electrical, mechanical and thermal characteristics of the generator is still under investigation.

3.3 Control systems

Control system is required to monitor and regulate the output power of the tidal current turbine. The control system helps to maintain an optimum speed of the generator so that maximum power can be harnessed from the turbine. According to Mbabazi for a particular water velocity there is always a rotor speed that produces maximum power. Few other researchers have also stated the same for wind and tidal turbines. The control system should be capable of protecting the system from over power. The effect of over power on the torque and rotational speed of the rotor for PMSG in case of wind turbines is discussed in [19]. The impact of over power in case of tidal turbine will also be similar as the same generator will be used.

The load characteristics regulate the torque characteristics for both the generator and turbine. Hence, the control of the load characteristics is important in order to optimize the operation of turbine and generator. One way of controlling the load characteristics is by controlling the current. The method of doing this is discussed in [19]. The results are also verified by simulating the same using Simulink.

3.4 Heating of winding

Tidal turbines which are installed for the commercial usage usually have large generators. The loading conditions of the turbine have an impact on its electrical, mechanical and thermal properties. The overloading of generator requires various measures as the cooling material is air which does not have heat transfer ability. Moreover, the windings of the generator rotor are confined in the steel slots and are subjected to temperature expansion problems. This is the reason that the tidal current turbine generator windings use various types of insulations. Hence, if overloading conditions are expected in a particular area which depends upon the power by tidal current various tests should be performed on the generator windings. These tests including with the results are discussed [20].

4 Proposed solution for the problems

The aforementioned problems are the few major challenges for the tidal current turbine generators. In this section few solutions are provided for the problems encountered.

4.1 Energy stabilisation

As explained in section 4.1 the power fluctuation is either small scale, large scale or both. Hence, there is a need to supply and absorb energy as per the requirement. The energy storage techniques used for marine current turbines have been explained [21]. The usage and the pros-cons of energy storage system by batteries and other methods have been done in table 2 and 3 respectively. The large scale fluctuations can be stabilized by battery storage systems, however, for small scale fluctuations the methods proposed in table 3 are more effective. Another method of stabilizing the power is increasing the power factor. A study has been done to analyze the power fluctuations in an established tidal current turbine in Seagen. The results show that the power fluctuation for Seagen was only because of the variation in tidal speed and the effect never exceed 1.25% for voltage and 0.65% for current. The study also says that the high power factor (0.99) helps to minimize the effect of power fluctuation. The problem of power fluctuation is common for all the renewable sources. As stated earlier the working principle of both wind and tidal turbine is the same. A novel technique of stabilizing the power for wind turbine was invented by Mohammad Lutfur Rahman and co-workers. The authors explained that the tidal turbine with an induction generator can be used as a motor or generator depending upon the requirement which will charge and discharge energy like a flywheel and this system will be connected after the wind turbine using a synchronous generator. The performance of this technique was verified by using a mathematical simulation on PSCAD. Mohammad Lutfur Rahman in the same study has also stated that the tidal power is more stable than wind energy and IGBT (insulated gate bipolar transistor) converters can be used for power stabilization. It is remarkable to note that IGBT's are used in removing the fluctuation in case of Seagen [21].

As the water of SoM does not have high variation in water velocity compared to oceans, hence, the use of IGBT converters can minimize the effect of power fluctuations.

4.2 Prevention of corrosion

As discussed in the previous sections the salinity of SoM is lower than the sea water. However, the presence of various chemicals increases the chances of bio-fouling and makes the environment harsh for tidal current turbines. As per Bahaj and co-workers this can be prevented by enclosing the generator parts in a nacelle and sealing it properly. Coating of the various components by paint and increasing the thickness of the parts also helps in the prevention of corrosion [14].

4.3 Control strategies

Control strategies are important for all the power generation systems. In case of tidal turbine no case of system failure has been reported due to poor generator control strategies. However, there is a case of whole turbine failure due to high speed tidal currents. This may have a serious impact on the working and efficiency of the generator. Depending upon the type of generator and the site velocity, various control strategies can be used. Including with this the connected load also plays a vital role to decide the appropriate control strategies. The deeper study on this has been made by various researchers [22].

4.4 Cooling method

The windings of generator are mainly heated up during the overloading conditions. Including with this for cost cutting the cross sectional area of the conductor, thickness of the insulator, and material in the steel core are reduced. This factors leads to the heating of the generator winding. The conditions for overloading also lead to the heating up of the windings. However, if proper measures are taken then this problem can be controlled to a great extent.

Table 2. Pros and Cons of various battery type energy storage system.

Battery Type	Pros	Cons
Lead acid [23]	easy installation low cost negligible maintenance low self-discharge rate (2-5 % the rated value) ideal for long term storage	low energy density (30wh/kg) short service life (1200-1800 cycles) Temperature dependent batteries not suitable for discharge over 20%
Nickel	longer life time (3000	high cost

based (Nickel Cadmium)	cycles) fully dischargeable	toxic components of Cadmium [24]
Nickel based (Nickel-metal hydride)	high energy density work over high temperature range	life cycle will be shorten by their repeatedly discharge on high load current [25]
Nickel based (Nickel-zinc)	deep cycle capabilities high energy density [26]	life cycle is hampered by the growth of the dendrites
Lithium ion	high energy density (90 to 190 Wh/kg) high power density (500 to 2000 Wh/kg) [27] low self- discharge	actual efficiency is lower than theoretical fragile difficult maintenance high cost
Sodium sulphur	high energy density (100 Wh/kg highly efficient at high temperature 100% depth of discharge low cost high efficiency deep discharge tolerance efficiency between 75% to 90% average cycle life of 2500 [28]	corrosive nature of sodium high operating temperature
Flow batteries [29]	power and energy ratings can be sized independently long service life (about 10,000 cycles at 75% depth of discharge) high safety negligible degradation for deep discharge negligible self-discharge	Complexity High cost

Table 3. Various types of energy storage systems.

Energy Storage System	Advantages	Disadvantages
Flywheel	Long service life (105 cycles with deep discharge) High power density High efficiency Low maintenance	High self-discharge High cost
Super Capacitor	High energy efficiency Long service life (500,00 cycles) Wide temperature range	High self-discharge Very low energy density (5 Wh/kg)
Pump Hydro Storage system	High efficiency (up to 75%) High power storage Favorable for load levelling	Appropriate geographic site High cost Long construction time
Compressed air energy storage	Cost effective High reliability Power smoothening	Geological dependence Use of fossil fuel Energy dissipation

4.5 Role of Simulation

Water turbines have complicated geometry. Thus, there are various factors which are considered before the selection and during the testing of generator used for a tidal current turbine. There are physical factors like marine current speed, electromagnetic fields and wakes. The other factors are the statistical factors like the voltage- current characteristics of the power harnessed, effect of the energy storage system on the output, deviation in power by using different generator to name a few. Study of the factors and situations can be done by using simulation software. Investigations of the physical factors (marine current speed, wakes, and electromagnetism) have been done by various researchers using a computational fluid dynamics software called OpenFOAM [30]. For the study of electromagnetism, partial discharge and losses a number of study has been done is still ongoing using the various software. The use of PSCAD, Matlab, and Opera is more common for the study of the response of electrical factors.

5 Conclusion

Marine current has the potential to generate electricity with a low negative impact on the surrounding environment. The sites of 'Straits of Malacca' like Pangkor Island and Malacca are capable of producing electricity which could supplement the energy needs of Malaysia and prevent the greenhouse gas emission. The velocity of the water is at these site is not high but with the use of direct drive permanent magnet synchronous generator the desired output can be achieved. High bio-fouling in the tropical water can be regulated by coating the generator parts and turbine with optimum materials. Simulation and numerical modelling will help in achieving better results without wasting the resources. Being a tropical water site the challenge of power fluctuation will be lower for SoM and can be controlled by using IGBT. The research going in the various parts of the world has great advantage regarding the slow velocity sites. The deployment and testing of a tidal current turbine at this site will serve as a milestone for the similar future projects.

References

1. Fergal RO, Boyle F and Reynolds A. Marine current energy devices: Current status and possible future applications in Ireland. Renewable and Sustainable Energy Reviews April 2010; 14 (3): 1026-1036.
2. Esteban M, Leary D. Current developments and future prospects of offshore wind and ocean energy. Applied Energy February 2012; 90 (1): 128-136.

3. Henrey J, Jay B, Winskel M. Accelerating the development of marine energy: Exploring the prospects, benefits and challenges. Technological Forecasting and Social Change April 2012; 23.
4. Julio V, McKesson C, Walczak M. Sustainable energy for the marine sector. Energy Policy October 2012; 49: 333-345.
5. Hassan HF, Karim OA, Al-Shafie E. Tidal current turbines glance at the past and look into future prospects in Malaysia. Renewable and Sustainable Energy Reviews October 2012. 16 (8): 5707-5717.
6. Chen WN. Renewable Energy Status in Malaysia. Sustainable Energy Development Authority Malaysia December 2012; Malaysia.
7. Andaya LY. Leaves of the Same Tree: Trade and Ethnicity in the Straits of Melaka. University of Hawaii Press 2008; USA.
8. Ramos V, Carballo R, Álvarez M, Sánchez M, Iglesias G. Assessment of the impacts of tidal stream energy through high-resolution numerical modelling. Energy 2013.
9. National Geospatial-Intelligence Agency. Sailing directions-Strait of Malacca and Sumatera 2013; Springfield, Virginia.
10. Frankel PL. Tidal Current Energy Technologies. International journal of avaian science- IBIS March 2006; 148 (s1): 145-151.
11. Looi LJ, Aris AZ, Johari WLW, Yusoff FM, Hashim Z. Article in Press. Baseline metals pollution profile of tropical estuaries and coastal waters of the Straits of Malacca. Marine Pollution Bulletin 2013.
12. Ng KW, Lam WH, Ng KC. 2002–2012: 10 Years of Research Progress in Horizontal-Axis Marine Current Turbine. Energies 2013; 6: 1497-1526.
13. Elghali B, Benbouzid SE, Benbouzid MEH, Charpentier JF. Generator Systems for Marine Current Turbine Applications: A Comparative Study. IEEE Journal of Oceanic Engineering 2012; 37 (3): 554-563.
14. Arief A, Dong ZY, Nappu MB, Gallagher M. Under voltage load shedding in power systems with wind turbine-driven doubly fed induction generators. Electric Power Systems Research March 2013; 96: 91-100.
15. Machowski J, Bialek JW, BumbyJR. Power System components in Power Sytem dynamics. John Wiley and Sons Ltd 19-25. Published Online 1997.
16. Kömürgöz G, Gündogdu T. Comparison of Salient Pole and Permanent Magnet Synchronous Machines Designed for wind turbines. Power Electronics and Machines in Wind Applications (PEMWA), 2012 IEEE.
17. Ghiţă C, Chirilă AI, Deaconu ID, Ilina DI. Wind turbine permanent magnet synchronous generator magnetic field study. International Conference on Renewable energy and Power Quality 2007; Santander, Spain.
18. Jadhav HT, Roy R. A comprehensive review on the grid integration of doubly fed induction generator. International Journal of electric Power and Energy System July 2013; 49: 8-18.
19. Chinchilla M, Arnaltes S, Burgos JC. Control of Permanent-Magnet Generators Applied to Variable-Speed Wind-Energy Systems Connected to the Grid. IEEE Transactions On Energy Conversion March 2006; 21 (1): 130-135.
20. Judendorfer T, Fletcher J, Hassanain N, Mueller M, Muhr M. Challenges to machine windings used in electrical generators in wave and tidal power plants. IEEE- Electrical Insulation and Dielectric Phenomena 2009; 238-241.
21. MacEnri J, Reed M, Thiringe T. Influence of Tidal Parameters on SeaGen Flicker Performance. Philosophical Transactions of the Royal Society A: Mathematical, Physical and Engineering Sciences 2013; 371 (1985).
22. Zhou Z, Scuiller F, Charpentier JF, Benbouzid M, Tang T. Power Limitation Control for a PMSG-Based Marine Current Turbine at High Tidal Speed and Strong Sea State. IEEE, 2013.
23. Culpin B. Secondary Batteries – LEAD– ACID SYSTEMS | Valve-Regulated Batteries: Oxygen Cycle. Reference Module in Chemistry, Molecular Sciences and Chemical Engineering, from Encyclopedia of Electrochemical Power Sources 2009; 705-714.
24. Espinosa DCR, Tenorio JAS. Recycling of nickle-cadmium batteries- Thermogravimetric behaviour of electrodes. Journal of Power Sources 2006; 160 (1): 644-751.
25. Notten PL, Latroche M. Secondary batteries- Nickel system Nickel Metal Hydride: Metal hydrides. Encyclopedia of Electrochemical Power Sources 2009; 502-521.
26. Pavlov AP, Grigorieva LK, Chizhik SP, Stankov VK. Nickel-Zinc batteries with long cycle life. Journal of Power Sources 1996; 62 (1): 113-116.
27. Khaligh A, Alireza, Zhihao L. Battery,ultra capacitor,fuel cell,and hybrid energy storage systems for electric,hybrid electric,fuel cell,and plug-in hybrid electric vehicles: state of the art. Vehicular Technology, IEEE transaction 2010; 59 (6): 2806-2814.
28. Chen H, Cong TN, Yang W, Tan C, Li Y, Ding Y. Progress in electrical energy storage system: A critical review. Progress in Natural Science March 2009, 19 (3): 291–312.
29. Dunn B, Kamath H, Tarascon JM. Electrical Energy Storage for the Grid: A Battery of Choices. Science 18 November 2011; 334 (6058): 928-935.
30. Higuera P, Lara JL, Losada IJ. Simulating coastal engineering processes with OpenFOAM®. Coastal Engineering, January 2013; 71: 119-134.

Numerical Analysis of the Adiabatic and Quasi Steady Model of Free Piston Stirling Engine

Ayodeji Sowale and Athanasios J. Kolios

Offshore Renewable Energy Engineering Centre, School of Water, Energy and Environment, Whittle Building 52, Cranfield University, Bedfordshire MK43 0AL

Abstract. This study presents the numerical simulation of the adiabatic and Quasi steady models of the free piston Stirling engine, the mathematical equations are presented, and design parameters are determined and used as input for the simulation. The simulations are computed under adiabatic and Quasi operating conditions, and their output results are compared. The similarities and differences in the model predictions in terms of the pressure to volume diagram, the amplitudes of the pistons and displacer, temperature, efficiency, power output and stable operation are observed and investigated. The models are validated against the experimental output and the results show a good agreement with the experiment. The adiabatic model predicted an output power of 862 W, while Quasi steady model predicted more accurate output power of 997W at frequency of 30 Hz in relation to the 1000 W of the experimental output. The effects of the variation of engine's parameters on the output power are also observed and presented.

1 Introduction

In recent years the consumption of energy and environmental pollution has been a major issue of controversy. The rapid increase in the cost of supply of energy and the continuous increase in environmental and noise pollution, have resulted to investigations into substitute machines that converts energy in the form of thermal, electrical or pressure to mechanical form. The free piston Stirling engine (FPSE) was introduced by Willian Beale in 1964. It is an external combustion engine with no mechanical linkage of the oscillatory components [1]. The disadvantage of the FPSE is the lack of rotating shaft, the engine uses springs and dampers in the oscillatory movements during operation, which can act as a hindrance in the stability and effective modelling of the FPSE. Therefore, accurate description and calculation is required to determine the operational performance and achieve a steady operation of the engine. Recent investigations have been conducted on the thermodynamic analysis of the free piston Stirling engine. Various approach have been carried out on the numerical model of first and second order models of the FPSE. A study was conducted by Formosa and Frechetter [2] on the scaling effects on the FPSE taking into consideration the heat exchangers geometry and the masses of the displacer and piston in relation to the heat losses in the engine. The output results showed that the indicated output power and efficiency of the miniaturized FPSE can be determined using the stipulated method. Begota et al [3] evaluated the stability analysis of the FPSE using the dynamic and thermodynamic method by employing the eigen values of state matrix and the model showed a good agreement with the experimental output. Riofrio et al [4] presented a dynamic model using the linear control method to design the FPSE to determine the stability of the engine. The influence of the temperature and damping load on the engine output was also observed and presented. Sowale [5] developed a numerical analysis of the Quasi, isothermal and adiabatic model of the FPSE by solving the motion equations of the displacer and piston simultaneously with the thermodynamic equations to determine the output performance of the engine. The results showed a good agreement with the experimental output. A mathematical model of the FPSE was developed by Jia et al [6] where the features of the pistons and displacer were defined including their natural frequency. The outputs showed the system was under forced excitation by varying the damping coefficient and spring stiffness. The piston and displacer's velocity and amplitude plotted against each other, formed a convergence which indicated the stable operation of the engine. Yang et al [7] developed thermodynamic analysis of a modified FPSE using the thermoacoustic theory. The results presented a good thermodynamic performance compared to the developed prototype. A numerical model of the FPSE was developed by Kim et al [8] to investigate the effects of the combustion phase on the motions of the displacer and piston. The output performance showed various spark timing for the piston velocity can be determined using efficiency conversion to work from combustion.

In this study, the adiabatic and Quasi steady model of the FPSE under the similar operating condition is

https://doi.org/10.1051/matecconf/201819804005

presented and results analysed to compare the performance of the two models. MATLAB is used to predict the performance of the free piston Stirling engine, where the characteristics of the engine are described by the linear dynamics theory. The modelling of the free piston Stirling engine requires an accurate dynamic equation of motion for the piston and displacer to predict the engine's performance. The analysis of the equations of motion for both the piston and displacer are solved in terms of the damping and stiffness coefficients of the engine. The aim of this paper is to predict the operating conditions of the FPSE at steady state and to generate the power output of 1kW. The numerical models are validated with experimental data of the Sunpower Re-1000 free piston Stirling engine prototype. The results of the output obtained from the proposed model using the standard description of the engine gives a close result to the experimental output.

2 Mechanism of FPSE

The mechanism of the FPSE is shown in Fig. 1. The engine consists of a displacer and piston, mechanical springs and dampers for both the displacer and piston and the engine casing. There is no mechanical linkage between the piston and displacer like the kinematic Stirling engines, the mechanical springs connect the displacer and piston the engine casing. The dampers act as an additional load to the piston and displace and they also helps to reduce vibration during oscillation. The pressure from the working gas volume in the engine acts on the piston and displacer and subject them to oscillatory motion in their cylinders. The displacer is designed to have a low mass compared to the piston so as to maintain necessary phase angle and produce more effective output. The working volume is made up of the expansion and compression spaces volume, and heat exchangers volume including hot, regenerative and cold volume. The working gas passes through a heating, regeneration and cooling process in the heat exchangers chambers.

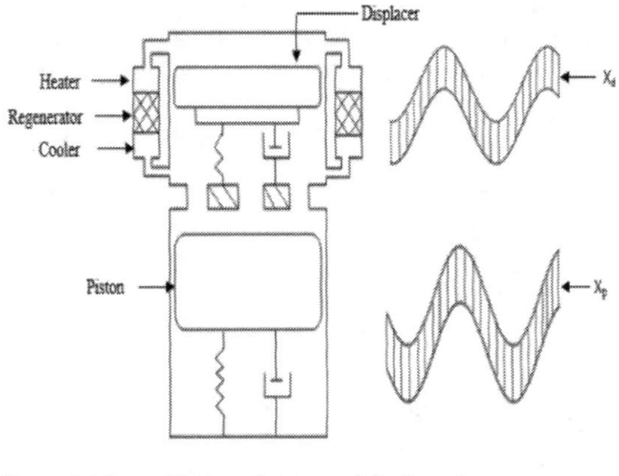

Figure 1. The oscillation of piston and displacer in correspondence with the position of the piston and displacer in the engine during operation.[5]

3 Methodology

The numerical model of the adiabatic and Quasi steady of the FPSE are developed in MATLAB program. The adiabatic was modelled under the adiabatic condition of the working fluid temperature within the walls of the expansion and compression space. The Pressure, volume and temperatures of the work spaces and heat exchangers were determined by making them functions of each parameter and was computed over the cycle until steady state was achieved. The Quasi steady model takes into consideration more details on the thermodynamic performance of the FPSE including the thermal losses, pressure drops. The operating conditions and assumptions were stated for the models. Input parameters used are that of the Sunpower RE-1000 FPSE and is presented in Table 1.

3.1 Assumptions for the ideal adiabatic model

- The total mass of gas in the engine is constant.
- There is no pressure drop in channels of the engine.
- The engine operates at steady state condition.
- The gas is an ideal gas.
- The spring's masses are not considered.
- The springs in the engine are linear.
- The motion of the engine casing is neglected.

3.2 Assumptions for the Quasi steady model.

The assumptions for the general operating conditions of the numerical models were determined and are stated below:

- The working volume pressure and block pressure is equal when the engine is in static equilibrium.
- The temperatures of the surrounding walls of the working gas volume changes with time.
- The masses of the piston and displacer springs are not considered.
- The springs are positioned in a linear order.
- No leakage is expected to occur between the mechanical components of the engine.
- The working fluid is an ideal gas.
- To maintain the required phase angle between the displacements of the moving elements the mass of the displacer is lighter than the piston.
- The regenerator temperature is required to be equal to the average of the cooler and heater temperature.
- The engine block is stationary.

The flow chart of the numerical procedure of the simulation of the FPSE is presented in Fig. 2.

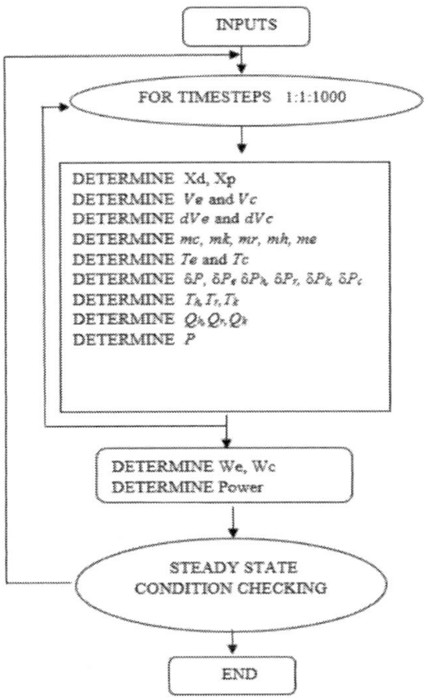

Figure 2. Flow chart of the numerical procedure

3.3 Equations used in the mathematical model

3.3.1 Generalized analysis

To describe the action of a free piston Stirling engine, it is essential to assume the harmonic motion of the displacer and piston. There are three forces that are applied during the operation, they are; applied force (force from the pressure of the working gas), spring force and damping force [9]. The equations of motion for both the piston and displacer are simplified to determine the operation of the engine. The outputs generated by the engine are the temperature, pressure, displacements and velocity of the displacer and piston. Using the Newton's second law that is employed to describe the motion of the piston and the displacer, it states:

$$\sum F = m\ddot{x} \tag{1}$$

$$F + F_{damping} + F_{spring} = M\ddot{x} \tag{2}$$

The motion of the displacer is a result of the spring force, damping force, area of the displacer and the pressure of the working fluid.

$$m_d\ddot{x} + K_d x_d + C_d\dot{x} = P(A_d) \tag{3}$$

Where m_d is the mass of the displacer. K_d displacer spring constant. C_d displacer damping constant. A_d cross sectional area of the displacer. The motion of the piston is a result of the damping force, spring force, area of piston, the bounce space pressure and the pressure of the working fluid.

$$m_p\ddot{x} + K_p x_p + C_p\dot{x} = A_p(P - P_b) \tag{4}$$

Where m_p is the mass of the piston. K_p piston spring constant C_p piston damping constant. A_p cross sectional area of the piston. P_b bounce space pressure. The equation for pressure in the engine for a closed cycle operation [1].

$$P = \frac{MR}{\left(\frac{Vc}{Tck}+\frac{Vk}{Tk}+\frac{Vr}{Tr}+\frac{Vh}{Th}+\frac{Ve}{The}\right)} \tag{5}$$

Boundary conditions for the conditional temperatures

If $m_{ck}{}' > 0$ *then* $T_{ck} = T_c$ *else* $T_{ck} = T_k$ *If* $m_{he}{}' > 0$ *then* $T_{he} = T_h$ *else* $T_{he} = T_e$ (6)

From the engine's geometry the positions of the piston and displacer x_p and x_d with the expansion and compression volumes Ve and Vc can be derived from the following equations:

$$Ve = (V_{cle} + x_d)A_d \tag{7}$$

$$Vc = (V_{clc} + x_p)A_p - (x_d + V_{cle})A_d \tag{8}$$

Where V_{cle} and V_{clc} are the expansion and compression space clearance volumes. By differentiating Ve and Vc above the new equations resulted to

$$dVe = \dot{x}\,(A_d) \tag{9}$$

$$dVc = \dot{x}(A_p) - \dot{x}(A_d) \tag{10}$$

The momentum equation for the displacer

$$m_d\ddot{x}_d + K_d x_d + C_d\dot{x}_d = P(A_d) \tag{11}$$

The momentum equation for the piston

$$m_p\ddot{x}_p + K_p x_p + C_p\dot{x}_p = A_p(P - P_b) \tag{12}$$

Work done in compression space and expansion spaces are

$$\frac{dW_c}{dt} = P_c\frac{dV_c}{dt};\ \frac{dW_e}{dt} = P_e\frac{dV_e}{dt} \tag{13}$$

The indicated work in the cycle is

$$W_{in} = \int_0^t \left(P_e\frac{dV_e}{dt} + P_c\frac{dV_c}{dt}\right) \tag{14}$$

The indicated power of the engine is

$$P_{in} = W_{in} f \tag{15}$$

3.4 Procedure for generalized simulation

MATLAB was employed to write the codes for the numerical simulation using the mathematical equations above. The initial input parameters for the engine are determined at the beginning of the program, some which are variable and constant. The working process of the FPSE is analysed using the mathematical equations listed above. The initial values of both the variable and constant parameters were defined, such as; the positions of the displacer and piston, the working spaces and heat exchangers temperature, then the equations of the motions of the piston and displacer are simplified using Runge-Kutta (rk4) solver in MATLAB that solves differential equations simultaneously using a trial step interval midpoints to remove errors in computation. For the Quasi model simulation, the initial values of the parameters are defined suh as; the temperatures of the heater and cooler, the dynamic pressure, positions of piston and displacer at equilibrium, the equations for pressure drop and thermal losses are determined. Also the initial temperature in the expansion and compression spaces are defined and Runge Kutta (rk4) was employed to solve all the equations using siimilar approach to the procedure in the adiabaic simulation. The simulation is then computed over 5 cycles in 1000 time steps to achieve steady state operation of the FPSE.

Table 1. Parameters Of The Sunpower Re-1000 Fpse [1]

Engine Data	Value		Value
General		**Geometric**	
Mean pressure	70 bar	Displacer frontal Area	0.018m^2
Heater temperature	780 K	Expansion space clearance volume	0.02m^3
Cooler temperature	350 K	Compression space clearance volume	0.016m^3
Oscillating frequency	30 Hz	**Masses**	
Phase angle	90 °	Mass of the piston	6.2kg
Geometric	0.054mm^3	Mass of the displacer	0.426kg
Dead volume	0.016mm^3	**Dynamic**	
Volume of heater	0.027mm^3	Piston spring stiffness	296000 N/m
Volume of regenerator	0.011mm^3	Displacer spring stiffness	25000 N/m
Volume of cooler	0.018m^2	Piston damping load	461.5 Nsm^{-1}
Piston frontal area	0.018m^2	Displacer damping load	35.34Nsm^{-1}

4 Results and Discussion

The output performance of the adiabatic and Quasi steady model of the free piston Stirling engine is presented in Figs. 3 to 11 to illustrate the comparison of the adiabatic and Quasi models of the FPSE. Their performance outputs are analysed and discussed.

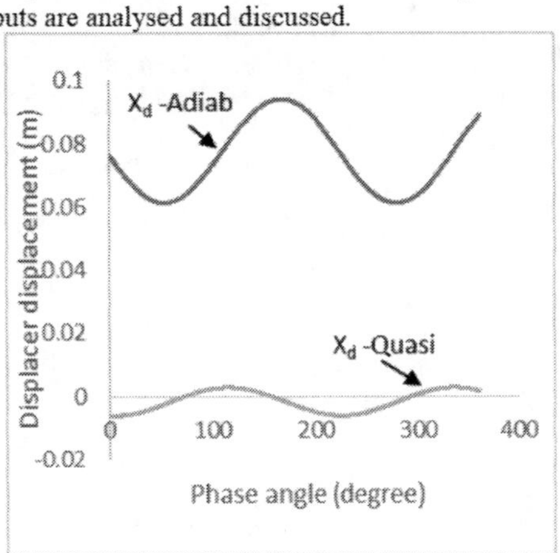

Figure 3. Displacer positions against phase angle

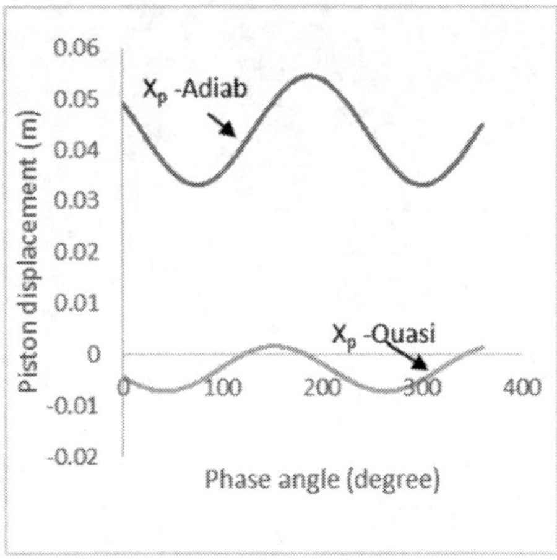

Figure 4. Piston positions against phase angle

The displacement of the displacer and piston amplitudes for the adiabatic and Quasi model against phase angle is presented in Fig.3, and Fig. 4. The displacer of the adiabatic model exhibited maximum displacement of 0.094 m at phase angle of 163°, while the displacer of the Quasi steady model exhibited maximum displacement of 0.002 m at 116°. This is due to the effect of the working gas pressure on the displacers, which also indicates position of the displacer at the top dead centre in the cylinder. In Fig. 4, the piston in the adiabatic model exhibited maximum displacement of 0.054m at phase angle of 186°, while the piston in the Quasi steady model exhibited maximum displacement of 0.001 m at 154°. The displacer and piston positions in the adiabatic model shows a considerable margin to the amplitudes of the Quasi steady model of the FPSE. This is due to the pressure function calculation in the numerical model and the effect of the damping load and spring stiffness on the oscillatory motions of the displacer and piston respectively in the two models. It can also be observed that the adiabatic models have an average piston amplitude of 0.03 m and displacer amplitude of 0.02m. While the Quasi model have an average piston and displacer amplitude of 0.003m and 0.006m. The variation is due to the magnitude of the working gas pressure exerted on the piston and displacer.

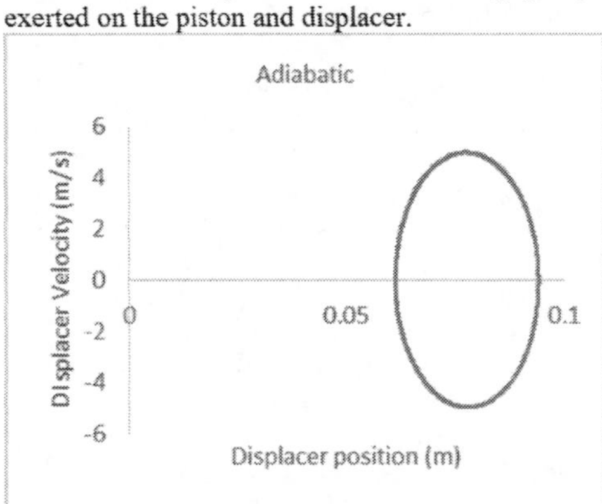

Figure 5. Displacer velocity against displacer position in adiabatic model

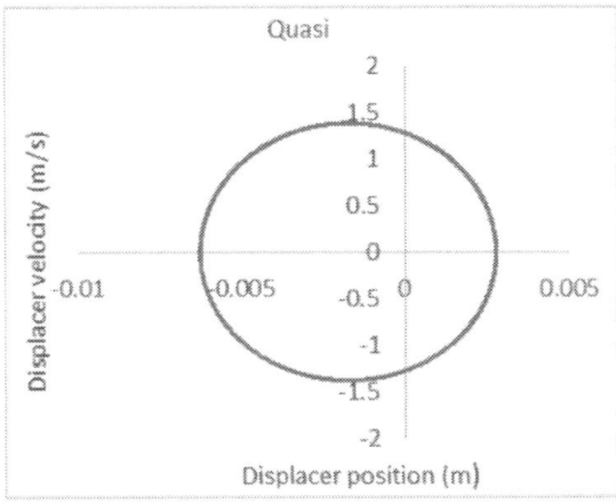

Fig 6. Displacer velocity against displacer position in the Quasi model.

Fig. 5 and Fig. 6, show the velocity of the displacer against the position of the displacer during engine operation, this represents the stability of the two models. It can be observed they form an ellipse each and do not overlap. This is also observed in the results presented in [6] and [10] on the stability check of the operation of the FPSE.

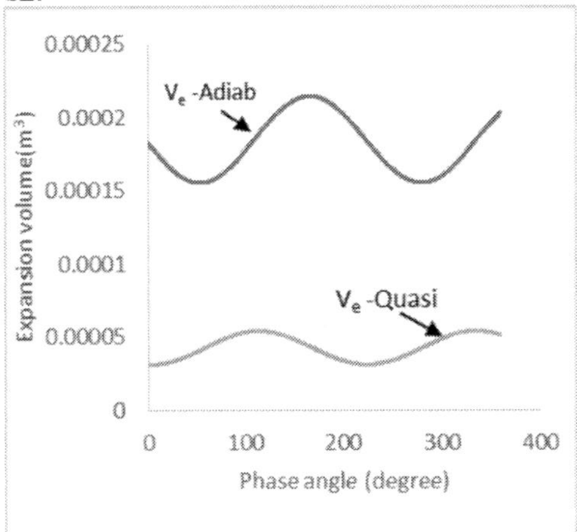

Figure 7. Expansion volume against phase angle

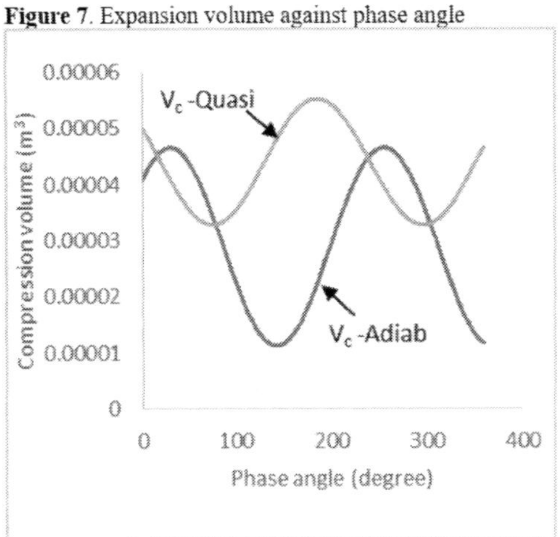

Figure 8. Compression volume against phase angle

The volumes of the working fluid in the expansion and compression spaces against phase angle for the adiabatic and Quasi steady models are illustrated in Fig. 7, and Fig. 8. The adiabatic model exhibited a higher volume difference compared to the Quasi steady model. The higher volume difference in the work spaces can relatively increase or decrease the output power of the FPSE. The adiabatic model predicted a lower output power compared to the Quasi model due to the volume difference in the work spaces. This is as a result of the average pressure and temperature over a complete cycle.

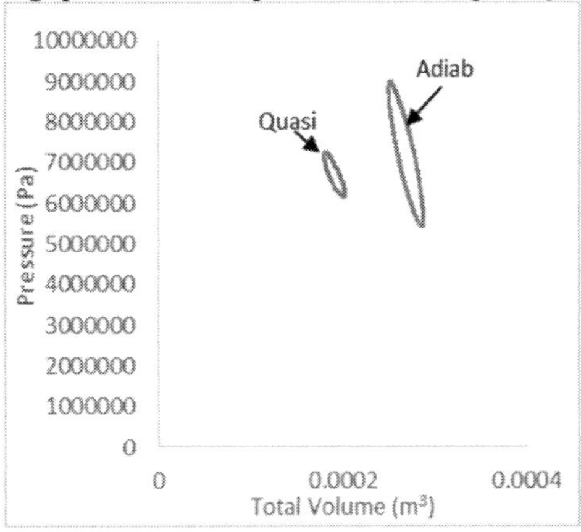

Figure 9. Pressure to Volume diagram

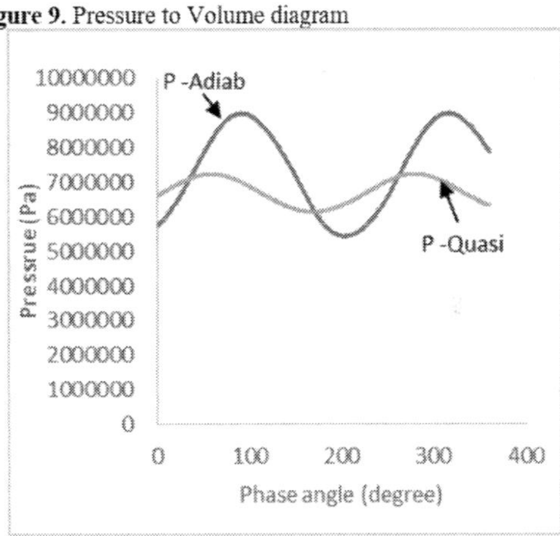

Figure 10. Pressure against phase angle

Fig. 9, represents the pressure to volume diagram of the two models in this study. They present identical elliptical shapes to the ideal Stirling cycle, but the area around the curvature indicates the dead volume not put to work in the expansion and compression spaces. The P-V diagram in the adiabatic and Quasi models have an average pressure of 70bar. This is due to the limited flexibility of the model and the pressure calculation in the numerical simulation. The adiabatic and Quasi steady models presented a more accurate prediction of the FPSE in the output power generated. Fig. 10, shows the pressure variation in the models against phase angle. The adiabatic model exhibited a higher pressure of 89.6 bar at 91° phase angle compared to the Quasi model that exhibited the maximum pressure of 72.3 bar at 65° phase angle. This is due to the temperature variation, volumes of the work spaces and the mass of the working fluid and gas constant. The consideration of thermal losses and pressure drop in pressure calculation of the Quasi model make a significant difference in the output performance of the models, as this is the major reason for the difference in model prediction. While in the adiabatic model, there was no account for the losses in the system, except the pressure was a function of the piston and displacer dynamics and temperature variation in the workspaces over a complete cycle.

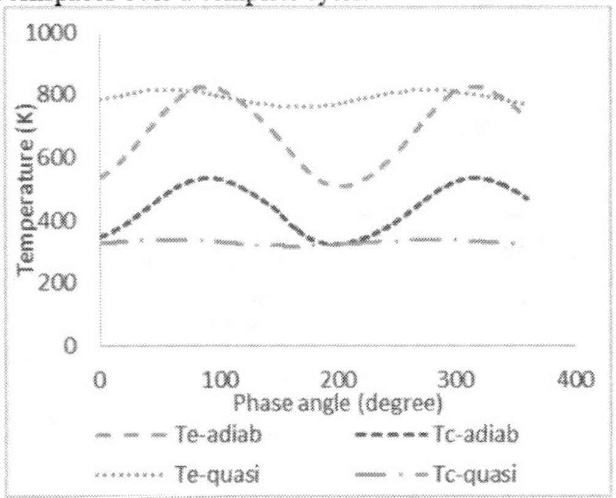

Figure 11. Expansion and compression space temperatures against phase angle

The comparison of the temperature in the expansion and compression spaces in the two models is presented in Fig. 11. In the adiabatic and Quasi steady model, the temperature in the work spaces vary with time in accordance to the assumptions relating to the real time approach of the FPSE. This gives a more accurate prediction of the engine performance. The start points of the temperature varies slightly in the two models due to different cycles each achieves steady state. The criteria for steady state condition checking for the adiabatic and Quasi model was determined by the expansion and compression space temperatures once the value at the beginning of the cycle is the same as the value at the end of the cycle, similar to the condition stated by Urieli and Berchowits [1].

Table 2. Comparison of the output power and efficiency of the models with experiment

Output	Experiment	Adiabatic	Quasi
Output power (W)	1000	862	997
Amplitude ratio	1.06	1.76	1.01
Frequency (Hz)	30	30	30

The difference in the output results of the two models of the FPSE is as a result of the thermal losses pressure drops and influence of the dampers and spring stiffness on the steady state operation of the engine. It can be observed in Table 2, that the Quasi steady state model predicted the closest result to the experimental output. The adiabatic model exhibited a lower value compared to the experimental output power due to the adiabatic condition of the expansion and compression work spaces and there was a slight difference in the displacer damping constant so as to achieve the steady state operation, during simulation at exact damping constant value of 35.34 ms-1 there was no convergence and the computation was altered after few completed cycles. The closest in output result to the experimental value is the Quasi steady state model. This model took into consideration the thermal losses and pressure drops and regenerator effectiveness. The temperature of the work spaces and the heat exchangers changes with time over the cycle similar to real time operation of the FPSE. Hence, the Quasi model predicted the most accurate result. Investigating the accuracy of the different models of the FPSE, It can be observed that the Quasi steady model predicted a power output of 997W, which is the most accurate result in comparison to the 1000 W of the RE-1000 FPSE experimental output. The adiabatic model predicted a power output of 862W which is not much different from the experimental output. The difference is due to the adiabatic condition of the simulation and the assumptions for the temperature in the expansion and compression spaces. The temperatures in the heat exchangers were assumed to be constant, hence the less accurate prediction of the output performance of the model.

4.1 Parametric sensitivity analysis

The influence of some parameters on the output power of the FPSE is presented and discussed in Tables 3 to Table 5.

Table 3. The effects of damping load on the engine performance

Damping load of piston	Damping load of displacer	Power output (Adiab) (W)	Power output (Quasi) (W)
465	35.34	862	997
470	45	884	1025
480	55	925	1076
490	65	973	1134

Table 4. The effects of spring stiffness on the engine performance

Spring stiffness (piston)	Spring stiffness (displacer)	Power output (Adiab) (W)	Power output (Quasi) (W)
297000	25000	862	997
300000	30000	785	932

https://doi.org/10.1051/matecconf/201819804005

350000	35000	733	893
400000	40000	691	825

Table 5. The effects of piston and displacer masses on the engine performance

Mass of Piston	Mass of displacer	Power output (Adiab) (W)	Power output (Quasi) (W)
6.2	0.426	862	997
7.2	0.5	954	1121
8.2	0.6	1540	1732
9.2	0.7	2265	2498

The parametric sensitivity analysis shows the influence of different parameters on the power output of the free piston Stirling engine. Table 3 shows the effect of the damping load on the engine output. It can be observed that the increase in the damping load of the piston and displacer increases the power output of the engine, this is due to the increase in the load on the masses of the piston and displacer which produces more work in the work spaces. The dampers are a form of additional load to the displacer and piston which is expected to add more work and increase the output power of the FPSE. In Table 4, the effects of spring stiffness on the power output of the engine is presented. It can be observed that the increase in the stiffness if the springs reduces the output power of the engine considerably. This is due to the limitation the stiffness result to on the allowance it gives to the displacer and piston during oscillation, this advertently reduces the displacement and velocity of the moving elements and thereby reducing the amount of the swept volume of gas converted to work in the expansion and compression spaces. The spring and damping constants are major determining factors for the stability and steady operation of the FPSE. The angular frequency depends solely on the mass and spring stiffness of moving elements. Due to complexity of the relation of these parameters to the motion of the piston and displacer in contrast to the crank mechanism of the kinematic Stirling engines, accurate determination of these parameters are required. In the Quasi steady model developed in this study, the damping load is calculated every time step with other parameters in the cycle as this is affected by the pressure variation, temperature change and volume difference in the work spaces. The accurate damping load is determined at the time step when the steady state of the FPSE is achieved. The spring stiffness is to add stability to the oscillatory motions of the moving elements. But on the contrary an adverse effect can be observed due to the increase in these factors and could cause instability in the engine. Hence, an accurate definition of these parameters is required for a steady operation of the engine. Table 5 shows the effects of piston and displacer masses on the engine's output power. It can be observed that the increase in the masses of the piston and displacer increases the output power of the engine respectively. This is due to increase in the swept volume of the working gas, and the increase of the forces exerted on the masses of piston and displacer which increases their acceleration. Therefore, from the observation of the effects of variation in the free piston Stirling engine's parameter, this indicates that careful consideration should be given to their selection during design stage of the engine.

5 Conclusion

In this present study numerical analysis is carried out on the adiabatic and Quasi models of the free piston Stirling engine in order to predict the satisfactory performance of the models and compare their outputs. The following conclusions are presented:

- The two models are developed and their output results analysed and the Quasi steady model predicted the most accurate results in terms of amplitude ratio and power output.
- Detailed and accurate analysis of the relation between the dynamic and thermodynamic characteristics of the FPSE is essential in order to understand better the performance characteristics of the engine.
- Parametric sensitivity analysis performed on the numerical models exhibited a considerable effect on the output power of the FPSE.

Therefore it can be concluded from this study that advanced algorithm and detailed description of the working condition of the FPSE is required in order to predict a satisfactory output performance of the engine.

NOMENCLATURE

A_d Cross sectional area of the piston (m^2)
A_p Cross sectional area of the displacer (m^2)
C_{ap} Specific heat at constant pressure (J/ kgK)
C_{av} Specific heat at constant volume (J/ kgK)
k Thermal conductivity (W/mk)
m_d Mass of the displacer (kg)
m_p Mass of the piston (kg)
T_k Temperature of the cooler (K)
T_r Temperature of the regenerator (K)
T_h Temperature of the heater (K)
T_{ck} Temperature of the compression space to cooler (K)
T_{he} Temperature of the heater to expansion space (K)
T_{rh} Temperature of the regenerator to heater (K)
T_{kr} Temperature of the cooler to regenerator (K)
V_b Bounce space volume(m^3)
V_r Volume of the regenerator(m^3)
V_h Volume of the heater(m^3)
V_k Volume of the cooler(m^3)
V_{clc} Compression space clearance volume(m^3)
V_{cle} Expansion space clearance volume(m^3)
x_p Displacement of piston (m)
$\dot{x}_p$ Velocity of piston (m/s)

$\ddot{x}_p$	Acceleration of piston (m/s^2)
x_d	Displacement of displacer (m)
$\dot{x}_d$	Velocity of displacer (m/s)
$\ddot{x}_d$	Acceleration of displacer (m/s^2)
R	Gas constant value
Q	Heat transfer rate (W)
P_e	Pressure in expansion space
P_c	Pressure in compression space
FPSE	Free piston Stirling engine
Adiab	Adiabatic

References

1. Urieli I. and Berchowitz, D. "Stirling Cycle Engine Analysis," *Altern. Sources Energy*, pp. 71–71, 1984.
2. Formosa F., and Frechette L.G., "Scaling laws for free piston Stirling engine design: Benefits and challenges of miniaturization," *Energy*, vol. 57, pp. 796–808, 2013.
3. Bégota S., Layes G., Lanzetta F., and Nika P., "Stability analysis of free piston Stirling engines," *Phys. Energy Transf. Convers. Storage*, vol. 61, no. Eur. Phys. J. Appl. Phys., 2013.
4. Riofrio J. A., Al-Dakkan K., Hofacker M. E, and Barth E. J., "Control-based design of free-piston stirling engines," *Proc. Am. Control Conf.*, pp. 1533–1538, 2008.
5. Sowale A. "Modelling and Optimisation of a Free Piston Stirling Engine for Micro-CHP Applications," Doctoral thesis Northumbria University; 2015.
6. Jia B., Tian G, Feng H., Zuo Z, and Roskilly A. P, "An experimental investigation into the starting process of free-piston engine generator," *Appl. Energy*, vol. 157, pp. 798–804, 2015
7. Yang Q, Luo E, Dai W, and Yu G, "Thermoacoustic model of a modified free piston Stirling engine with a thermal buffer tube," *Appl. Energy*, vol. 90, pp. 266–270, 2012.
8. Kim J., Choongsik B, and Kim G., "Simulation on the effect of the combustion parameters on the piston dynamics and engine performance using the Wiebe function in a free piston engine," *Appl. Energy*, vol. 107, pp. 446–455, 2013.
9. Walker G., Senft J.R. (1985) Free-Piston Stirling Engines. In: Free Piston Stirling Engines. Lecture Notes in Engineering, vol 12. Springer, Berlin, Heidelberg.
10. Rogdakis E.D, Bormpilas N.A, and.Koniakos I.K, "A thermodynamic study for the optimization of stable operation of free piston Stirling engines," *Energy Convers. Manag.*, vol. 45, pp. 575–593, 2004.

The Application of System identification method to characterize the performance of NiMH batteries in hybrid vehicles

Karin Kandananond

Valaya Alongkorn Rajabhat University, Department of Industrial Management, 13180 Prathumthani, Thailand

Abstract. Nickel-metal hydride (Ni-MH) battery is one of the electric sources which is widely used in hybrid electric vehicles. As a result, it is important to understand the characteristics of Ni-MH battery which is connected to direct current machine in the vehicle. However, the crucial problem is the complexity of the vehicle system which deals with the charging and discharging process of battery in order to maintain the designated speed. The system is considered as a black box and the system identification method is utilized in to characterize the dynamic behavior of the system. The system inputs are battery voltage, armature current and state of charge (SOC) while the output is the speed of DC machine. However, the system identification method will not work properly if the available data available has played an importable role on the determination of the model. As a result, the data regarding all parameters were collected and transmitted to the data logger and used to construct different models. The results from the system identification method indicate that the autoregressive model with exogenous input (ARX) is the most appropriate model to explain the relationship between inputs and output. Therefore, the performance of hybrid vehicle related to the characteristics of Ni-MH batteries is elaborately characterized and this study leads to the effective maneuver.

1 Submitting the manuscript

The characteristics of batteries are the critical factors affecting the performance of the electric vehicle which is driven by the direct current (DC) motor. DC motor is the main drive mechanism of the electric vehicle. Basically, the speed and torque of DC motor are adjusted by varying the amount of supply voltage and input current at the armature component. If the armature is supplied with the line voltage, the capability to control the speed of the motor is accurate and efficient since the value of line voltage is stable over the time and is considered as the infinite source. However, electric vehicles mainly rely on the power from batteries so the supplied voltage that DC motor receives is different than the one from line voltage. This is because the battery pack is the limited source of power supply because, at the supplying stage, they discharge the current to the sink. On the other hand, they are required to be charged in order to maintain the capability of supplying power. As a result, there is another factor which plays an important role on the supply of power to the motor, i.e., state of charge (SOC) or the gauge which measures the amount of battery charge. Moreover, since the charging characteristic of battery packs depends on the chemical reaction of different compounds, the study also focuses on one type of batteries, Nickel-metal hydride (NiMH). For Lithium-ion batteries, they work on the basis that Lithium-ion moves from one electrode to another. The advantages of Lithium-ion batteries are small memory effect, low self-discharge, and high energy density. For this reason, they are popular and commonly used for portable electronics. However, NiMH is based on the chemical reaction of the positive electrode (Nickel Oxide Hydroxide: NiOH) and a negative electrode (alloy). Because of its energy density equivalent to the one of Li-ion, it is also widely used in the hybrid cars and portable electronic devices. The spotlight of this study has two folds: one is the characteristic study of these two types of batteries. Another focus is the utilization of system identification method to mechanize the relationship between input factors, e.g., supply voltage, armature current, SOC and the output, i.e., motor speed.

2 Literature Review

There are several studies with the concentration on the characteristic of battery packs. Hu, Yurkovich, and Yurkovich [1] utilized the system identification for identifying the dynamic model of the electro-thermal model of Li-ion batteries. The factors concerning this model include state-of-charge, temperature, and current direction. Juang, Kollmeyer, Jahns, and Lorenz [2] had estimated the output energy of the batteries of electric vehicles by considering the voltage and current measurement of the batteries. The system identification was also deployed by Sabatier et al. [3] in order to estimate the state of charge for the lead acid battery and the result shows that the accuracy of estimation is up to the error value of five percent. The system identification algorithm was developed by Sitterly, Wang, Yin, and Wang [4]. The highlight of this algorithm is the capability to self-update the identification model in the real-time model. Moreover, this study also focuses on other factors such as aging conditions and the chemical reactions of the batteries. The dynamic characteristic adaptation of the model was also discussed by Hu, Yurkovich, and Yurkovich [5]. The proposed algorithm was tested with two types of batteries, Li-ion and NiMH and the genetic algorithm were utilized to optimize the system identification model.

Besides the linear model, the application of non-linear model was introduced by Juang et al. [6] to identify the voltage and current relationship of lead-acid batteries in the electric vehicle. The research results show that the estimation accuracy was improved significantly after the application of the new model.

3 System Identification

System identification method is the utilization of different statistical methods to construct a mathematical model of a dynamic system by focusing on inputs and outputs. The general equation of dynamic model is the discrete linear time-invariant, where input (y(t)) and output (u(t)) is shown as follows:

$$y(t)=G(q)u(t)+H(q)e(t) \quad (1)$$

, where G(q) is the relationship between input and output and it is represented in the form of the following transfer function or parameter of the dynamic system. The pole of the dynamic system is determined in F(q).

$$G(q)=B(q)/F(q) \quad (2)$$

On the contrary, H(q) is the relationship between noise and output or parameters of noise system as shown in (3).D(q) determines the noise of the system.

$$H(q)=C(q)/D(q) \quad (3)$$

, where B(q), F(q), C(q) and D(q) are polynomials with

$$B(q)=b_1\, q^{-1}+\cdots+b_n\, q^{-n} \quad (4)$$

$$F(q)=f_1 q^{-1}+\cdots+f_n q^{-n} \quad (5)$$

$$C(q)=c_1 q^{-1}+\cdots+c_n q^{-n} \quad (6)$$

$$D(q)=d_1 q^{-1}+\cdots+d_n q^{-n} \quad (7)$$

, where q-n is the shift operator so $q^{-n}u(t)$ is the output signal at time t = t-n or u(t-n). According to (1)-(7), (8) is Box-Jenkins (BJ) model.

$$y(t)=[B(q)/F(q)]u(t)+[C(q)/D(q)] \quad (8)$$

BJ model has the capability to estimate both dynamic and noise parameters. Therefore, the construction of a model is based on the measurement noise, not the input noise. As a result, the structure of BJ model provides high flexibility for modeling noise. Moreover, the above model can be extended into different subclasses of how to model dynamic and noise parameters. The dynamic model can be extended into a number of models. Among these models is the output error (OE) model which focuses on the estimation of the dynamic model only while a noise model is set at the constant level and equal to 1. The OE model is shown in (9).

$$y(t)=[B(q)/F(q)]u(t)+e(t) \quad (9)$$

These models are utilized to construct a mathematical model to explain the relationship between the output and the inputs.

4 Battery Packs and Hybrid Vehicles

In this study, the experiment and data collection is based on the Ni-MH batteries used in the hybrid vehicles (Figure 1 and 2). The drive shaft of a hybrid vehicle is driven by DC motor and engine (Figure 3) which are connected in parallel through the coupling. As a result, the vehicle can run on either engine or DC motor alone or both together. Moreover, not only the battery can supply voltage to DC motor but also it can act as a generator to charge the battery when it is run backward to stop the vehicle (regeneration braking).

Figure 1. Ni-MH battery packs in a hybrid vehicle.

Figure 2. Ni-MH battery packs in a hybrid vehicle.

Figure 3. Engine.

5 Data Processing

To compare the performance of batteries for hybrid vehicles, the focused parameters are a voltage supplied from the battery, state of charge (SOC), armature current (A) and motor speed. The rated voltage of the battery is 220 V with the energy charge of 7 Ah. The battery is connected in parallel to a DC machine which acts as the motor and generator. The charge and the discharge of battery depend on the value of the state of charge (SOC). When the SOC us below 40 percent, a negative load torque is applied to the machine which will generate the electricity and recharge the battery. On the other hand, when SOC is above 80 percent, the load torque will be taken off so the machine will behave as a motor. All related parameters, i.e., voltage (V), SOC (percent), armature current (A) and motor speed (radian per second) will be measured and transmitted to a data logger. The make of data logger used is HOBO UX120 which has four channels.

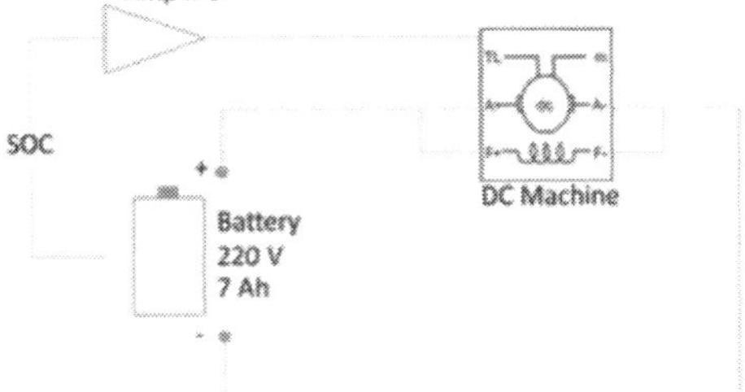

Figure 4. Schematic diagram of the battery and machine.

The real-time data of three factors, the output voltage of the battery, state of charging (SOC), and the armature current, are collected and shown in Figure 5. Due to Figure 5, the motor tends to gradually climb up when the voltage, current, and SOC increase.

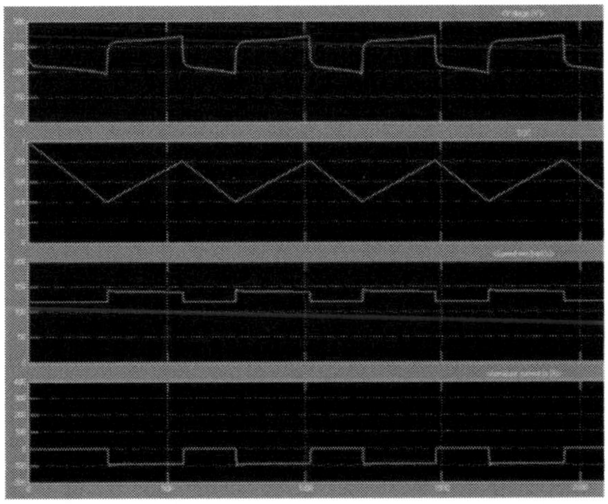

Figure 5. Outputs from Ni-MH battery and response motor speed.

6 Model Analysis

The system identification toolbox in MATLAB is deployed to analyze data and construct the most appropriate model to estimate the speed when the inputs are voltage, armature current, and SOC. The chosen model is the autoregressive model with exogenous input (ARX) and this result explicitly shows that both the respondent output and the input are used to derive the predicting model.

6.1 SOC VS Speed

The relationship between SOC, represented by u(t) and speed, y(t), is illustrated by the autoregressive model with exogenous input (ARX) which is chosen as the predicting model as follows:

```
Discrete-time IDPOLY model: A(q)y(t) = B(q)u(t) + e(t)
A(q) = 1 - 1.466 q^-1 + 0.3727 q^-2 + 0.08055 q^-3 + 0.004258 q^-4
B(q) = -995.9 q^-1 + 1811 q^-2 - 708.3 q^-3 - 109.1 q^-4
Estimated using ARX from data set eDat
Loss function 2.0744 and FPE 2.09283
Sampling interval: 1
```

Figure 6. ARX model for SOC VS speed.

The ARX model is the combination of the utilization of the time series (historical data) or autoregressive and the external input. According to the result, the equation of A(q) and B(q) indicates that the history data of speed and SOC up to the lag of four (the highest order of lag operator is q^{-4}). After the model is constructed, the simulated value of speed based on the SOC is shown in Figure 7.

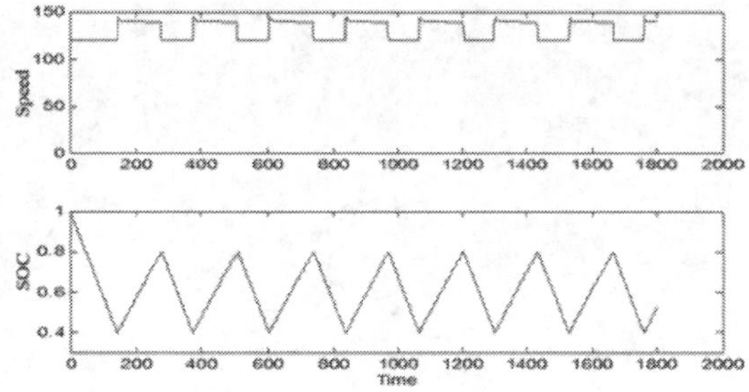

Figure 7. Simulated values of speed based on SOC.

6.2 Voltage VS Speed

The relationship between SOC, represented by u(t) and speed, y(t), is illustrated by the autoregressive model with exogenous input (ARX) which is chosen as the predicting model as follows (Figure 8):

Discrete-time IDPOLY model: A(q)y(t) = B(q)u(t) + e(t)
A(q) = 1 - 1.466 q^-1 + 0.3727 q^-2 + 0.08055 q^-3 + 0.004258 q^-4
B(q) = -995.9 q^-1 + 1811 q^-2 - 708.3 q^-3 - 109.1 q^-4
Estimated using ARX from data set eDat
Loss function 2.0744 and FPE 2.09283
Sampling interval: 1

Figure 8. ARX model for voltage VS speed.

As a result, the simulated output due to the constructed model with the applied voltage is depicted in Figure 9.

6.3 Current VS Speed

On the other hand, the relationship between the current measured at the armature part of the motor and the speed is determined and shown in Figure 10.

Discrete-time IDPOLY model: A(q)y(t) = B(q)u(t) + e(t)
A(q) = 1 - 1.466 q^-1 + 0.3727 q^-2 + 0.08055 q^-3 + 0.004258 q^-4
B(q) = -995.9 q^-1 + 1811 q^-2 - 708.3 q^-3 - 109.1 q^-4
Estimated using ARX from data set eDat
Loss function 2.0744 and FPE 2.09283
Sampling interval: 1

Figure 10. ARX model for armature current VS speed.

The motor speed corresponding to the armature current based on the system identification model is shown in Figure 11.

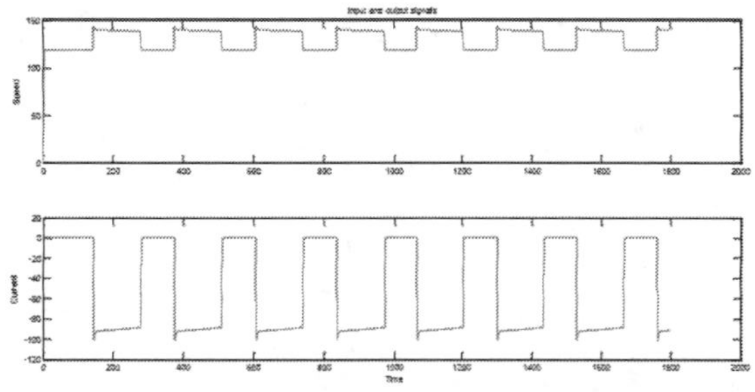

Figure 11. Simulated values of speed based on armature current.

After the system identification method is utilized in the form of ARX model to explain the relationship is these paired inputs and outputs: SOC VS Speed, Voltage VS speed and Current VS speed, the results show that the output speed can be efficiently predicted by a single input (SOC, voltage and current) and their historical data. The difference between the actual value of the motor speed based on the system identification model and the actual value is shown in Table 1. According to the results in Table 1, there is no significant difference between the actual speed and the predicting value based on the ARX model. According to the results in Table 1, there is no significant difference between the actual speed and the predicting value based on the ARX model since the difference is ranged from 1.05 to only 2.36 percent. The lowest is from the prediction of SOC by ARX model (1.05 percent) while the highest is the voltage prediction (2.36 percent).

Table 1. Difference between actual and predicting values.

Input of ARX	% Difference between actual speed and predicting value
SOC	1.05
Voltage	2.36
current	1.42

6 Conclusions and Discussions

The characteristics of the Ni-MH battery pack in hybrid vehicles, i.e., voltage, current and SOC, and the motor speed are studied to determine the relationship between these inputs and the output speed. All data is measured and transmitted to the data logger. Afterwards, the system identification method is utilized to characterize the statistical model explaining the relationship. The chosen model is the autoregressive model with exogenous input (ARX). After comparing the actual and simulated scenario, the result indicates that the constructed model can simulate the output speed accurately. The results show that the empirical model can be utilized without the need to derive the mathematical model between the input and the output. Therefore, the proposed model is capable to characterize the performance of the motor of hybrid vehicles when there are different inputs and their values are varied.

References

1. Y. Hu, S. Yurkovich, and B.J. Yurkovich, J. Power Sources **196**, 449-457 (2011).
2. L. W. Juang, P. J. Kollmeyer, T. M. Jahns, and R. D. Lorenz, *Proc. IEEE Energy Conversion Congress and Exposition (ECCE)* 3903-3910 (2010).
3. J. Sabatier, M. Aoun, A. Oustaloup, G. Gregoire, F. Ragot, and P. Roy, Signal Process., **86**, 2645-2657(2006).
4. M. Sitterly, L. Y. Wang, G. G. Yin, and C. Wang, IEEE Trans. Sustain. Energy **2**, 300-308(2011).
5. Y. Hu, S. Yurkovich, and B.J. Yurkovich, Control Eng. Pract. **17**, 1190-1201(2009).

6. L. W. Juang, P. J. Kollmeyer, T. M. Jahns, and R. D. Lorenz, IEEE Trans. Ind. Appl. **49**, 1480-1483(2013).
7. O. Tremblay, L. –A. Dessaint, A. –I. Dekkiche, *proc. IEEE Vehicle Power and Propulsion Conference* (2007).

Embedded Position Control of Permanent Magnet Synchronous Motor Using Model Predictive Control

Agus Ramelan[1], Arief Syaichu Rohman[1], and Allen Kelana[1]

[1]*School of Electrical Engineering and Informatics, Institut Teknologi Bandung, Indonesia*

Abstract. This paper presents an implementation embedded system for position control of Permanent Magnet Synchronous Motor (PMSM). The control system consists of raspberry pi 3 as a microcontroller, ASDA-A2 servo drive, and Delta Servo ECMA type. The software design includes simulation tool and Python included on Raspbian OS. Communication between Raspberry Pi 3 and ASDA-A2 drivers using the ASCII Modbus communication protocol. Raspberry Pi 3 processes the reference data and the actual reading result and calculates the resulting error. The control algorithm used in this research is Model Predictive Control (MPC). As a Linear Quadratic Regulator, MPC aims to design and generate an optimal control signal with the ability to anticipate saturation, receding horizon, future event and take control accordingly In the design of the MPC technique to adjust the speed of the PMSM to take action of reference tracking, performance index optimization is done by adjusting the value of weighting horizon N, Q and R. The simulation and implementation results showed that the PMSM can reach the stability point on each desired setpoint and result in a near-zero steady-state error.

1 Introduction

Electric drives have an important role in the industry. One type of electric drive that is often used is a permanent magnet synchronous motor (PMSM)[1]. PMSM provides reliable speed control for acceleration and deceleration. As an actuator, there is often a saturation problem at the input. This will degrade system performance if it is not embedded in the design. With such non-linearity nature, it is difficult to obtain a stability guarantee [2]. For that, we need the method of MPC (Model Predictive Control) as one way to solve the problem. Example of implementation of MPC on drive system has been done before. That paper gives an exhaustive description of the design procedure of MPC applied to the combined control of the motor speed and current[3]. The implementation of an MPC on microcontroller also has been done. An example is the implementation of MPC on an Arduino and an FPGA board. MPC had implemented to control the speed of BLDC motor using Arduino Mega 2560. MPC gives better performance than PID and it is shown that MPC capable to handle input saturation of the system[4]. Besides that, there is a paper that discusses the MPC application for TITO (Two Input Two Output) system and its implementation onboard SPARTAN 6 FPGA SP605 Evaluation Kit with Microblaze MCS. The implemented MPC using Algorithm-3 on board can be used by a different plant with fast responses[5]. The differences that the author did with the previous one, will be discussed at the end of this chapter. MPC is one of the control methods that calculate the current control input by solving an optimal control problem of a limited open-loop horizontal and using the first element of the optimum control sequence at each sampling time. This method has the ability to overcome saturation on both the control input, system input, and system output. The saturation element is included as an additional requirement that must be calculated in resolving the problem of optimizing the objective function of MPC continuously. The process of optimizing quadratic objective function with restrictions on its variables is called quadratic programming (QP). An algebraic loop that involves saturation equivalent to a QP problem that has its variable constraints. Iterative of QP with an algebraic loop which produces MPC convergent solution may become slow, so we need a fast QP iteration algorithm for tracking such as Algorithm-2 and Algorithm-3[6].

In this paper, we will discuss the performance of MPC by using the algorithms-2[6] which then evaluated its performance. The new thing in this study is the microcontroller, motor type, driver, and communication system used. The system used in the implementation is a position control system PMSM controlled by Raspberry Pi 3. Type of motor used is ECMA K11820SS with its driver type ASDA-A2-2 43-M. The communication series used is RS-232 Modbus, where ASDA driver as master and Raspberry Pi 3 as the slave.

2 System Design

The PMSM motor control system model used is PMSM 2 KW 3 Phase and Driver Motor with 24 Volt supply. Mathematical modeling of PMSM has been done by previous research, for example in the paper [7]. In this study, the modeling is done by identification system tool. This study was conducted as one of the stages of rail simulator making. It consists of four integrated controlled PMSM motors. Before real implementation, a controller test is required for a quarter of the model. We use raspberry pi because it has GPIO port, HDMI, and USB. So it will be easier in programming and interconnect with other devices.

The ASDA-A2 driver is a PMSM motor controller driver which is a production of Delta Company. In this driver is equipped with features that are complete enough for the purposes of motor control. These features are input-output channels, communication ports with master/slave controllers and other devices, encoder sensors, writable and readable registers with other devices, and complete built-in HMI software [8]. Figure 1 is a block diagram of the PMSM motor position control system module with Raspberry Pi 3 based Raspberry Algorithm.

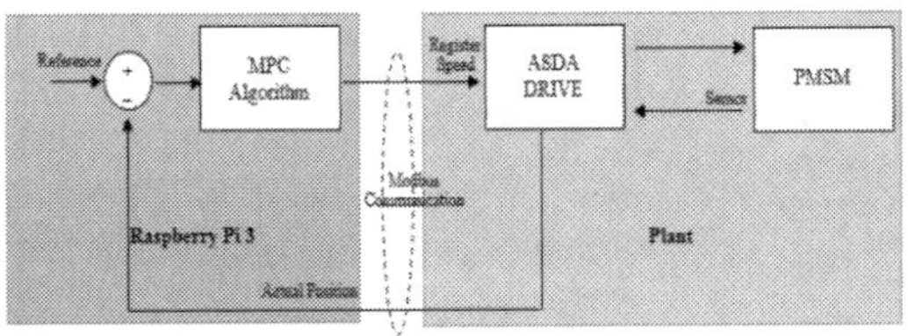

Figure 1. Block diagram of system design.

Raspberry Pi 3 processes the reference data and the actual reading result and calculates the resulting error. This error is the input for predictive model control. The result of predictive model control calculation is a control signal in the form of the input value of speed to be given to actuator that is PMSM motor. In the implementation of this control signal is limited in accordance with the specificity of the minimal-maximal values.

3 Controller Design

3.1 MPC Design

MPC or also known as receding horizon control is a control technique in the form of discrete time and has a purpose to predict the future behavior of the system process to obtain optimal closed-loop control that minimizes an objective function on-line to a certain time horizon with certain limits [9]. Broadly speaking, the controlling stages of MPC, in general, can be explained as follows,

a. Calculate the output to come for predefined horizons,
b. Calculates the control input sequence that will come by optimizing the objective function,
c. From the set of control signals, only the first $u(k)$ input is assigned to the system and back to step a.

In order to achieve zero tracking errors for a position control, adding integrator is not needed. The open loop system can be written as the following equation.

$$x_{k+1} = A_d x_k + B_d u_k \quad (1)$$

$$y_k = C_d x_k + D_d u_k \quad (2)$$

The problem with the MPC strategy is to design a control by finding U, within the constraint of U_{sat}, that minimizes the objective function J along the horizon length, N, at each sampling time.

$$J = e_{k+N}^T \overline{Q}_N e_{k+N} + \sum_{j=0}^{N-1}\left(e_{k+j}^T \overline{Q} e_{k+j} + u_{k+j}^T \overline{R} u_{k+j}\right) \quad (3)$$

$$J = e_k^T \overline{Q} e_k + E^T Q E + U^T R U \quad (4)$$

$$J = J_x + X^T\left(\overline{\Phi}^T Q \overline{\Phi}\right)X + U^T M U + 2U^T F_X X \quad (5)$$

where

$$\overline{Q} = \overline{Q}^T > 0, \overline{R} = \overline{R}^T > 0 \quad (6)$$

$$X = \begin{bmatrix} x_k \\ -\Omega \end{bmatrix}; Y = \begin{bmatrix} y_{k+1} \\ y_{k+2} \\ \vdots \\ y_{k+N} \end{bmatrix}; \Omega = \begin{bmatrix} \omega_{k+1} \\ \omega_{k+2} \\ \vdots \\ \omega_{k+N} \end{bmatrix}; E = \begin{bmatrix} u_{k+1} \\ u_{k+2} \\ \vdots \\ u_{k+N} \end{bmatrix} \quad (7)$$

minimizing J is equivalent to minimizing J_R within the constraint of U_{sat}.

$$\frac{1}{2}(U + M^{-1}F_X X)^T M (U + M^{-1}F_X X) = J_R \quad (8)$$

According to [6], by comparing equation [8] and [13]

$$-U_0 = M^{-1}F_k x_k \quad (9)$$

$$U_0 = M^{-1}F_k x_k \quad (10)$$

The objective function optimization algorithm used is QP and can be solved by a class of algebra loop [6]. The results stated that a class of algebra loop could be used to complete an on-line QP. One of the block diagrams which is an alternative QP solution can be explained on the Figure 2.

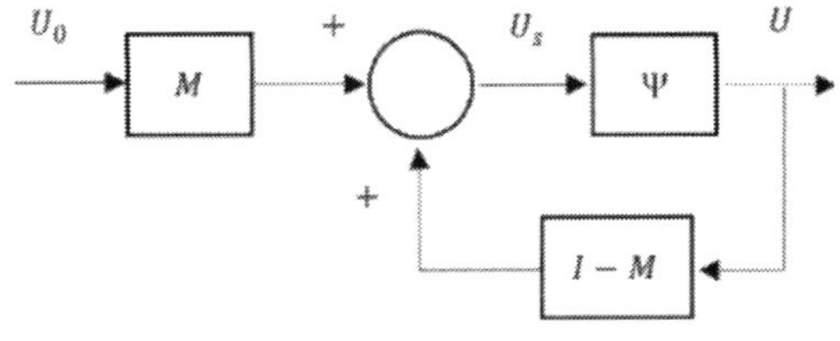

Figure 2. Nonlinear algebraic loop.

The loop algebra shown in Figure 2 is equivalent to solving the following QP problem based on the objective function:

$$U = \arg\min (J),\ U \in U_{sat} \tag{11}$$

$$U_{sat} = U_{min}, \leq U_i \leq U_{max}, ,\ i = 1,2, \dots N \tag{12}$$

and

$$J = \frac{1}{2}(U - U0)^T M(U - U0) \tag{13}$$

$$M = M^T > 0 \tag{14}$$

$$\overline{\Phi} = [\Phi \quad I_N]; \overline{\Gamma} = \Gamma\,; \tag{15}$$

$$\overline{\Phi} = \left[\begin{bmatrix} C_dA_d \\ C_dA_d{}^2 \\ \vdots \\ C_dA_d{}^n \end{bmatrix}\right];$$

$$\Gamma = \begin{matrix} C_dB_d & 0 & \dots & 0 \\ C_dA_dB_d & C_dB_d & \dots & 0 \\ \vdots & \vdots & \ddots & \vdots \\ C_dA_d{}^{n-1}B_d & C_dA_d{}^{n-2}B_d & \dots & C_dB_d \end{matrix}\ ; \tag{16}$$

Finally, the first u_k control vector in U is used as the current control input for the control.

$$u_k = [I_m\ 0 \dots 0]U \tag{17}$$

$$u_k = E_1^T U \tag{18}$$

3.2 Modelling

3.2.1 Plant Model

Modeling is done with the Identification System tool. The transfer function is obtained by giving varying inputs to the system. Input and output are the reference values of speed and actual read speed. The transfer function is approximated by order 2 with 1 zero in the identification system tool. The result of the transfer function is then integral to get the transfer function in position. From the modeling results obtained the function of continuous transfer of PMSM motor system is as follows.

$$G(s) = \frac{653.5\,s + 15600}{s(s^2 + 266.6\,s + 15600\,)} \tag{19}$$

Sampling time Ts is chosen with the longest computation time in the input stage is 0.03. Ts is chosen based on the consideration of the average speed of data transmission and command execution is 25 ms. So we get the transfer function in the form of discrete as follows.

$$G(z) = \frac{0.05061\,z^2 - 0.02072\,z - 0.00224}{z^3 - 1.079\,z^2 + 0.07901\,z - 0.0003361} \tag{20}$$

The model of the control must be detectable and stabilizable. One more step in the prediction can be used recursively to find the next prediction. The prediction can be specified in length based on the horizon selection. The output of the MPC is expected to follow a reference input $\omega_k \in \mathbb{R}^n$.

3.2.2 Implementation

Implementation of MPC consists of several steps, such as calculate state estimation, calculate the output to come for predefined horizons, calculates the control input sequence that will come by optimizing the objective function. Finally, from the set of control signals, only the first *u(k)* input is assigned to the system and back to the first step. The block diagram of MPC implementation can be explained on the figure 3.

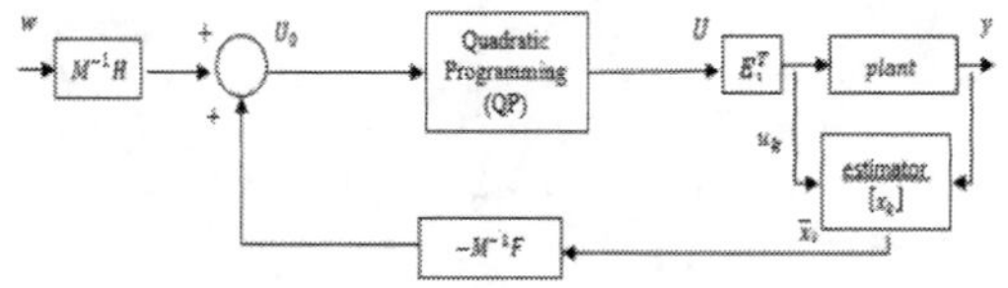

Figure 3. Block diagram of MPC implementation.

3.2.3 Quadratic Programming

The quadratic programming to be used is algorithm-2 which is quadratic iteration programming algorithm with block diagram as figure 4.

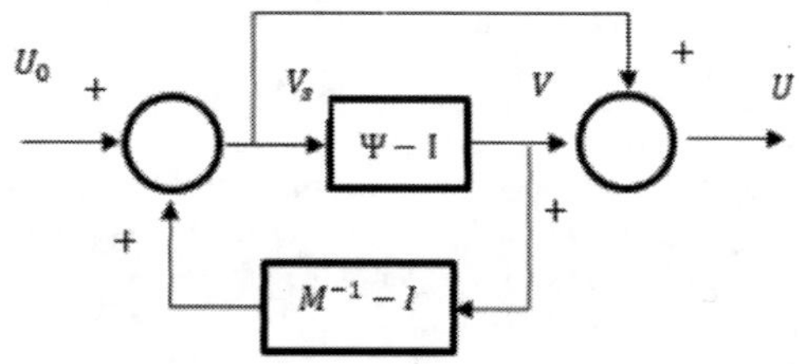

Figure 4. Algorithm-2 block diagram.

$$\Psi_a = \Psi_a - I \tag{21}$$

Ψ is diagonal saturation that representing U_{sat}. The loop equation will be as follows:

$$V = \Psi_a(V_s) = \Psi_a(U_0 + (M^{-1} - I)V) \tag{22}$$

$$U = V_s + V = M^{-1}V + U_0 \tag{23}$$

That iteratively becomes

$$V_{k+1} = \Psi_a(U_0 + (M^{-1} - I)V_k) \tag{24}$$

If iterating convergent, then the solution U^* is obtained as follows.

$$U^* = V^* + (M^{-1} - I)V^* + U_0 = M^{-1}V^* + U_0 \tag{25}$$

3.2.4 State Estimator

Given that the built model is not known the relationship in detail through the mathematical model, and the model obtained is done with tools identification system. Then all state is estimated by Luenberger approach of full order state estimation as shown in figure 5.

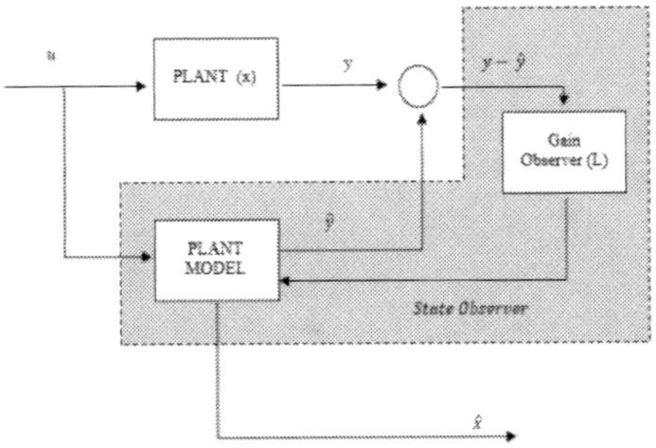

Figure 5. Algorithm-2 block diagram.

The equation of full order state estimator (observer) as shown above is as follows

$$\hat{x}\,(k+1) = A\,\hat{x}\,(k) + Bu(k) + L(y(k)\text{-}\hat{y}(k)) \qquad (26)$$

$$\hat{Y}(k) = C\,\hat{x}\,(k) + Du(k) \qquad (27)$$

4 Simulation and Experimental Result

Both in simulation and implementation, the selection of the same parameters. These parameters include the time sampling, the horizon, Q, R, and the length of the iteration. In this study, selected time sampling = 0.03 s, N = 33, Q = 185, R = 1, and the length of iteration. The horizon is chosen with considering the desired closed-loop response time is T and the control interval is Ts, try p such that T ≈ pTs[10]. The QP iteration length is not limited to convergent because each input signal requires a different length of iteration.The simulation is done with simulation software and the implementation is done with Python programming language.

4.1 Simulations Result

In this simulation, the horizon value is chosen based on the reference to the MATLAB library[10] where the horizon is the comparison between the targeted time response (1 s) with the sampling time (0.03 s). Other parameters are selected by trial to produce the best results. In the first simulation is done rounding the nearest value of control signal according to the requirement on ASDA system that requires integer value. MPC requires the state value of the system to perform computations but only known y (k) system outputs. From this it is estimated the state value with full order state estimator. Gain estimator L can be searched by Pole Placement (Ackerman). The iteration termination criterion is performed with a convergent error <1%. The simulation results are shown in the following figure 6 and figure 7.

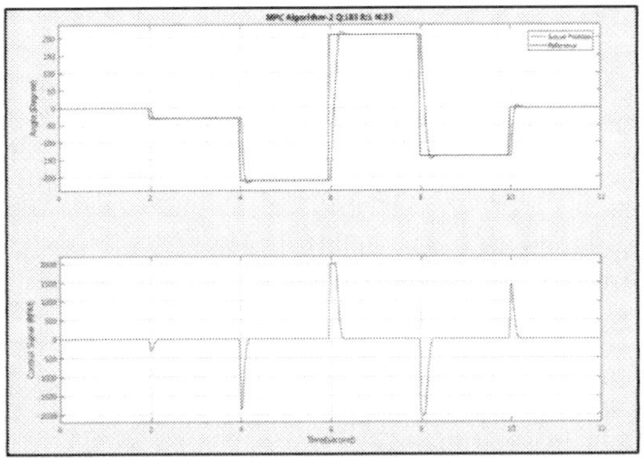

Figure 6. Respone system for tracking trajectory.

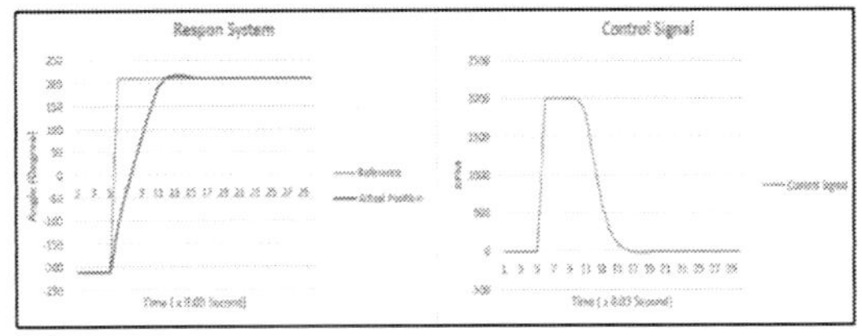

Figure 7. Respone System for Tracking Trajectory at 6 seconds $\leq t \leq 8$ seconds.

Settling time and steady state error are reviewed from the reference input and output when the output is steady at 6 seconds $\leq t \leq 8$ seconds, ie -210° to 210°. From the results it can be seen that the settling timenya is 0.45 seconds and the steady state error is 0.03 degree. So the simulation results can be concluded that the system response meets the desired specification of settling time less than 1 second and steady state error close to equal 0 degree.

4.2 Experimental Result

Similarly, with simulations, experiments are performed by giving variable reference inputs in a clockwise and counterclockwise direction from the shortest to the farthest angle. The farthest reference value is -210° to 210°. The experimental results are shown in the following figure 9 and figure 10.

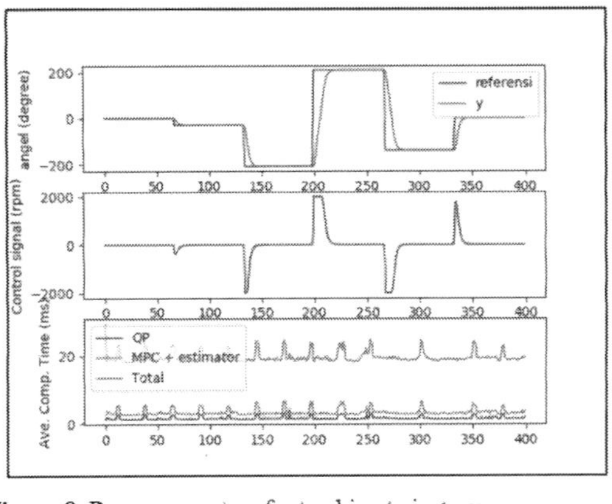

Figure 8. Response system for tracking trajectory.

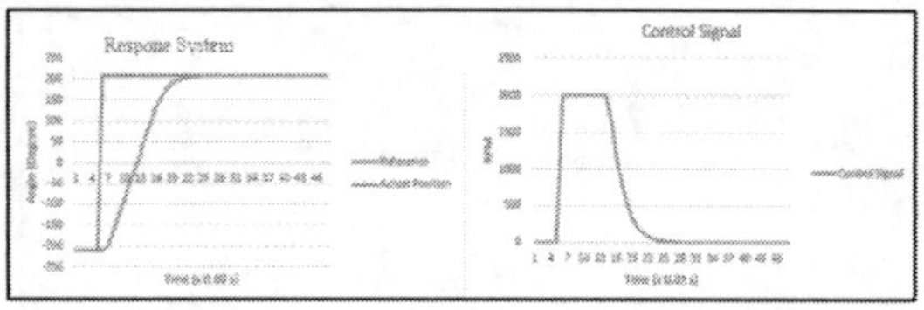

Figure 10. Respone system for tracking trajectory at 6 seconds ≤ t ≤ 8 seconds.

Settling time and steady state error are reviewed from the reference input and output when the output is steady at 6 seconds ≤ t ≤ 8 seconds, ie -210° to 210°. From the results it can be seen that the settling timenya is 0.66 seconds and the steady state error is 0.0096 degree. In addition, Figure 9 also shows the average calculation time QP = 1.62 ms, MPC + Estimator = 3.49 ms, total time for execution 1 signal reference = 20.1050 ms, and the maximal of total time for execution 1 signal reference = 28.31 ms. So the experimental results can be concluded that the system response meets the desired specification of settling time less than 1 second and steady state error close to equal 0 degree.

5 Conclusion

The Model Predictive Control has been successfully simulated and implemented as a Permanent Magnet Synchronous Motor (PMSM) position controller. MPC is computed inside Raspberry Pi 3 and the resulting control signal is configured with ASDA-Servo with serial communication rs-232 Modbus. The use of raspberry memory for this MPC implementation is about 25%. The simulation and experimental results can be concluded that the system response meets the desired specification of settling time less than 1 second and steady state error close to equal 0 degree. Their results show similarities. In the simulation occurs a small overshot while the implementation is not. However, this difference is not so significant.

6 Acknowledgments

This work was supported by LPDP Scholarship from Indonesian Ministry of Finance, School of Electrical Engineering and Informatics, Institut Teknologi Bandung, especially Control and Intelligence System Research Group.

References

1. F. De Lillo, F. Cecconi, G. Lacorata, A. Vulpiani, EPL, 84 (2008) [1] Aamir Hashim O.A. "Optimal Speed Control for Direct Current Motors Using Linear Quadratic Regulator". Journal of Science and Technology - Engineering and Computer Sciences, Vol. 14, No. 3, June 2013.
2. Prasetyo, Hanif. (2017): Quadratic Programming Implementation in Model Predictive Control using Arduino Mega2560 for Speed Control of BLDC Motor , Master's Program Thesis, Institut Teknologi Bandung.
3. Saverio Bolognani, Silverio Bolognani, Luca Peretti, and Mauro Zigliotto. Design and Implementation of Model Predictive Control for Electrical Motor Drives. IEEE TRANSACTIONS ON INDUSTRIAL ELECTRONICS, VOL. 56, NO. 6, JUNE 2009.
4. Hanif Fauzan Prasetyo; Arief Syaichu Rohman; M. R. A. R. Santabudi. Implementation of model predictive control using Algorithm-3 on Arduino Mega 2560 for speed control of BLDC motor. 2017 3rd International Conference on Science in Information Technology (ICSITech).
5. Cahyantari Ekaputri; Arief Syaichu-Rohman. Model predictive control (MPC) design and implementation using algorithm-3 on board SPARTAN 6 FPGA SP605 evaluation kit. 2013 3rd International Conference on Instrumentation Control and Automation (ICA).
6. Syaichu-Rohman, Arief, R.H. Middleton, and M.M Seron. "A Multivariable Nonlinear Algebraic Loop as a QP with Application to MPC." The Proceedings of the European Control Conference. Cambridge, 2003.
7. A. B. Dehkordi, A. M. Gole, and T. L. Maguire. "Permanent Magnet Synchronous Machine Model for Real- Time Simulation". International Conference on Power Systems Transients (IPST'05) in Montreal, Canada on June 19-23, 2005 Paper No. IPST05 - 159.
8. Delta Electronics Inc. "Delta High Resolution AC Servo Drive for Network Communication Applications ASDA-A2 Series". User Manual. 2017.
9. Dede Irawan Saputra. "Electric Speed Control In Electric Vehicle Simulator Using Model Predictive Control". Thesis Magister Electrical Engineering STEI Bandung Institute of Technology. 2017
10. MATLAB R2017b Help. Avaible on : https://www.mathworks.com/help/mpc/ug/choosing-sample-time-and-horizons.html

MATEC Web of Conferences **198**, 04008 (2018) https://doi.org/10.1051/matecconf/201819804008
MEAE 2018

Condition Monitoring of Wind Turbine Based on Copula Function and Autoregressive Neural Network

Zhongshan Huang[1,2], Ling Tian[1,2], Dong Xiang[1,2], Sichao Liu[1,2] and Yaozhong Wei[1,2]

[1]*Department of Mechanical Engineering, Tsinghua University, Beijing 100084, China*

[2]*Beijing Key Lab of Precision/Ultra- precision Manufacturing Equipment and Control, Beijing 100084, China*

Abstract. The traditional wind turbine fault monitoring is often based on a single monitoring signal without considering the overall correlation between signals. A global condition monitoring method based on Copula function and autoregressive neural network is proposed for this problem. Firstly, the Copula function was used to construct the binary joint probability density function of the power and wind speed in the fault-free state of the wind turbine. The function was used as the data fusion model to output the fusion data, and a fault-free condition monitoring model based on the auto-regressive neural network in the faultless state was established. The monitoring model makes a single-step prediction of wind speed and power, and statistical analysis of the residual values of the prediction determines whether the value is abnormal, and then establishes a fault warning mechanism. The experimental results show that this method can provide early warning and effectively realize the monitoring of wind turbine condition.

1 Introduction

Wind energy has played an increasingly important role in the world energy structure in recent years, which results in the rapid development of wind power equipment. Statistically, the installed capacity of global wind turbines has increased from 24 GW in 2001 to 433 GW in 2015. According to the current growth trend, it is expected that the capacity will reach 800 GW by 2020[1]. Wind farms are generally located in remote areas where the working environment is quite complex; therefore, the wind turbines have a high probability of failure. If the key components of the turbine fail, the equipment will be damaged, and even the whole unit may be shut down, causing huge economic losses. Condition monitoring can effectively diagnose the internal failure of wind turbines, which makes it possible to optimize maintenance strategy, reduce unplanned shutdowns, and save operation and repair costs. It can be seen that it is necessary to monitor the operating status of the gearbox and realize fault early warning.

As technology advances, there are diverse statuses monitoring methods for different parts of wind turbines [2]. BP neural network was applied to turbine fault forecasting, and a multi-agent method was proposed to give comprehensive evaluation of the overall operating status [3]. Wavelet analysis was adopted to process vibration signals for fault detection [4]. Nonlinear State Estimate Technology (NEST) was deployed to detect and early-warn the temperature of gearbox bearings in wind turbines [5]. An artificial neural network approach was applied to the gearbox bearing and cooling oil temperature condition monitoring modelling [6]. Condition monitoring technology has developed greatly, but it is still based on single parameters (vibration, temperature, etc.) at present to detect turbine status and carry out fault early warning. Therefore, it is significant to use the overall correlation between the monitoring parameters of wind turbines to conduct state monitoring research.

There is a close relationship between the power curve of the wind turbine and the faults. At the same time, the power curve is greatly influenced by the wind speed. Therefore, a condition monitoring model based on power and wind speed can be considered as a whole. This paper proposes a wind turbine condition monitoring method based on Copula function and auto-regressive neural network, and uses Copula function to construct a joint probability density function data fusion model of wind speed and power binary monitoring parameters.

2 Theoretical basis

2.1 Copula function

Copula is a mathematical tool that connects the edge distribution and joint distribution of random variables. Copula function described in Eq.1 is defined as a connection function of standard uniform random variables [7].

$$C(u_1, \mathrm{L}\ , u_n) = P_r\{U_1 \le u_1, \mathrm{L}\ , U_n \le u_n\} \qquad (1)$$

where $U_i : U(0,1), i=1,2,\text{L} ,n$ is a standard uniform distribution.

The Copula function obeys Sklar's theorem [8]: Let F be the joint cumulative distribution function of an n-dimensional variable. For the edge cumulative distribution F_1,F_2,L ,F_n , there is an n-dimensional Copula function C such that

$$F(x_1,x_2,\text{L} ,x_n)=C(F_1(x_1),F_2(x_2),\text{L} ,F_n(x_n)) \quad (2)$$

If F_1,F_2,L ,F_n are continuous functions, then function C exists uniquely. Otherwise, the Copula function C is uniquely determined only within the range of the edge cumulative distribution function. The density function in Eq.3 can be obtained through the derivation of Eq.2

$$f(x_1,\text{L} ,x_n)=c(F_1(x_1),\text{L} ,F_n(x_n))\prod_{i=1}^{n} f_i(x_i) \quad (3)$$

where $c(u_1,u_2,\text{L} ,u_n)=\dfrac{\partial^n C(u_1,u_2,\text{L} ,u_n)}{\partial u_1 \partial u_2 \text{L} \ \partial u_n}$, $u_i=F_i(x_i)$, and $f_i=(x_i)$ is the edge probability density function of x_i .

Archimedes Copula is widely used because of its simple structure and good nature. The commonly used Archimedes Copula functions have Gumbel Copula functions, Clayton Copula functions, and Frank Copula functions. Gumbel Copula is sensitive to the tail-tailed correlation, Clayton Copula is sensitive to the tail-tail correlation, and Frank Copula is sensitive to the symmetry of the variable. Let u, v be a random variable and be a related parameter, then the analytical formulas of the commonly used three types of binary Archimedean Copula are shown in Table 1.

Table 1. Archimedean copula functions.

Copula function	$C(u,v\|\theta)$
Clayton	$(u^{-\theta}+v^{-\theta})^{-1/\theta}$
Frank	$-\frac{1}{\theta}\ln\{1+[(e^{-\theta u}-1)(e^{-\theta v}-1)]/(e^{-\theta}-1)\}$
Gumbel	$\exp\{-[(-\ln u)^{\theta}+(-\ln v)^{\theta}]^{1/\theta}\}$

2.2 Autoregressive neural network

Some signals have strong autocorrelation, and their future time values can be predicted by learning signal autocorrelation features. Commonly used time series modelling methods include auto-regressive moving average (ARMA) models, artificial neural networks (RNN) and other methods. In this paper, a non-linear auto-regression (NAR) dynamic neural network is introduced to model the monitoring signal time series. Generally, the static neural network processes input signals from the input layer and then goes through each layer in order, and finally outputs the output results through the neurons in the output layer. The NAR and other dynamic networks can feed back the output signal to the input end, so that the output signal can participate in the next iterative training with memory function, so it can better describe the characteristics of time-varying systems with non-stationary, nonlinear and other complex mapping relationships [9], and overcome the shortcomings of the ARMA model that can only be modelled for stationary linear signals.

The NAR neural network can be defined as shown in Eq.4

$$\hat{y}(t+1)=f[y(t),y(t-1),\text{L} ,y(t-n_y)] \quad (4)$$

where $f(g)$ is a nonlinear mapping, $y(t),y(t-1),\text{L} ,y(t-n_y)$ is the historical time series of the monitoring signal, and $\hat{y}(t+1)$ is a prediction value.

2.3 Establishing fault-free monitoring model

2.3.1 Determination and selection of Copula functions

Determining the edge distribution function of each variable is the first step in determining the Copula function, so firstly an edge distribution function for wind speed and power is built. In most cases, the monitoring signal does not completely obey the Gaussian distribution, so the Gaussian distribution cannot be used to fit the data. This paper uses the kernel density estimation (KDE) nonparametric estimation method to fit the edge distribution. Subsequently, the maximum likelihood estimation method was used to solve the parameters of the Copula function to determine the Archimedean Copula function expression. Then select the appropriate Copula function type based on the binary frequency histogram.

2.3.2 Establishing condition monitoring model based on NAR network

First determine the input data for the NAR neural network. In this paper, based on the joint probability density function of wind speed and power, it is used as a data fusion mapping model. The model is used for data fusion of multivariate parameters to obtain a new fusion data vector, which are denoted as $\boldsymbol{y}_t=[y(1),y(2),\text{L} \ y(t)]$. Using the new data vector $\boldsymbol{y}_t$ as the input of the NAR neural network, the operational status of the system is described in terms of wind speed and power.

After using the Copula function to fuse the wind speed and power data in the fault-free state, the fault-free fusion data can be obtained, which is denoted as $\boldsymbol{y}_{t,norm}$, then it is used as the input vector of the NAR network. After training, a single step prediction model under normal conditions can be established. This model is called a faultless condition monitoring model.

Furthermore, the single-step prediction residuals of training sample were obtained using the no-fault monitoring model, and the distribution of residual samples was analysed using statistical principles. After training, after the NAR network fully extracts the information of the training sample, the residual sample generally approximately obeys the normal distribution. After training, the NAR network can generally fully extract the effective information of the training sample, so the residual sample generally approximately obeys the normal distribution. Consider the residual as a random variable A, and then in the no-fault state, there should be $X_e : N(\mu,\sigma)$, where and are the mean and standard deviation of the residual sample.

Finally, using the residual distribution law to monitor the running status of the system in real time, it is judged whether the single-step prediction residuals of the non-fault state monitoring model obey the residual distribution law. If the single-step prediction residual is deviating from the distribution of residual, it is considered that the system has an early fault signal, the monitoring model will carry out fault early warning.

The above process is the so-called state monitoring process based on Copula function and autoregressive neural network. The process can be described as

$$\begin{cases} \boldsymbol{u}_t = F(\boldsymbol{x}_1), \boldsymbol{v}_t = F(\boldsymbol{x}_2) \\ \boldsymbol{y}_t = C[\boldsymbol{u}_t, \boldsymbol{v}_t] \\ \hat{y}(t+1) = f(\boldsymbol{y}_t) \\ e(t+1) = y(t+1) - \hat{y}(t+1) \\ X_e = \{e_1, e_2 \text{L}\ e_N\} \\ X_e : N(\mu,\sigma) \end{cases} \tag{5}$$

where $F(\boldsymbol{x}_1)$ and $F(\boldsymbol{x}_2)$ are the edge cumulative distribution of wind speed and power, C is the binary Copula joint probability density function, $\boldsymbol{y}_t$ is the output vector of Copula function, and f is the NAR neural network mapping.

3 Application verification and analysis

This paper uses the wind speed and power data collected by the wind turbine Supervisory Control and Data Acquisition (SCADA) system as an application verification to verify the effectiveness of the proposed method. The sampling interval for wind speed and power is 10 s. The acquisition time is from 1:00:10 on November 15, 2015 to 11:59:00 on December 2, 2015. Among them, at the time of 11:45:00 on December 2, 2015, the SCADA system showed that a right yaw reaching limit fault occurred in the system. This article will use 8640 data sets from 11:59:00 on December 1st, 2015, to 11:59:00 on December 2nd, 2015 as a test sample. This data contains failure data; use the remaining data as a training sample training model. This article will use a total of 8640 × 2 sets of data from 11:59:00 on December 1st, 2015 to 11:59:00 on December 2nd, 2015 as a test sample, which contains fault data of a right yaw reaching limit fault. The remaining data are used as a training sample to train the NAR model.

The frequency histograms of the two datasets are shown in Figure 1(a) and Figure 1(b), obviously, neither of them fully obeys the normal distribution. The wind speed and power are converted into uniform distributions u and v on [0, 1]. The binary frequency histogram is shown in Figure 2. It can be seen that the graph has obvious symmetry, so the Frank Copula function was chosen to fit the model.

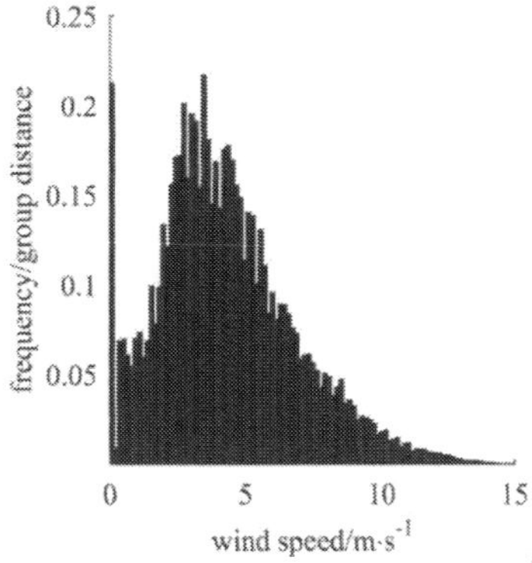

(a) Wind speed

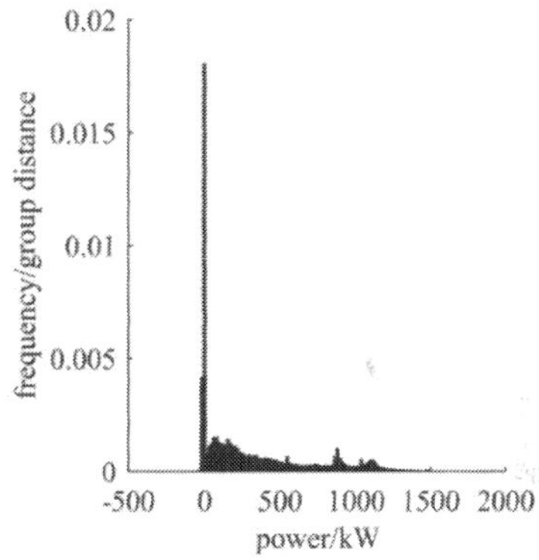

(b) Power

Figure 1. Wind speed and power frequency histogram.

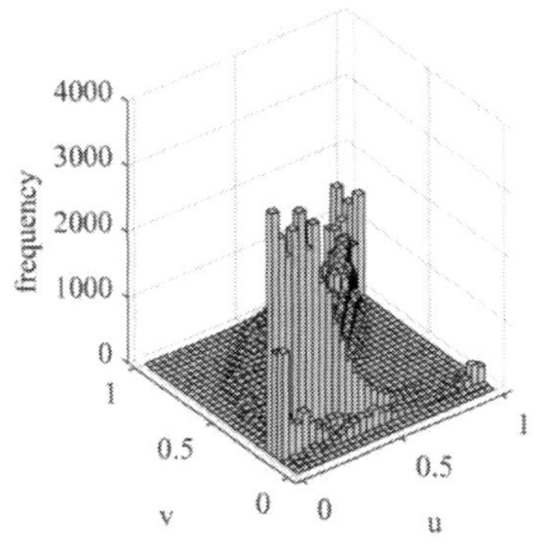

Figure 2. Binary frequency histogram.

Using KDE non-parametric estimation method to fit the edge distribution of wind speed and power data, and then using the maximum likelihood estimation method to obtain the parameter of the Frank Copula function as $\theta = 11.706$, then the joint probability density function of the binary Frank Copula is

$$C(u,v;\theta) = -\frac{\ln\left[1+\frac{(e^{-11.706u}-1)(e^{-11.706v}-1)}{e^{-11.706}-1}\right]}{11.706} \quad (6)$$

Enter u, v into the joint probability density function of Eq. 6 to find the time series y_t of fusion sample points based on the Copula function. The time series of wind speed, power, and fusion data are shown in Figure 3.

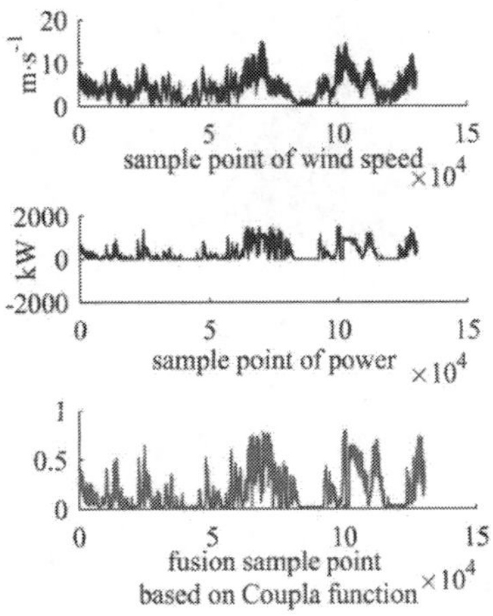

Figure 3. Wind speed, power and fusion data.

The fusion data sample is used as a training set, and further divided into sub-training sets, sub-validation sets, and sub-test sets, respectively accounting for 70%, 15%, and 15% of the total fusion data samples, and the NAR model is trained using the data set. The training error goal is set as 5×10^{-4}. After several parameter optimizations, the hidden layer of the NAR neural network selected in this paper is 30, and the delay order is 10. The training process shows that the convergence step is up to 4 steps, which satisfies the error requirements. The model establishment speed is fast and all parameters of the network meet the requirements.

The statistical analysis of the single-step prediction residual random variables of the training samples in the no-fault state can be considered as approximately obeying the normal distribution, and there is $X_e : N(-1.89\times10^{-4}, 0.0211)$.

Through the above training, a fault-free monitoring model based on the NAR neural network in a fault-free state can be obtained. The model is used to perform online monitoring of the test set containing the fault signal. By judging whether the residual of the single-step prediction value obeys the normal distribution, it can be judged whether or not the failure has occurred. The data of wind speed and power in the test set are input into the fault-free condition monitoring model to obtain the residual sequence of the test set. The time series of the monitoring parameters and the residuals are displayed synchronously as shown in Figure 4.

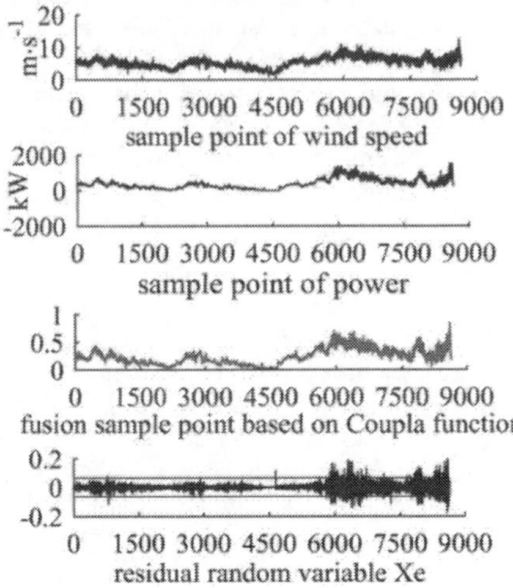

Figure 4. Parameters simultaneously displayed.

In Figure 4, the residual values of the single-step prediction clearly exceeds the confidence interval of $[\mu-3\sigma, \mu+3\sigma]$. In order to overcome the effects of extreme errors, an abnormal signal is considered to be present only if there is a continuous period of time exceeding the threshold and the sample point satisfying this condition appear at 3:40 on December 2, 2015. By comparison, it can be found that the fault early warning time is 8 hours earlier than the fault recording time of the SCADA system. The results show that the fault-free monitoring model proposed in this paper can effectively monitor the early abnormal signal before the failure occurs compared with SCADA system.

In order to fully illustrate the effectiveness of the method, it is applied to the actual monitoring to observe whether more faults can be monitored. After a period of testing, it is found that the method can effectively monitor the following faults: the left yaw reaching limit fault, the gearbox oil temperature overrun fault, and the generator bearing temperature overrun fault. Compared with the fault alarm time of the SCADA system, the prediction time of the three kinds of faults in advance is shown in Table 2. The residual error sequences of the above three kinds of faults are shown in Figure 5.

Table 2. Various types of fault early warning time

Type of fault detected	Early warning time / h (compared to SCADA system)
Left yaw reaching limit	4.3
Gearbox oil temperature overrun	16.2
Generator bearing temperature overrun	1.3

 https://doi.org/10.1051/matecconf/201819804008

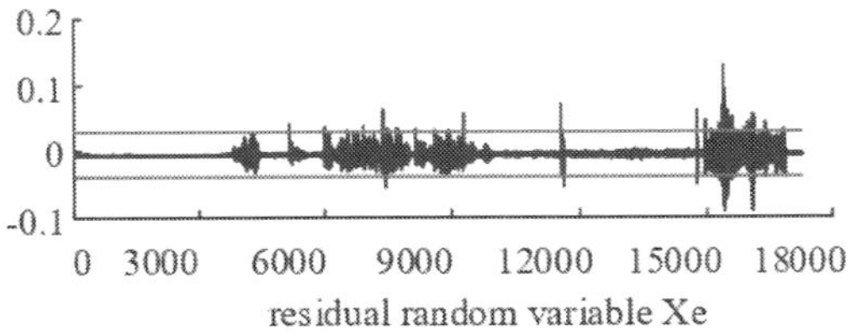

(a) Left yaw reaching limit fault

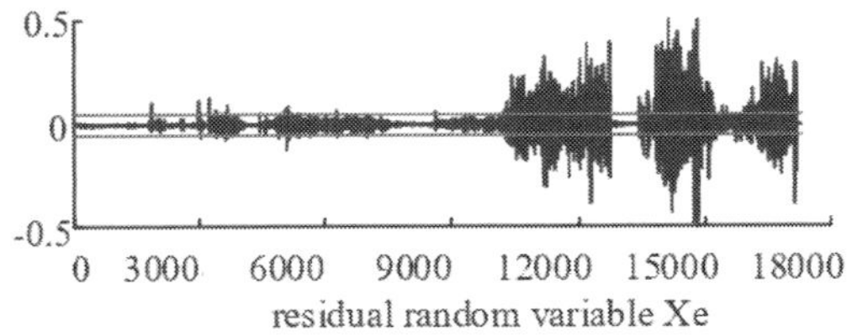

(b) Gearbox oil temperature overrun fault

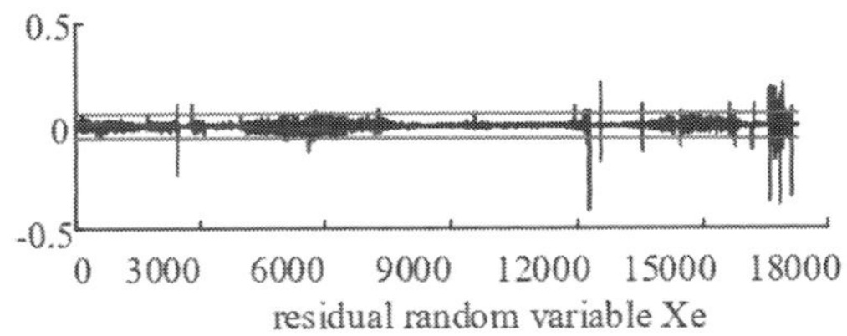

(c) Generator bearing temperature overrun fault

Figure 5. Various types of fault residual sequences.

As can be seen from the Table 2, this method can give an alarm earlier than the SCADA system, which is of great significance for the condition monitoring activities of wind turbines. The analysis proves that the method proposed in this paper has a high state monitoring and early warning capability.

4 Conclusions

This paper makes use of the wind speed and power parameters of wind turbines to monitor the status of the system. A condition monitoring method based on Copula function and autoregressive neural network is proposed. The final application verification shows that the warning time of this method is 8 hours earlier than the SCADA monitoring system, which proves that this method can effectively perform state monitoring and abnormal alarm for wind turbines.

References

1. Global Wind Report 2015: Annual Market Update [R]. Global Wind Energy Council, 2016.
2. CJ Crabtree, Y Feng, PJ Tavner. Detecting Incipient Wind Turbine Gearbox Failure: A Signal Analysis Method for On-line Condition Monitoring [J]. Journal of Organic Chemistry, 2010, **75**(18): 6122-6140.
3. A Zaher, SDJ Mcarthur, DG Infield, et al. Online wind turbine fault detection through automated SCADA data analysis [J]. Wind Energy, 2009, **12**(6): 574-593.
4. W Yang, PJ Tavner, CJ Crabtree, M Wilkinson. Cost-Effective Condition Monitoring for Wind Turbines [J]. IEEE Transactions on Industrial Electronics, 2010, **57**(1): 263-271.
5. P Guo, D Infild, X Y Yang. Wind turbine gearbox condition monitoring using temperature trend analysis [J]. Proceedings of the CSEE, 2011, **31**(32): 129-136.
6. A Zaher, SDJ Mcarthur, DG Infield, et al. Online wind turbine fault detection through automated SCADA data analysis [J]. Wind Energy,2009, **12**(6): 574-593.
7. RB Nelsen. An Introduction to Copulas (Springer Series in Statistics) [M]. 2006: xx,315.
8. M Sklar. Fonctions de repartition a n dimensions et leurs marges [J]. Publinststatistunivparis, 1959, 8.
9. M Mordjaoui, S Haddad, A Medoued, A Laouafi. Electric load forecasting by using dynamic neural network [J]. International Journal of Hydrogen Energy, 2017.

Deep sub-micron ESD GGNMOS layout design and optimization

Shi Jun

College of Mechanical and Electronic Engineering, Shanghai Jianqiao University, Shanghai, P.R. China

Abstract. In the field of integrated circuits, ESD (Electro Static Discharge) has always been a rather serious problem of reliability. Enhanced ESD tolerance of IC chips became a focus of research on IC failure protection design. The thesis is better to solve the multi-fingered non-uniform conduction of ESD devices under electrostatic pulse. Layout parameters DCGS (Drain-Contact to Gate Spacing), SCGS (Source-Contact to Gate Spacing) and BS (Substrate-source spacing) size in the paper can be used as reference for ESD GGNMOS (Gated Ground NMOS) layout design. Also this paper provides setting the DRC (Design Rule Check) command to check the distance between the N+ diffusion regions of different potentials so that ESD failure is prevented effectively. TLP (Transmission Line Pulse) current pulse signal is adopted to measure characteristics of the GGNMOS. The thesis descripts a ESD Optimal layout design from five aspects of introduction, Key elements of ESD circuits layout design, ESD layout optimization, a ESD GGNMOS layout instance and conclusion.

1 Introduction

There are two kinds of failures caused by ESD: One is the thermal failure caused by excessive current; the other is due to the high voltage which causes the electric effect to cause breakdown of the medium (such as gate oxide). The thermal failure is, because the local current excessive concentration, causing a large number of heat accumulation, resulting in metal interconnection or semiconductor melt, further lead to two breakdown, the damage mainly concentrated in vias, diffusion resistance, poly-silicon resistors and metal wires. The electrical failure is the ESD voltage applied directly to the vulnerable thin oxide layer, leading to dielectric breakdown or surface breakdown. A variety of reasons have led to the failure of the chip under ESD pulse. These reasons may be due to the inappropriate layout design [1], parasitic ESD devices, the internal ESD sensitive circuit structure and the incorrect placement of the power pins.

2 Key elements of ESD circuit layout design

In deep submicron CMOS (Complementary Metal-Oxide Semiconductor) technology, ESD protection circuits protect Integrated circuit from high current or high voltage attacks, usually are very larger, in order to achieve the I/O input and output ESD protection, power clamp protection, and rail to rail ESD protection. GGNMOS with large W/L ratio is generally designed with multi-finger layout. Under ESD pulse, the biggest problem is the non-uniform conduction of large size multi-finger transistors [2]. Secondly, in the high density unit chip, the parasitic NPN transistor exists, Under ESD pulses, parasitic NPN transistors may be triggered by substrate current, resulting in ESD failure [1]. Therefore, in the layout design of large size GGNMOS, we must solve the key multi-finger transistor layout, as far as possible to ensure the GGNMOS uniform conduction, or through the layout structure design to improve it the breakdown voltage of two V_{t2}, avoided premature thermal breakdown; At the same time, by checking DRC validation command in the layout design rule, the space between the N+ diffusion regions of different potentials in the I/O unit is examined to avoid it too small to induce NPN turn-on.

3 ESD layout optimization design

3.1 GGNMOS multi-finger structure and optimization of layout design parameters

The channel length L (Length), channel width W (Width) and even multi-finger number n of the MOS device have been determined in the GGNMOS layout design. Therefore, it is the key to solve the layout design of GGNMOS multi-finger structure, DCGS, SCGS and BS, these GGNMOS structural parameters as shown in Figure 1.

In deep sub-micron conditions, increasing the channel length L does not improve the withstand voltage of GGNMOS leakage voltage [3-5]. Therefore, the channel length L is usually designed to be 0.5μm-0.8μm in length. When designing $W=nW_{UNIT}$ with multi finger, the Fingered-Width W_{UNIT} usually cannot exceed 50μm, and the number of fingers n uses even value [1], [4], [6].

Therefore, the key in design to solve the layout design of GGNMOS multi-fingered structure is DCGS, SCGS and BS, as shown in Figure 1 GGNMOS structural parameters [2, 7].

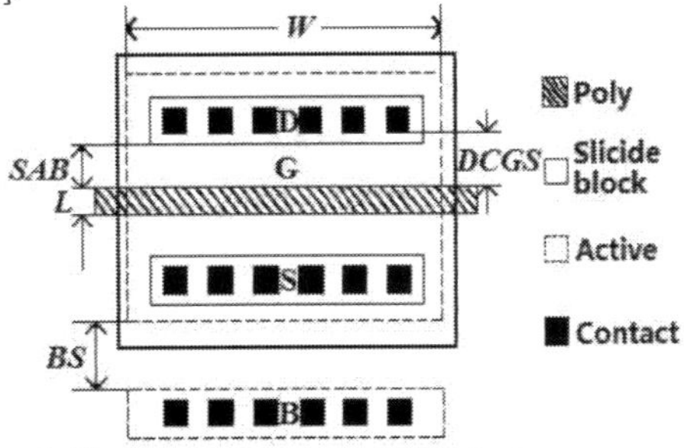

Figure 1. The layout parameters of MOS device.

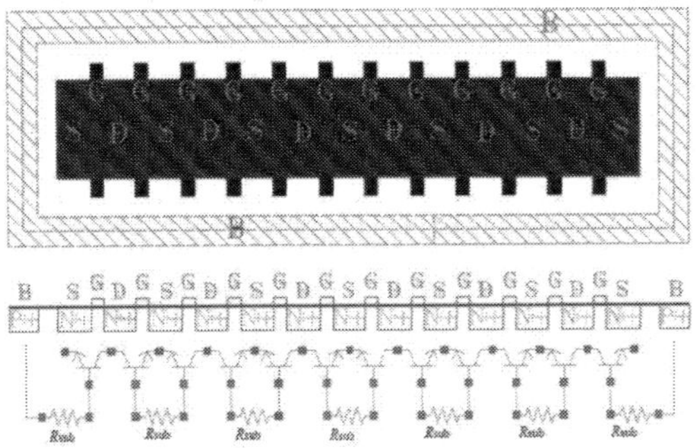

(a) BSGDGSGDGSGDGSGDGSGDGSGDGSB

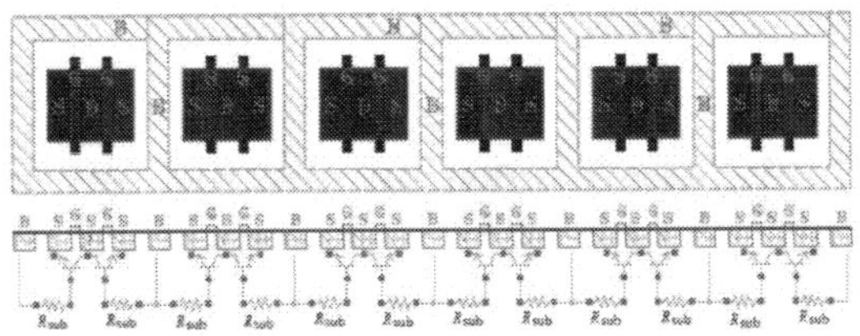

(b) BSGDGSBSGDGSBSGDGSBSGDGSBSGDGSBSGDGSB
Figure 2. GGNMOS multi-fingered structure.

The GGNMOS multi-fingered structure is usually designed with S (Source) - S merging and D (Drain) -D merging. And then the source and drain metal lining leads a multi-fingered structure, which is compact, but the problem of non-uniform conduction is serious, as shown in Figure 2(a). In deep submicron process, in order to solve the inhomogeneous conduction problem, instead of using BSGDGSGDGSGDGSGDGSGDGSGDGSB layout as shown in Figure 2(a), GGNMOS is designed with BSGDGSBSGDGSBSGDGSBSGDGSBSGDGSBSGDGSB layout as shown in Figure 2(b), which has better uniform conduction performance. In Figure 2(b), the substrate resistance R_{sub} of each parasitic NPN composed of NMOS Drain, Source and P-sub substrate is the same, which ensures uniform conduction under ESD pulse; Whereas As shown in Figure 2(a), the substrate resistance R_{sub} is larger due to the resistance of the substrate R_{sub} on both sides of the substrate is half of the adjacent substrate resistance. And the intermediate substrate resistance R_{sub} is largest. On the ESD pulse, it cause inhomogeneous conduction that the middle parasitic NPN conducting but side parasitic NPNs not, substrate resistance is bigger, NPN parasitic is more easy to conduct [8, 9].

Because the GGNMOS device turn-on trigger voltage V_{t1} decreases with the increase of DCGS [4], When the DCGS is 5μm, the maximum secondary reverse breakdown current I_{t2} and the smaller GGNMOS turn-on voltage V_{t1} can be obtained. The size of DCGS is a key factor that can affect the performance of GGNMOS.

With the decrease of SCGS, the V_{t1} and I_{t2} show an increasing trend, which is not conducive to the effective triggering of LNPN so that SCGS can consider the minimum design size [4].

As the V_{t1} decreases with the increase of BS, this is beneficial to the effective triggering of the LNPN structure of GGNMOS devices, The I_{t2} shows the lowest peak value at SCGS=4.0μm. Therefore, the design as shown in Figure 2(b) is available in GGNMOS layout.

3.2 Verification of different N+ potential intervals

Because of the N type region of GGNMOS, the N type region with different potential nearby and P-substrate, there will be parasitic NPN transistor. The most simply practical and effective way to avoid parasitic NPN is to increase the spacing of N type diffusion regions with different potentials. By increasing the Base thickness of parasitic NPN transistor, the distance between parasitic NPN emitter junction and collector junction is increased. The minority carrier in Base region cannot be collected in collector region, and parasitic NPN cannot be conducted. Alternatively, the heavily doped P-substrate enhances the threshold voltage of parasitic NPN.

The command can be set in DRC (Design Rule Check) to check the spacing of the N diffusion region with different potentials as follow:

```
ESD.sn.chk.1{
  @enimp of different potential space must >= 4
  EXT      enimp<  4      ABUT<90SINGULAR
REGION  NOT  CONNECT
  }
```

Among them, enimp is the N type diffusion region with different potentials; NOT CONNECT represents a N type region that is not connected by a metal or other connecting layer; REGION indicates that when the error is found, the range is displayed when the error is displayed; SINGULAR means that a touch or a touch is wrong; ABUT<90 indicates the intersection angle less than 90 degree error. Here, the different potential N type diffusion region minimum spacing is 4μm.

4 A ESD GGNMOS layout optimization model

4.1 ESD GGNMOS optimized layout structure

Layout structure as shown in Figure 3 is part of layout refer of the serials shown in Figure 2(b). The ESD device consists of GGNMOS in which drain D and source S form multi-fingered structure. The gate G is connected with the source S through the metal-contact-poly at the top and bottom, and the drain D is from the up line. The gate G and the source S are contacted with the substrate, and the B is formed by assembling the lower finger to line below the device.

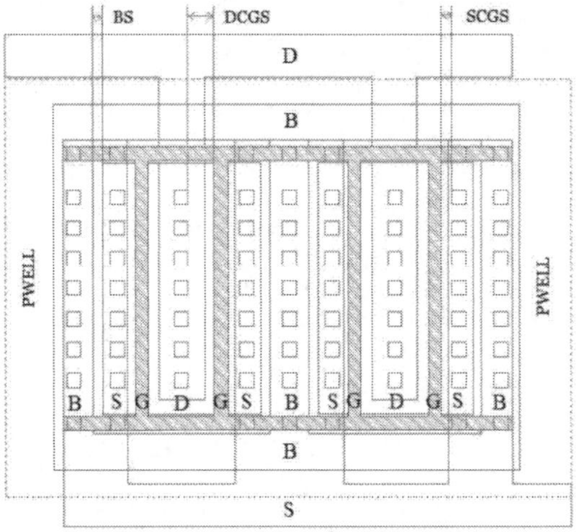

Figure 3. GGNMOS layout structure optimization.

4.2 GGNMOS layout based on 0.13μm process

In this paper, the GGNMOS ESD layout of LDO (Low Drop-Out) regulator chip is completed based on the 0.13μm as shown in Figure 4. And the Figure 4 is the layout of the GGNMOS and bonding-pad. The input and output signals are connected to the PADs, which is formed from the Input PAD into the internal IC circuit and from the internal IC circuit to the Output PADs.

According to the above optimization principle, the GGNMOS layout is optimized that parameters of it are shown in Table 1.

Figure 4. GGNOMS in ESD design.

Table 1. GGNMOS layout design parameters

W_{UNIT}	L_{UNIT}	*Multiple*	DCGS	SCGS	BS
50μm	0.5μm	12	1.8μm	0.63μm	2μm

W_{UNIT} is the channel width of the unit GGNMOS, L_{UNIT} is the channel length of unit GGNMOS, and ***Multiple*** is the number of multi-fingered.

5 Performance after ESD layout optimization

TLP [10-14] current pulse signal, which is 100ns pulse width, 10ns rising time, 10ns falling time and 40mA currents, is added to the drain of the GGNMOS device, at the same time the gate, source and substrate of it are grounded. As shown in Figure 5, the V_D - I_D characteristic is the result of the simulation of the GGNMOS device. It is shown in the Figure 4 that the V_D - I_D characteristic coincides with the snapback obtained by theoretical analysis after GGNMOS layout optimization. It is uniform conduction within the GGNMOS device, because the drain secondary breakdown voltage V_{t2} is greater than the trigger voltage V_{t1} that is about $8V$. Meanwhile, it is beneficial to GGNMOS triggered conduction, and the drain holding voltage V_h in the diagram is also reasonable. Figure 6(a) is the hole carrier current density, Figure 6(b) is the electron carrier current density. They are shown the internal states near the channel when the GGNMOS is triggered.

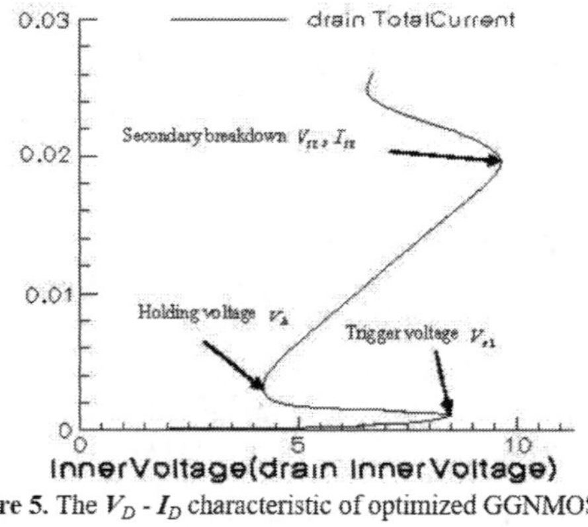

Figure 5. The V_D - I_D characteristic of optimized GGNMOS.

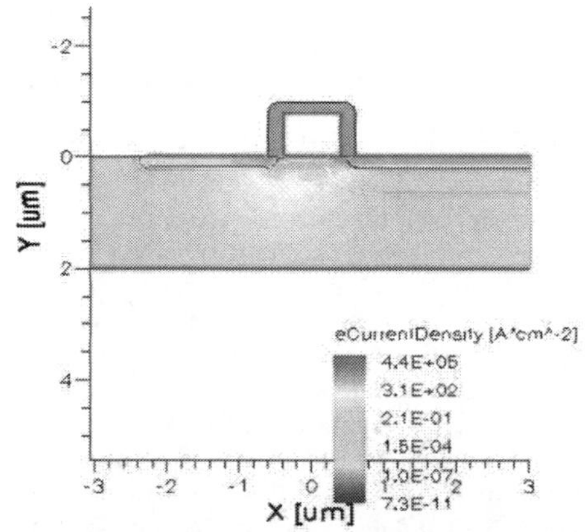

(a) the hole carrier current density

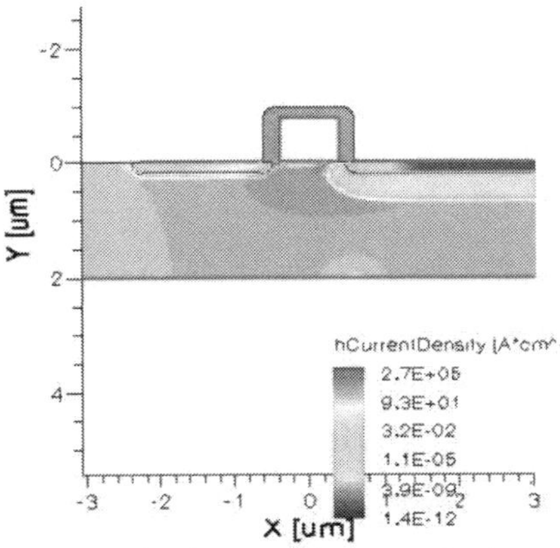

(b) the electron carrier current density

Figure 6. The current density of optimized GGNMOS.

6 Conclusion

In this paper, the GGNMOS layout optimization design of ESD devices is presented. It is better to solve the multi-fingered non-uniform conduction of ESD devices under electrostatic pulse. Layout parameters DCGS, SCGS and BS size in the paper can be used as reference for ESD GGNMOS layout design. The parasitic NPN around the GGNMOS could be found by Setting the DRC command to check the distance between the N+ diffusion regions of different potentials so that ESD failure is prevented e effectively. TLP current pulse signal is added to the drain of the GGNMOS device, at the same time the gate, source and substrate of it are grounded. As the result, the V_D - I_D characteristic coincides with the snapback obtained by theoretical analysis after GGNMOS layout optimization.

7 Relation to prior work

Recently a LDO (Low Dropout Linear Regulator) layout of SOC is designed based on 0.13μm. The ESD protective device is using the GGNMOS designed according to this work. At last we will research for ESD characteristics biased on nan-meter process.

Acknowledgment

Some simulations and experiments have conducted in Shanghai Institute of Microsystem, CAS (The Chinese Academy of Sciences).

References

1. Tao Jian-lei, Fang Pei-yuan, Wang Jia-qi. "Study of Layout-Induced Failure in ESD Protection Circuit in CMOS IC," Semiconductor Technology. Vol. **32**, No. 11, pp. 1003-1006, (2007)
2. T.-Y. Chen and M.-D. Ker, "Analysis on the dependence of layout parameters on ESD robustness of CMOS devices for manufacturing in deep-submicron CMOS process," IEEE Trans. Semiconductor Manufacturing, vol. **16**, no. 3, pp. 486-500, (Aug. 2003)
3. Li Ruo-yu, Li Bin, Luo Hong-wei, "The dependence of GGNMOS ESD robustness on its gate length," Journal of Circuits and Systems. Papers **10**(10), (2005). J. Clerk Maxwell, A Treatise on Electricity and Magnetism, 3rd ed., vol. **2**. Oxford: Clarendon, 1892, pp. 68–73
4. Shi Jun, "ESD characteristics of GGNMOS Device in deep sub-micron CMOS technology," The 5th International Conference on Audio, Language and Image Processing: ICALIP'2016, Shanghai: IEEE, pp. 367-372
5. Li Ruo-yu, Li Bin, LUO Hong-wei, "The dependence of GGNMOS ESD robustness on its gate length," Journal of Circuits and Systems. Vol. **10**, No. 5, pp. 93-96, (Oct. 2007)
6. Jiang Yu-xi, "Influence of layout parameters on snapback characteristic for a gate-grounded NMOS device in 0.13-μm silicide CMOS technology," Journal of Semiconductors. Papers (8), (2009)
7. K.-H. Oh, C. Duvvury, K. Banerjee, and W. Dutton, "Impact of gate-to-contact spacing on ESD performance of salicided deep submicron NMOS transistors," IEEE Trans. Electron Devices, vol. **49**, pp. 2183-2192, (Dec. 2002)
8. X.F. Gao, J.J. Liou, A. OrtizConde, J. Bemier, G.Croft, "A physics based model for the substrate resistance of MOSFETs," SolidState Electronics. Papers **6**(46), pp. 853-857 (2002)
9. Alan Hastings, [The Art of Analog Layout], Beijing: Publishing House of Electronics Industry, (2011)
10. M.-D. Ker, J.-J. Peng, and H.-C. Jiang, "ESD test methods on integrated circuits: an overview," in Proc. IEEE Int. Conf. Electronics, Circuits and Systems, vol. **2**, pp.1011-1014, (2001)
11. G. Notermans, P. de Jong, and F. Kuper, "Pitfalls when correlating TLP, HBM, and MM testing," in Proc. EOS/ESD Symp., pp. 170-176, (1998)
12. M.-D. Ker, H.-C. Hsu, and J.-J. Peng, "ESD implantation for sub-quarter-micron CMOS technology to enhance ESD robustness," IEEE Trans. Electron Devices, vol. **50**, no. 10, pp. 2126-2134, (Oct. 2003)
13. "Electrostatic Discharge Sensitivity Testing—Human Body Model (HBM)—Component Level," ESD Association, ESD Association Standard Test Method ESD STM-5.1, (1998)
14. "Electrostatic Discharge (ESD) Sensitivity Testing—Machine Model (MM)," EIA/JEDEC, EIA/JEDEC Standard Test Method A115-A, (1997)

A flexible motion tracking system based on inertial sensors

Zhonghao Han, Lei Hu, Na Guo, Biao Yang, Hongsheng Liu and Yuhan Wang

School of Mechanical Engineering and Automation, Beihang University, 100191 Beijing, China

Abstract. As a newly emerging human-computer interaction, motion tracking technology offers a way to extract human motion data. This paper presents a series of techniques to improve the flexibility of the motion tracking system based on the inertial measurement units (IMUs). First, we built a most miniatured wireless tracking node by integrating an IMU, a Wi-Fi module and a power supply. Then, the data transfer rate was optimized using an asynchronous query method. Finally, to simplify the setup and make the interchangeability of all nodes possible, we designed a calibration procedure and trained a support vector machine (SVM) model to determine the binding relation between the body segments and the tracking nodes after setup. The evaluations of the whole system justify the effectiveness of proposed methods and demonstrate its advantages compared to other commercial motion tracking system.

1 Introduction

Human motion tracking extracts joints data from human in motion. A whole range of applications become possible due to the motion tracking technique gets cheap, reliable and convenient. These applications can roughly be categorized under two titles: control and analysis [1].

In control applications, the human motion data are used to control something. These things can be physical things like manipulators or other in general exoskeletal parts, which human can control remotely so avoiding physical risks [2]. The thing can also be virtual thing, like applications in the film, game industry, virtual reality and so on [3].

Analysis applications are most used in medical and sports fields. For example, the rehabilitation described in [4] used a full-body motion tracking system to analyse the patient's exercise in order to develop customized treatment strategies.

There are many techniques to implement the motion tracking, such as optical, image-based, inertial measurement units (IMUs) based, mechanical, magnetic, acoustic, and hybrid systems [5-7]. Among these methods, the systems based on IMUs are most promising due to the versatility, robustness and almost unlimited range of motion [8]. Thanks to recent advances in the technology of micro-electromechanical systems, a miniature circuit board can integrate a three-axis accelerometer, a three-axis gyroscope, and a three-axis magnetometer simultaneously [9]. The accelerations, angular velocity, and magnetic data can be fused to estimate the orientation of the IMUs using a complementary filter or extend Kalman filter (EKF) [10-12]. Many motion tracking systems have been developed based on the inertial sensors, the MVN of the Xsens Technologies B.V. is most outstanding one, which is usually used as gold standard to evaluate the performance of other systems. Its sensor fusion technologies enable a seamless interaction between the physical and the digital world. However, the limitation of commercial tracking systems based on IMUs lies in many aspects other than the high cost.

The first one is the size and weight of the module. There are many efforts in reducing the size of the IMUs [13-14], but this is an endless way because a smaller IMU can usually be mounted on the body segments more easily and causes fewer obstacles to human movement, which means more accurate and convenient.

Other drawbacks depend on the architecture of the motion tracking system. In general, there are two wireless structures based on IMUs. The first one is hub-based. In this architecture, all the tracking nodes connect to a central hub directly or relayed by other nodes. The hub services as a power supply and data exchange center. Due to the centralized design, this architect usually features a higher data exchange rate and easy recharge. But the setup procedure can be onerous even integrating all the tracking nodes into a tracksuit, which generally take more than ten minutes to complete with the help of the others. The entire equipment can be cumbersome when the wire around body hinders the movement of the subject. Besides, this method is only friendly to the whole-body motion capture and is not appropriate when we just want to track the part motion of our body.

The second architecture is hub-free, means that individual measurement module is independent, holding its own battery and communicating with PC station directly. In this method, the nodes usually mount the segments by the strap, so the movement of the subject would be freer. The number of sensors can be customized according to the applications. However, the data transfer rate remains fixed in these system, means you cannot get a higher data transfer rate by using fewer nodes. Besides, in all current designs, the binding relation between the tracking nodes and the body segments is confirmed before the setup procedure. So, these systems are characterized by a configurable ID in the firmware of node and a label with the name of the body segment it mounts. In this way, the tracking nodes are not interchangeable, which complicates the setup procedure and increases the cost of mass production and maintenance.

Based on this, we propose a flexible motion tracking system based on IMUs, which has features as follows: Firstly, the tracking node is ultra-small with low power consumption and can be easily mounted on the human body. Secondly, the number of the measurement modules used in motion tracking can be arbitrary according to the applications. An asynchronous query method will always optimize the data transfer rate between the tracking nodes and PC station. Finally, the binding relation between the body segments and tracking nodes is determined using a trained support vector machine (SVM) model after setup procedure, which simplifies the setup procedure and makes the interchangeability of all nodes possible.

2 System architecture

2.1 Hardware design

The system in this paper is hub-free (see Figure 1), which means all tracking nodes are independent, containing its own IMU, wireless communication component, and power supply. The IMU focuses on determining the orientation of each individual node, and orientation data are transferred to a PC station via a WI-FI module and wireless router. To maximize the transmission speed, an ethernet cable is recommended between the PC station and router.

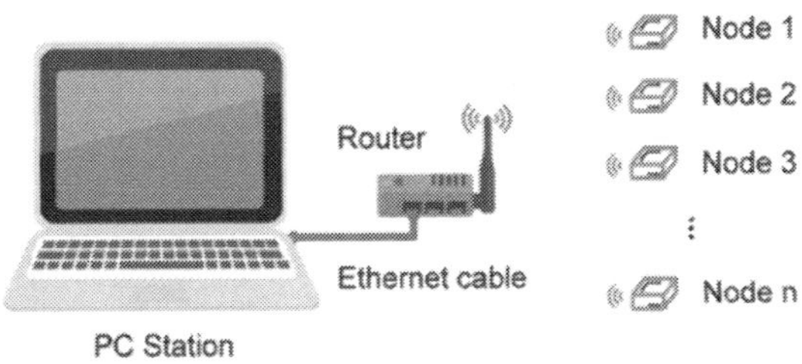

Figure 1. The hub-free architecture of the motion tracking system

The IMU is the core of the tracking node, for that it determines the accuracy of the system directly. One IMU usually consists of a three-axis accelerometer, a three-axis gyroscope, and a three-axis magnetometer. The orientation of the IMU can be estimated from the accelerations, angular velocity and magnetic data using a complementary filter or EKF. In this paper, JY901 made by Wit-Motion was chosen as the integrated IMU. Consisting of an MPU9250 and a 32-bit high performance MCU, the JY901 features a miniaturized 15.24 × 15.24 × 2mm package size. The output data can be customized in a range of acceleration, angular velocity, magnetic field component, Euler angle and quaternion fused by EKF. The output frequency can be configured between 0.1Hz and 1kHz, which satisfies the requirements of most situations. The other specifications of JY901 are shown in Table 1.

Table 1. The specifications of JY901

Name	Value
Dynamic angle error	0.05°
Accelerator range	±16g
Angular velocity	±2000°/s
Angle	±180°
Operating voltage	3.3V
Operating current	10mA
Interface	UART, IIC

In the motion tracking field, the ZigBee and Wi-Fi protocol are usually used for wireless data communication [15]. The former provides a low-power network connection with a great deal of flexibility [16]. The theoretical maximum transfer rate is 250 Kbps [17], but in the actual test, this number is close to only 25Kbps, which may not be suitable for data-intense whole-body motion tracking. Though the power consuming of Wi-Fi protocol is relative high, the data transfer rate can be improved significantly [18]. In this system, an EPS-12S is chosen for the purpose of high-speed communication. This Wi-Fi module integrates a low power 32-bit MCU, a radio frequency (RF) component, and several communication interfaces.

When connecting the UART interface of EPS-12S to the JY901, the orientation data can be transferred to the PC station using a wireless method. To save the space, the two components were attached on the offside of the printed circle board (see Figure 2). The highly integrated design features an ultra-small size of 28×32×10mm (weight: 14g) which is the smallest tracking node in all these systems proposed in [3], [13-14]. The whole power consumption is only 330mW.

Figure 2. The structure of the tracking node

2.2 Software design

In the original design, all tracking nodes sent the data with a fixed interval time plus a small random delay but the data transfer rate reduced significantly when more modules join the network. Besides, due to the fixed transmit frequency, we cannot get a higher data transfer rate by using fewer nodes. So, an asynchronous query strategy was adopted to guarantee the data transfer rate. In this method, the EPS-12S is configured to send the orientation data from the JY901 after receiving one byte of data. For the PC station, there are three sub-program (thread).

The first is the device detecting program. When the main program starts, this program scans all the devices that connect to the router and builds the inquiry table (see Figure 3). One term of the table includes a state flag, an IP address, and a quaternion data buffer. In the process of data collecting, the program deletes the items of devices that don't return data for a period and adds the new devices that connect to the router dynamically. This program executes with a frequency of 1Hz.

The second program is the data collecting program. This program loops through the inquiry table and send one-byte data to the tracking nodes designated by the IP address with an interval of 1ms. In most instances of motion tracking, the real-time of the data is important than reliability, so the UDP protocol was adopted to maximize the data transfer rate. If the program receives one data, it will check the flag of the corresponding inquiry table term. The return quaternion data are saved in the data buffer.

The third program is the main program. This program displays the animation model after processing the quaternion in the data buffer with an update frequency of 30Hz.

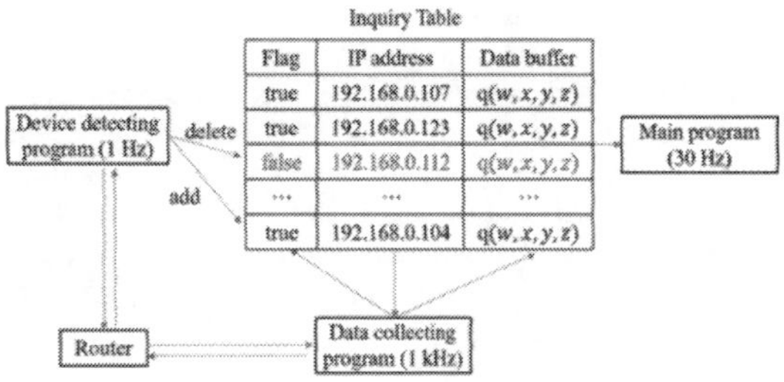

Figure 3. The program architecture and inquiry table

3 Calibration method

3.1 Theoretical analysis

For one specific IMU, the quaternion when subjects pose into a standard posture (T-Pose in this paper) is Q_P and the quaternion changes to Q_C after calibration gesture. Then, the transform quaternion from Q_P to Q_C in the world coordinate system (JY901 uses an "East North Up" (ENU) coordinate system) can be formulated as:

$$Q_T = Q_C \otimes Q_P^* \tag{1}$$

where $\otimes$ denotes a quaternion multiplication and * the complex conjugate of the quaternion.

The result can be expressed as:

$$Q_T = (w_T, x_T, y_T, z_T) \tag{2}$$

Transfer the quaternion to axis-angle representation:

$$\theta = 2\arccos w_T$$

$$\boldsymbol{e} = \left(\frac{x_T}{\sin\frac{\theta}{2}}, \frac{y_T}{\sin\frac{\theta}{2}}, \frac{z_T}{\sin\frac{\theta}{2}}\right), \text{ if } \theta = 0,\ \boldsymbol{e} = (0,0,0) \tag{3}$$

The rotation vector $\boldsymbol{\theta}$ is given by:

$$\boldsymbol{\theta} = \theta\boldsymbol{e} = \left(\frac{2x_T \arccos w_T}{\sin(\arccos w_T)}, \frac{2y_T \arccos w_T}{\sin(\arccos w_T)}, \frac{2z_T \arccos w_T}{\sin(\arccos w_T)}\right) \tag{4}$$

We represent rotation vector as:

$$\boldsymbol{\theta} = (R_x, R_y, R_z) \tag{5}$$

In the process of calibration, the subjects always keep perpendicular to the ground but can face any horizontal orientations. That means the rotation vector $\boldsymbol{\theta}$ will rotate around the Z axis for a specific calibration gesture. However, Figure 4 shows that $R_x^2 + R_y^2$ remains constant when rotation vector $\boldsymbol{\theta}$ rotates. As a consequence, a transform quaternion of an IMU for one calibration gesture can be determined by two critical components R_z and $R_x^2 + R_y^2$.

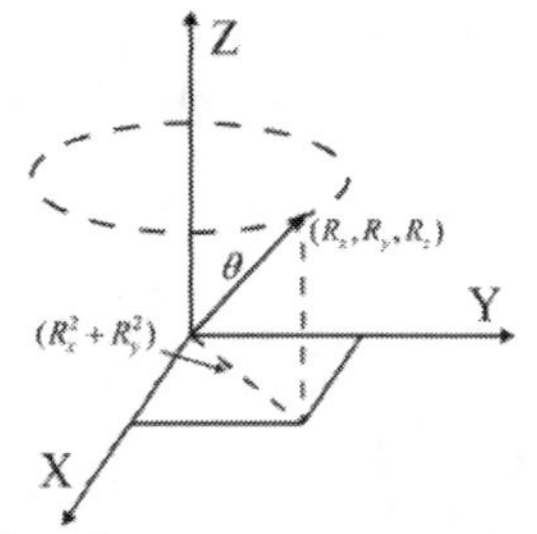

Figure 4. The $x^2 + y^2$ remains constant when rotation vector $\boldsymbol{\theta}$ rotates around Z axis

3.2 Design calibration procedure

There are three guidelines for designing calibration positions:

(1) The calibration gestures should be easily performed that the subjects can make a standard pose just following the guide of tutorial or program.

(2) For one calibration gesture, the R_z or $R_x^2 + R_y^2$ components of each node should be different enough to guarantee the nodes can be classified rightly.

(3) Each single calibration gesture should identify as many nodes as possible.

Based on this, we have designed a whole calibration procedure consists of four gestures. Figure 5 shows the three views of each gesture.

Figure 5a) is the T-Pose calibration, in which the subject keeps upright with arms horizontally and thumbs forward. This pose, determining the initial quaternion of the calibration, is the basic pose of all the calibrations.

The gesture in Figure 5b) is used to determine the binding relation between the tracking nodes and upper limbs including the head. In this gesture, the subject lowers head and puts hand together toward the front. The angle between forearm and arm should be greater than 90° (145° is recommended for the accuracy of recognition). All the upper limbs should keep horizontal.

Figure 5c) and d) show the gestures to recognize the nodes bound to the left lower limb and right lower limb respectively. The two gestures are mirror symmetry. The thigh and foot should be as horizontal as possible, and the calf keeps inward. The whole leg turns out.

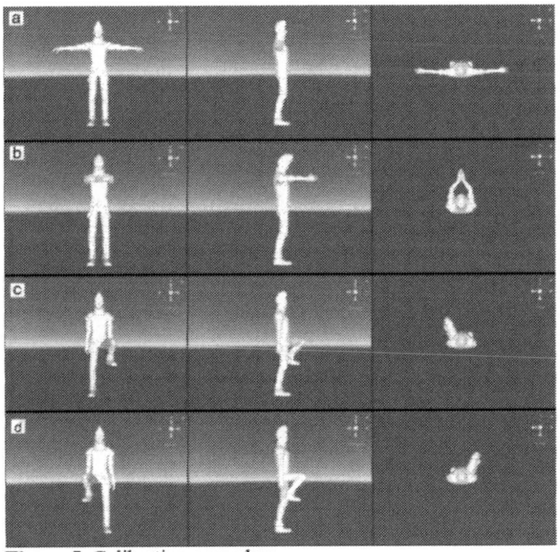

Figure 5. Calibration procedure

3.3 Training SVM model

During the model training phase, three subjects were chosen to perform 3-4 calibration procedures respectively. Figure 6 shows the points extracted from the transform quaternion of all nodes with $R_x^2 + R_y^2$ as abscissa and R_z as ordinate.

Note all these points are two-dimensional linear separable. An SVM was used to classify these points [19].

$$\max_{\alpha} \sum_{i=1}^{n} \alpha_i - \frac{1}{2} \sum_{i=1}^{n} \sum_{j=1}^{n} \alpha_i \alpha_j y_i y_j (\kappa(\boldsymbol{x}_i, \boldsymbol{x}_j))$$

$$s.t. \sum_{i=1}^{n} \alpha_i y_i = 0 \ , \quad \alpha_i \geq 0, \ i = 1, 2, \ldots, n \qquad (6)$$

where $\boldsymbol{x}$ is the vector represent of the points, y is the classification label, the kernel function we used is $\kappa(\boldsymbol{x}_i, \boldsymbol{x}_j) = \boldsymbol{x}_i^T \cdot \boldsymbol{x}_j$ and α_i can be found by using SVC learning algorithm [20].

Then, the dividing line $w^* \boldsymbol{x} + b^* = 0$ can be got by:

$$w^* = \sum_{i=1}^{N} \alpha_i^* y_i \boldsymbol{x}_i \qquad (7)$$

$$b^* = y_i - \sum_{i=1}^{N} \alpha_i^* y_i (\boldsymbol{x}_i^{\mathrm{T}} \cdot \boldsymbol{x}_j) \qquad (8)$$

The classification decision function is given by:

$$f(\boldsymbol{x}) = sign(w^* \boldsymbol{x} + b^*) \qquad (9)$$

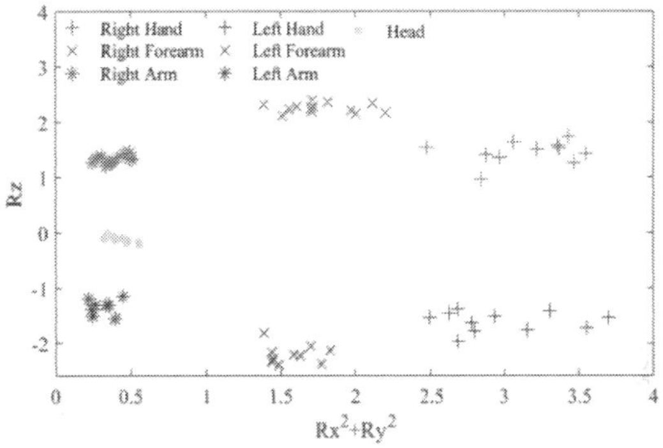

(a)

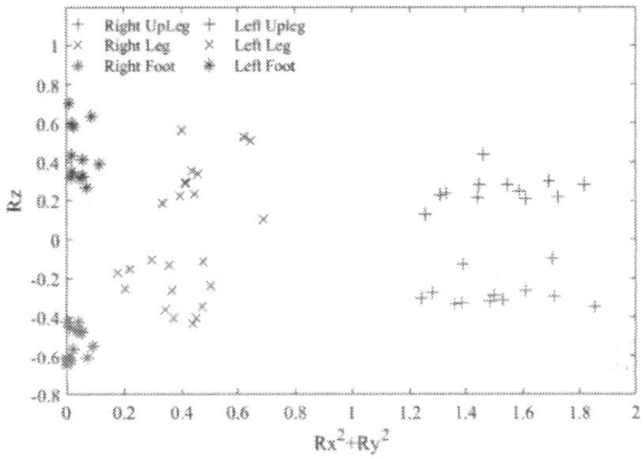

(b)

Figure 6. a) Extracted points of upper limbs and head; b) Extracted points from lower limbs

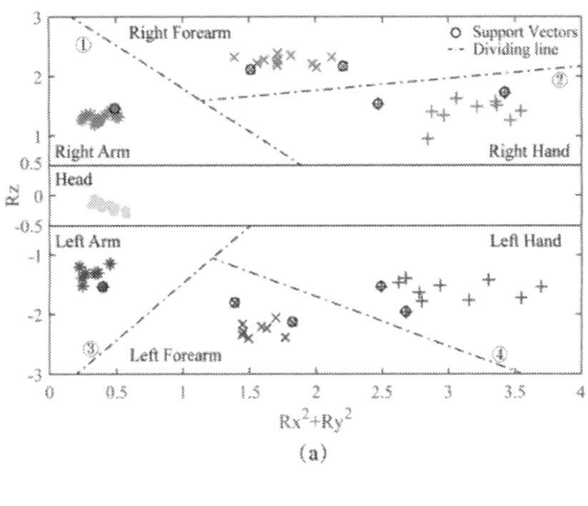

(a)

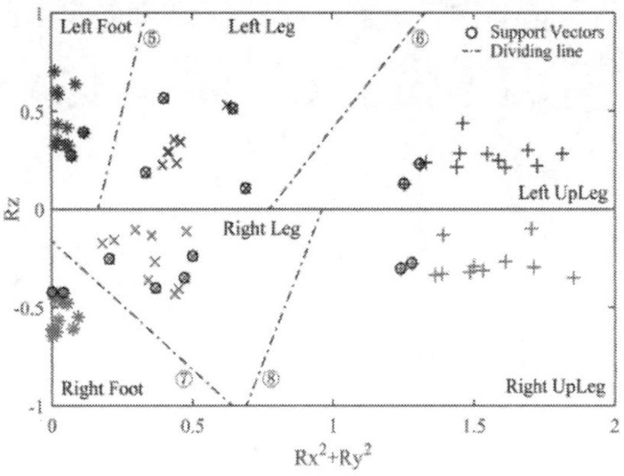

(b)

Figure 7. a) The SVM classification results of upper limbs; b) The SVM classification results of lower limbs

According to the distribution of points, we built two lines to separate the data of each limb, and the head is distinguished by the limitation of z. Compared to the traditional one versus one method for multi-classification problem, this can guarantee one point will belong to one specific class. Figure 7a) and b) show the eight dividing lines of all body segments. Table 2 shows the parameters of each dividing line represented by $y = kx + b$, which can be used as a reference.

Table 2. The parameters of each separated lines in SVM model

	L1	L2	L3	L4	L5	L6	L7	L8
k	-1.49	0.20	1.91	-0.83	5.80	1.80	-0.91	3.69
b	3.25	1.38	-3.38	-0.04	-0.96	-1.39	-0.24	-3.55

4 Experimental evaluation

For the evaluation, the system was built following the architecture presented in the Section II. The router is TL-WR886N made by TP-Link and the network card of the PC station is RTL8168/8111/8112 Gigabit Ethernet Controller made by Realtek. An ethernet cable is used to connect the PC station to the router.

The data transfer rate using asynchronous inquiry method was investigated when different numbers of nodes access the network. A specific program was designed which can find out how many frames of data received from each node in a given period. In the experiment, the max number of tracking nodes was 17, a standard value for whole body motion tracking. The average distance between the nodes and router was about 4m. The data collection program executed with a fixed frequency of 1kHz. As a comparison, the relation between the receive frequency and transmit frequency using auto sending method was also investigated when there were 17 tracking nodes in the network.

To construct the data set for analysis, the tracking nodes were snugly strapped to the subjects' body using straps. Then, the subjects performed the calibration procedure described in the Section III under the tips of the motion tracking program. Three subjects each performed 3 calibrations and tried to face the different direction in each calibration. The calibration time and nodes transformation quaternion were recorded.

5 Result

Figure 8a) shows relation the between data transfer rate and the number of nodes in the network using asynchronous inquiry method. The receive frequency is the average number of received frame data of all joined nodes in a period. The receive frequency is 815Hz when only one node in the network and the transfer rate decreases when more node joined the network. The average receive frequency of communicating with 17 nodes is 49 Hz. Figure 8b) shows the relation between the receive frequency and transmit frequency of 17 nodes using auto sending strategy, the maxim receive frequency is 32Hz when each node sends data with a transmit frequency of 40Hz. The results confirm the effectiveness of the asynchronous inquiry method.

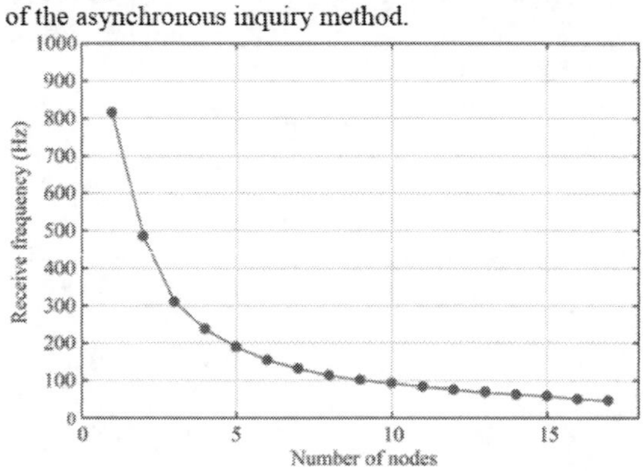

(a)

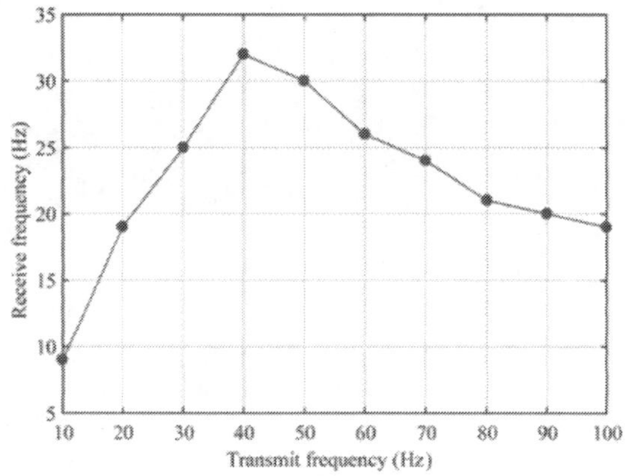

(b)

Figure 8. a) The receive frequency according to the nodes number; b) The receive frequency according to transmit frequency

Figure 9 shows the classification results of test data. Although the points distribution of a specific gesture is relatively decentralized, the SVM can distinguish these points with 100% accuracy. But we can find some points (for example, left foot) are close to the separated line, so, the model can be optimized by training more data further. For the reason that tracking nodes can be deployed arbitrarily, the average setup time is reduced to 6 minutes, and the average calibration time is 20.3s (see Table 3). These results verify the feasibility and reliability of the calibration method and the trained SVM model. Furthermore, a comparison of the main features of proposed motion tracking system with other commercial systems is shown in Table 4.

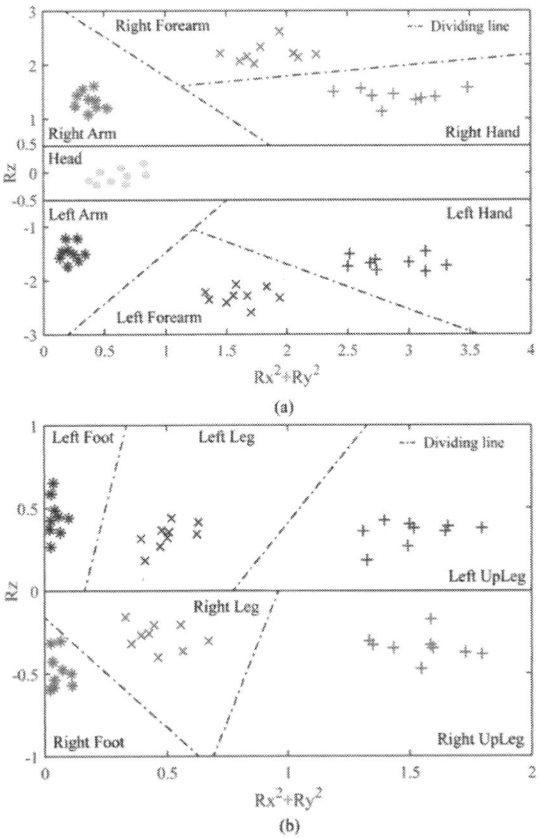

Figure 9. a) The classification results of the upper limbs and head using SVM model; b) The classification results of the lower limbs using SVM model.

Table 3. The time of setup and calibration

	Setup(mins)	**Cali. 1(s)**	**Cali. 2(s)**	**Cali. 3(s)**
s1	5.6	19	23	21
s2	5.8	18	19	17
s3	5.6	19	23	21

Table 4. Comparison among MVN AWINDA, MVN link and system in this paper

Parameters	**MVN Link**	**MVN Awinda**	**This Paper**
Architecture	Hub-based	Hub-free	Hub-free
Node Size	33×26×10mm	47×30×13mm	32×28×10mm
Battery life	10 hours	6 hours	7 hours
Node number	17 (fixed)	17 (fixed)	17 (adjustable)
Way of fixed	Tracking Suit	Strap	Strap
Setup Time	10 minutes	8 minutes	6 minutes
Node Interchangeabilit	No	No	Yes
Sample Rate	240 (fixed)	60 (fixed)	49 (17 nodes)
Hardware Price	$15380	$12820	$2600

6 Conclusion

In this paper, we have presented a wireless motion tracking system based on a hub-free architecture. An integrated IMU, a Wi-Fi module and a power supply made up a most simplified tracking node which features a minimize size and significant low power consumption. An asynchronous query strategy was proposed to monitor the network connection and optimize the data transfer rate between PC station and tracking nodes. Besides, given the particularity of calibration procedure, a transformation quaternion of one tracking node can be mapped to a single two-dimensional point. Base on this, we designed a specific calibration procedure and trained an SVM model using pre-collected data to classify the tracking nodes.

With all these techniques, this motion tracking system is extremely flexible. Due to the miniature size and weight, the tracking nodes can be easily mounted on the human body which reduces the resistance and improves the accuracy. The number of nodes used in one tracking task can be customized or adjusted dynamically according to the real applications, and the data transfer rate will always remain optimized. The experiments demonstrated that the average transfer rate for 17 tracking nodes is 49 Hz, which is high enough for some demanding applications. Moreover, compared to the commercial systems, the setup time was reduced to 6mins by a factor of the arbitrary arrangement of nodes. After calibration procedure, the trained SVM model can map the body segments and the nodes mounting on with 100% accuracy. Last but not least, considering individual node in this system is identical and burned same firmware, the cost of mass production and maintenance will be reduced significantly.

Acknowledgement

This work was supported by National Hi-Tech Research and Development Program of China ("863" Project) (Grant Nos.2015AA043204).

References

1. O T. Moeslund, A. Hilton and V. Krüger, "A survey of advances in vision-based human motion capture and analysis," Comput. Vis. Image Und., vol. **104**, no. 2-3, pp. 90-126 (2006)
2. S. Won, F. Golnaraghi and W. Melek, "A Fastening Tool Tracking System Using an IMU and a Position

Sensor With Kalman Filters and a Fuzzy Expert System," IEEE Trans. Ind. Electron., vol. **56**, no. 5, pp. 1782-1792 (2009)
3. D. Roetenberg, H. Luinge and P. Slycke, "Xsens MVN: full 6DOF human motion tracking using miniature inertial sensors," Xsens Motion Technologies BV, Tech. Rep. (2009)
4. S. Komada, Y. Hashimoto, N. Okuyama, T. Hisada and J. Hirai, "Development of a Biofeedback Therapeutic-Exercise-Supporting Manipulator," IEEE Trans. Ind. Electron.vol. **56**, no. 10, pp. 3914-3920 (2009)
5. R. Zhu and Z. Zhou, "A Real-Time Articulated Human Motion Tracking Using Tri-Axis Inertial/Magnetic Sensors Package," IEEE Trans. Neur. Sys. Reh., vol. **12**, no. 2, pp. 295-302 (2004)
6. Y. Nakamura, K. Yamane, Y. Fujita and I. Suzuki, "Somatosensory computation for man-machine interface from motion-capture data and musculoskeletal human model," IEEE Trans. Robot., vol. **21**, no. 1, pp. 58-66 (2005)
7. A. Sabatini, "Estimating Three-Dimensional Orientation of Human Body Parts by Inertial/Magnetic Sensing," Sensors, vol. **11**, no. 12, pp. 1489-1525 (2011)
8. J. K. Lee and E. Park, "A Fast Quaternion-Based Orientation Optimizer via Virtual Rotation for Human Motion Tracking," IEEE Trans. Biomed. Eng., vol. **56**, no. 5, pp. 1574-1582 (2009)
9. A. Kim and M. Golnaraghi, "A quaternion-based orientation estimation algorithm using an inertial measurement unit," in PLANS., pp. 268-272 (2004)
10. SP. Tseng, WL. Li, CY. Sheng, JW. Hsu and CS. Chen. "Motion and attitude estimation using inertial measurements with complementary filter," in ASCC., Kaohsiung, Taiwan, pp. 863-868 (2011)
11. AM. Sabatini, "Quaternion-based extended Kalman filter for determining orientation by inertial and magnetic sensing," IEEE Trans. Biomed. Eng., vol. **53**, no. 7, pp. 1346-1356 (2006)
12. X. Yun and ER. Bachmann, "Design, implementation, and experimental results of a quaternion-based Kalman filter for human body motion tracking," IEEE Trans. Robot., vol. **22**, no. 6, pp. 1216-1227 (2006)
13. H. Harms, O. Amft, R. Winkler, J. Schumm, M. Kusserow and G. Tröster, "Ethos: Miniature orientation sensor for wearable human motion analysis." Sensors, pp. 1037-1042 (2010)
14. C. Brigante, N. Abbate, A. Basile, A. Faulisi and S. Sessa, "Towards Miniaturization of a MEMS-Based Wearable Motion Capture System," IEEE Trans. Ind. Electron., vol. **58**, no. 8, pp. 3234-3241 (2011)
15. N. Abbate, A. Basile, C. Brigante and A. Faulisi, "Development of a MEMS based wearable motion capture system," in HSI, Catania, pp. 255-259 (2009)
16. C. Cifuentes, A. Braidot, L. Rodríguez, M. Frisoli, A. Santiago and A. Frizera, "Development of a wearable ZigBee sensor system for upper limb rehabilitation robotics," in BioRob, Rome, pp. 1989-1994 (2012)
17. P. Kinney, "Zigbee technology: Wireless control that simply works," Communications design conference. vol. **2**, pp. 1-7 (2003)
18. J. Lee, Y. Su and C. Shen, "A Comparative Study of Wireless Protocols: Bluetooth, UWB, ZigBee, and Wi-Fi," in IECON, pp. 46-51 (2007)
19. J. Suykens and J. Vandewalle, "Least squares support vector machine classifiers," Neural Processing Letters, vol. **9**, no. 3, pp. 293-300 (1999)
20. VN. Vapnik, and V. Vlamimir, Statistical learning theory, New York: Wiley, pp123-129.0 (1998)

Conceptual sizing of next supersonic passenger aircraft from regression of the limited existing designs

Keith F. Joiner[1], Jordan Zahra[1] and Obaid Rehman[1]

[1]*School of Engineering and Information Technology, University of New South Wales (NSW), Australian Defence Force Academy, Canberra, 2600, Australia*

Abstract. Despite previous versions, there are no current supersonic passenger transport aircraft. Much aircraft research is focused on hypersonic flight and the new technologies therein and is therefore unlikely to add to commercial versions anytime soon. This study re-examines conceptual sizing of a supersonic transport aircraft based on extant supersonic designs in order to ignite research into whether a commercially-viable design might exist. Key metrics are developed using distances between likely airport network nodes, an assumed number of passengers, and a reduction in transport time to one-third of current journeys. The study uses multiple response regression of known designs to develop key performance formulae, which are then optimized to set performance values so as to estimate an initial aircraft size, including an expected value analysis to guide the next conceptual design iteration. Twenty years ago a NASA Langley Research optimization system was used to examine non-linear regression of supersonic aircraft designs and to optimize such a design around similar performance criteria. In contrast, this work is the first supersonic transport aircraft sizing to use commercially-available Excel add-on software and standard design-for-six-sigma analysis techniques; notably for the sensitivity analyses to guide the next design iteration.

1 Nomenclature

b =	aircraft wing span (ft)
h =	aircraft height (ft)
l =	aircraft length (ft)
L/D =	lift to drag ratio
$MTOW$ =	maximum take-off weight (lb)
S =	aircraft wing area (sq ft)
T/W =	thrust to weight ratio
W_E =	empty weight (lb)
W/S =	wing loading (lb/sq ft)

2 Introduction

Current airliner technology features large-capacity, turbine-powered aircraft travelling at subsonic speeds in order to travel relatively large distances as quickly as these aircraft can. There have been supersonic aircraft such as the Concorde and Tupolev TU-144, however since Concorde stopped flying in 2003, all current and planned commercial passenger transport aircraft remain subsonic. Much of the airline design activity is focused on making these subsonic aircraft longer-range to decrease stop-overs, but the future of air transportation needs to commercially re-cross the sonic 'barrier' to substantially get intercontinental aircraft travel times down. Just as very-fast trains have leveraged increasing travel demand to create high-speed network cores in Europe and Japan, the increasing demand for air travel should create the opportunity for a supersonic air transport network.

Much of the major aircraft research is into hypersonic flight and the technological benefits that ensue if this can be achieved. Examples are the European Union's Long-Term Advanced Propulsion Concepts and Technologies (LAPCAT) program [1] and the U.S. military experimental aircraft like the X-51 [2]. Such hypersonic research aircraft use advances in technology that are unlikely to achieve commercially-viable designs for the airline industry for many decades.

What is more likely to realize reduced intercontinental airline times in the short-term is if some aircraft research is devoted to revisiting supersonic aircraft design, using proven technologies focused on optimizing commercial viability. Such multidiscipline aircraft design optimization was seminally characterised in 1997 [3] and research at that time using the NASA Langley Research Center [4] sought to minimize the gross take-off weight of a supersonic transport aircraft designed to carry 305 passengers at a cruise speed of Mach 2.4 for a range of 5,000 nautical miles (nm). This optimization research used both non-linear regression and neural network techniques with a great many design criteria, importantly including considerations like the take-off and landing criteria and aircraft engine performance. For relevance to this study, it had a solved aircraft take-off weight of around 666,665 lb and wing size of 8,162 sq ft. At a more graduate level, such a supersonic conceptual design would conventionally use the textbook pre-solved regression methods of [5], which involve usually two factors at a time and which is illustrated well for supersonic aircraft in [6].

This study revisits the supersonic transport conceptual design problem and optimizes with a method between the two extremes of the NASA optimization techniques and the Raymer textbook method. The study does so by exploiting new Excel software add-ons available to most graduates and several experimental analysis techniques used in Design For Six Sigma (DFSS) techniques [7]. As such, this research exploits improved probabilistic modelling and optimization packages to perform a high-level initial aircraft sizing based on prior design data. Use of DFSS methods for robust optimization was first reviewed extensively in [8] and formally this study is the first possible iteration in the process of 'Identify, Design, Optimize and Validate' as described and exemplified in [9], however in this case applied to aircraft conceptual design. The publication of such examples is hoped will promote wider use of DFSS capabilities outside the U.S. and Europe [10], especially sensitivity analyses in conceptual design.

The analyse packages used in this research were DOE PRO XL™ and Quantum XL™, both of which are Excel add-ons copyrighted by Air Academy Associates, LLC, and SigmaZone.com.

3 Historical Data Regression Set-up

The inputs to the study were initially as vast as possible whilst maintaining enough degrees of freedom to analyse interactions between the input parameters. The available degrees of freedom for analysis was constrained by the limited number of candidate aircraft able to be analysed; namely the 16 in Table 1. Hence a candidate list of different input design parameters was gathered (Table 2) so as to then screen the most important factors. Four different design output parameters were gathered: Maximum Speed, Maximum Mach Number, Service Ceiling, and Range. Note that for variable parameters, such as the wing area of many variable-sweep aircraft or the ability to use afterburners for several aircraft, the parameter with the highest velocity was always taken (i.e., often resulting in wings swept back and wet thrust using afterburners).

A screening analysis was conducted in DOE PRO XL™ using the 16 independent supersonic aircraft for 15 input factors but with no statistically significant factors and saturated degrees of freedom. To rectify this problem, a much wider screening study would be needed with more different aircraft, however, there is simply not enough unique examples of supersonic aircraft. Using different variants of the same aircraft would have unfairly skewed the results, while putting multiple military fighter aircraft into the study would not have embodied the high take-off weights needed for comparison against supersonic aircraft concepts. Instead, classical aircraft texts [5, 11] were used to inform reducing the analyses to just the eight key factors shown in Figure 1.

4 Regression Analysis

The regression analysis used a practical DFSS method [7, pp. 135-150] and ran easily for the effects of the eight first-order input factors. However, the analysis required careful repeat coding within the seven degrees of freedoms to garner any of the 28 possible two-way interactions and eight quadratic terms that could be significant. Also, each of these 44 terms are different for each of the four output parameters. The analysis excluded higher-order interactions such as three-way interactions and cubic terms as there was insufficient degrees of freedom. Insignificant factors were then removed progressively from the regression model based on their level-of-significance. Such refining of factors significant for all output responses is '*inverse screening*' [12, p. 224] and is a robust way to simplify. The final model for each output parameter was checked for its adjusted R-squared value, multi-collinearity and overall model significance [7]. Only the multi-collinearity was poor, however, this is frequently the case in historical data analysis. While results were otherwise statistically adequate, a diagnostic residual analysis revealed high leverage and the Cook's D statistic on two runs: namely the X-15 and X-2 aircraft. Both these aircraft are experimental and very short range when compared to the other aircraft, so they were removed and the process repeated with 14 aircraft types and five degrees of freedom.

Table 1 Supersonic Aircraft Designs used

Tupolev Tu-160	XB-70 Valkyrie	B-1B Lancer	Tupolev Tu-144
Concorde	Sukhoi T-4	Tupolev Tu-22M	Convair B-58
Lockheed SR-71	BAC TSR-2	Mikoyan MiG-31	F-111G Aardvark
MiG-25P Foxbat	F-14 Tomcat	X-15	X-2 Starbuster

Table 2. Candidate Input Parameters to Supersonic Aircraft Regression

Year Commissioned	Empty Weight (W_E)	Loaded Weight	Engine Type
Length (l)	Wingspan (b)	Height (h)	Wing Area (S)
Wing Type	Wing Loading (W/S)	Number of Engines	Lift to Drag (L/D) Ratio
Maximum Take Off Weight (MTOW)	Maximum Engine Thrust	Thrust to Weight (T/W) Ratio	

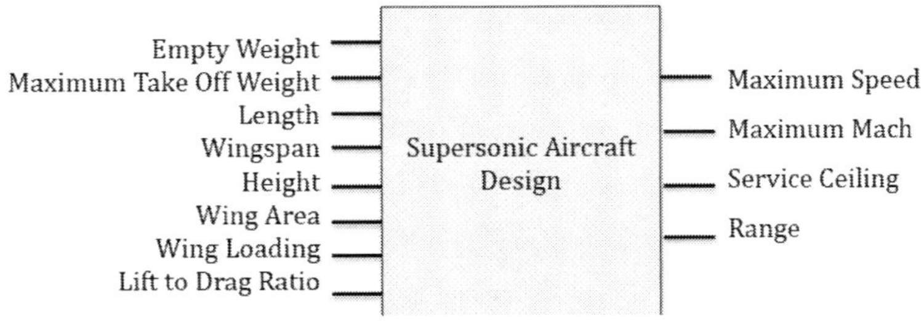

Figure 1. Illustration of final multi-factor inputs (8 of) and multi-responses (4 of)

The analysis yielded the regression formulae in equations 1-4 and associated Pareto charts of significant terms. Only the Pareto for maximum speed is given as an example of the process used. The formulae are un-coded and the units are in pounds and feet, while key statistical regression information are in the footnotes. The use of these Pareto charts to visualize effects on each of the output responses aided enormously in the inverse screening as forecast in [12, pp. 225-226].

$$Maximum\ Speed\ \text{(knots)} = -3288.7693 - 0.01658 \times W_E - 0.01551 \times MTOW + 74.210 \times l + 133.665 \times b - 122.186 \times h - 0.1032 \times S + 5.337 \times {}^{W}/_{S} + 192.235 \times {}^{L}/_{D} + (0.000391 \times MTOW \times h) + (0.00000221 \times MTOW \times S) - (1.0593 \times l \times b) \quad (1)^{b}$$

The most significant effect on maximum speed from equation (1) and Figure 2 are of course geometrical, leading with the interaction of aircraft length (factor C) and wingspan (factor D), shown as interaction CD. Other key factors are wing area, empty and take-off weights and the interactions of take-off weight (factor B) with both wing area (factor F) and with height (factor E) (i.e., BE & BF).

$$Maximum\ Mach = -5.6143 - 0.00002914 \times W_E - 0.0000293 \times MTOW + 0.139 \times l + 0.2392 \times b - 0.2358 \times h - 0.0001942 \times S + 0.008393 \times {}^{W}/_{S} + 0.2991 \times {}^{L}/_{D} + (0.000000773 \times \text{MTOW} \times h) + (4.151 \times \text{h} - 09 \times MTOW \times S) - (0.001991 \times l \times b) \quad (2)^{c}$$

The most significant effect on maximum Mach Number mirrored maximum speed, since these outputs are closely linked, being the same both proportionally and in order. The only difference to maximum speed was that more of the lesser terms achieved significance such as *MTOW* and wingspan.

$$Service\ Ceiling\ \text{(ft)} = 45490.25 - 2.0881 \times W_E + 0.5237 \times MTOW + 758.075 \times l + 1962.076 \times b - 4730.7083 \times h + 31.646 \times S - 119.996 \times {}^{W}/_{S} - (18.625 \times b^2) + (10.1235 \times b \times {}^{W}/_{S}) \quad (3)^{d}$$

The service ceiling equation was weak: unfortunately, all relationship in this formula collapses if the regression is forced to try to improve by rigorously removing terms with only a likelihood of significance. This weak relationship was persisted with so as to retain some related constraint in this dimension for the optimization. The most significant effect on service ceiling were likely to be empty and take-off weights, followed by wing area.

$$Maximum\ Range\ \text{(nm)} = 8041.429 - 0.11526 \times W_E + 0.08326 \times MTOW - 28.185 \times l - 227.287 \times b + 89.661 \times h - 0.64534 \times S - 7.6520 \times {}^{W}/_{S} + (115.9715 \times {}^{L}/_{D}) - (0.0000002481 \times W_E \times MTOW) + (0.002516 \times W_E \times b) - (0.0003904 \times MTOW \times b) \quad (4)^{e}$$

The most significant effects on range were not surprisingly from weight, leading with interactions of empty weight (factor A) with wingspan (factor D) (i.e., AD) and with maximum take-off weight (factor B) — the latter (i.e., AB) being a weight squared.

Residual diagnostics of each output response regression revealed a number of aircraft in the analysis

[b] R-sq. = 0.9969; Adj. R-sq. = 0.9797; $p < 0.0444$ all terms except *MTOW* at p = 0.0715 (likely sig.) and *l*, *b* and *h* all kept due to hierarchy; F = 58.05, Sig. F = 0.0171; Multi-coll. poor with tol. < 0.23.

[c] R-sq. = 0.9992; Adj. R-sq. = 0.9948; $p < 0.0253$ all terms except *h* at p = 0.0826 (likely sig.); F = 228.00, Sig. F = 0.0044; Multi-coll. poor with tol. < 0.24.

[d] R-sq. = 0.9190; Adj. R-sq. = 0.6489; $p < 0.05$ all terms except W_E, *S* and *W/S* where $p < 0.1$ (likely sig.) and *L/D* and one interaction are weak sig. ($p < 0.15$); F = 3.4028, Sig. F = 0.1710; Multi-coll. poor with tol. < 0.2.

[e] R-sq. = 0.9980; Adj. R-sq. = 0.9912; $p < 0.0146$ all terms except *W/S* at p = 0.0627 (likely sig.) and *b* kept due to hierarchy; F = 147.11, Sig. F = 0.0008; Multi-coll. poor with tol. < 0.1687.

https://doi.org/10.1051/matecconf/201819805001

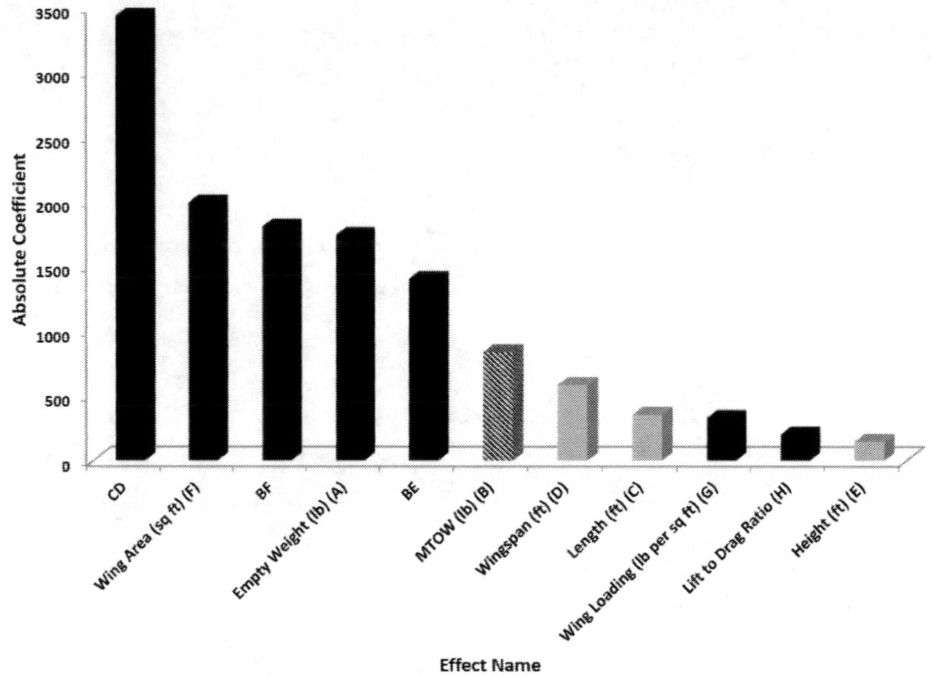

Figure 2. Pareto of Coefficients for Maximum Speed (knots) with black indicating significant ($p < 0.05$), hatched likely to be significant ($0.05 < p < 0.1$) and grey non-significant

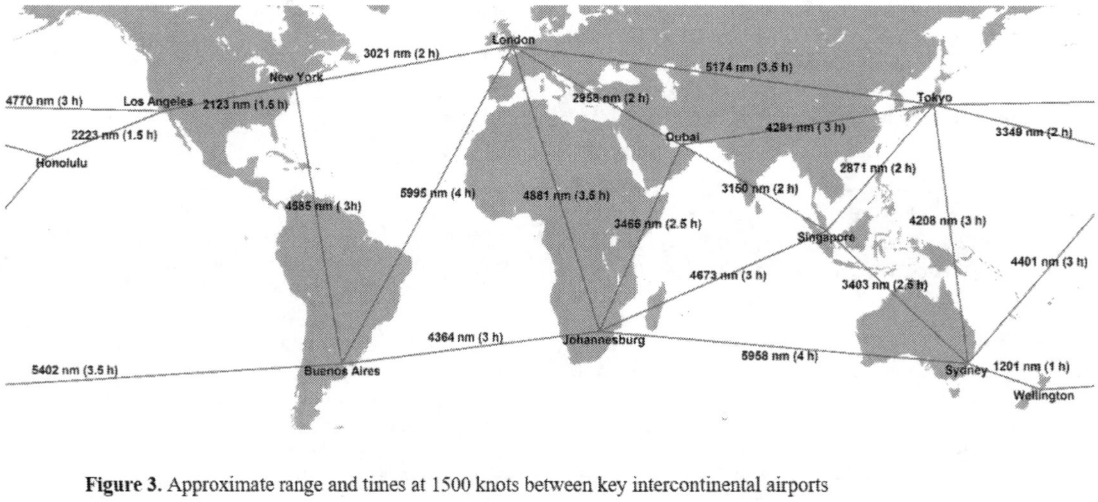

Figure 3. Approximate range and times at 1500 knots between key intercontinental airports

still produce high values of the *Cook's D* statistic (> 2.0), indicating high leverage on the subject formula and a high standardized residual. This is likely to be because the aircraft, while useful for this analysis, were optimized for different missions and involved different technologies.

5 Optimising Criteria for Sizing

The three main criteria for the initial size of a new supersonic transport aircraft are range, speed and the number of passengers. Range was explored as shown on the intercontinental network map shown in Figure 3. There are very long routes in current use, in excess of 8500 nm and taking around 16 to 17 hours, such as Sydney to London. Some of these, like Australia to South America are still relatively low volume. Hence, the range decision needed to be a trade-off that includes time-of-flight.

Noting the aim of commercial viability, the possible speed of a supersonic transport was kept to realistically achievable values of Mach 3 and 1,500 knots, because:

(1) this is high enough to drop travel times substantially to a third of current values;

(2) is still where the fuel consumption of conventional turbojets is better than an ideal ramjet, which are as yet unproven commercially [11, p. 767]; and
(3) is still within the historical cases on which the regression is based.

With speed set at 1,500 knots and Mach Number at 3, range then becomes an issue of the tolerance for stop-overs, versus the challenge of a high fuel consumption (at such speed) and the difficulties of accommodating a high fuel fraction of overall take-off weight within a necessarily slender supersonic body. Passengers travelling London to Sydney are primarily likely to want to spend less time travelling and the only current option is taking out stop-overs. Consider the Sydney to Los Angeles route, which currently takes about 14 hours and is direct. At 1,500 knots a supersonic aircraft can achieve this in three hours to Honolulu and 1.5 hours to Los Angeles, which even with an hour stop-over, would still reduce the journey to between a half and a third. Similarly, London to Sydney can be achieved in three short flights of around two hours each, or an overall time of about nine hours. Hence the trade-off range suggested is 6,000 nm with the travel times and possible supersonic network nodes as shown in Figure 3.

an expectation that this may need to be as low as 150 passengers.

6 Optimizing Initial Size Solution

The eight dimensional input space of the formulae from the historical regression needed more than the three main criteria to converge on a sensible solution. Initial attempts sought to go underground with a negative ceiling, or would drive take-off weight to very high proportions relative to the empty weight, which from conventional methods [5] was known to be unrealistic. So the following additional criteria were imposed:

(1) Service ceiling was limited to between 20,000 and 50,000 feet, where the challenges of engine intakes, air traffic control and passenger safety are well-known.

(2) The ratio of take-off weight to empty weight was constrained to 2.5, which is realistic with four of the historical aircraft achieving this ratio or better.

(3) The lift-to-drag ratio was allowed to go up to 3.5, which is outside any of the historical designs (i.e., > 3.18), based in part on the modern ability for computational fluid dynamics to drive superior aerodynamic conformal

Table 3. EVA inputs and output constraints with initial settings and final solution

EVA Inputs				
Factor	Distribution Type	Initial Value (mid or set)	Standard Deviation	Optimised Solution
Empty Weight (lb)	Normal	148,667	1,000	223,960.4
MTOW (lb)	Normal	340,305	1,000	518,771
Length (ft)	Normal	200	3	200
Wingspan (ft)	Normal	74.415	3	120.314
Height (ft)	Normal	29.5	1	42.516
Wing area (sq ft)	Normal	3,223.88	9	6,568.05
Wing loading (lb / sq ft)	Normal	105.5	0.6	56.614
Lift to drag ratio	Normal	2.14	0.2	2.715

EVA Output Constraints		
Name	LSL	USL
Maximum Speed (knots)	1,300	1,600
Maximum Mach No	2	3
Service Ceiling (ft)	20,000	50,000
Range (nm)	6,000	7,000
Ratio MTOW to Empty Weight	1	2.5
Lift to Drag Ratio	2	3.5

Note: LSL & USL are lower & upper specification limits

The subsonic Airbus 380 is 238 feet long and seats comfortably about 484 passengers. The Concorde was 202 feet long and carried 128 passengers and the Tu-144 was 215 feet long and carried 140 passengers. Both of these supersonic historical aircraft targeted the high-end executive traveller and thus perhaps more passengers could be fitted if similar space constraints are applied as current short-time economy flights. The slenderness requirements of fuselages in supersonic flight is going to be a design challenge. Rather than try to set a number of passengers, this conceptual sizing used aircraft length as the substitute, setting a minimum length of 200 feet with the target of this holding up to 200 passengers, but with

as robust design [7, pp. 205-212]. The input variations and output specification limits for the EVA and robust design optimization are in Table 3, where the initial values to begin the EVA were the mid-range values for the 14 historical aircraft, except for length, which was set to 200 feet as discussed earlier. The input variations were set around estimates of initial design limits, such as a 1,000 pounds standard deviation on empty weight, or three feet in length; however, these estimates should at some stage be confirmed with a fully representative design team.

The solution was optimized using the criteria of minimizing the observed defects per million in each of 1,000 Monte Carlo simulations, where all six outputs

https://doi.org/10.1051/mateconf/201819805001

were weighted equally. The solution compares to that outlined earlier [4] as lighter (518,771 c.f. 666,665 lbs) and smaller in wing area (6,568 c.f. 8,162 sq ft), however that solution was based on 305 passengers vice 150-200 passengers, was for a range of 5,000 nm rather than 6,000 nm, and included the greater fidelity of a dedicated aircraft optimization computer. The aircraft shape is still to be determined, but examples of what these dimensions would look like are in Figure 4.

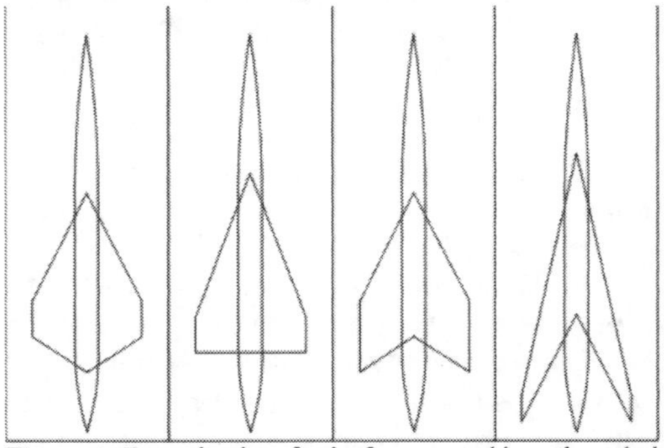

Figure 4. Example aircraft planforms to achieve the solution dimensions in Table 3

The achieved output values are shown probabilistically with their specification limits overlaid in Figures 5-6. The ratios of *MTOW* to empty weight and lift-to-drag are not shown, as these were both at values well within the specified limits, with means and standard deviations of {2.316, 0.0118} and {2.717, 0.198} respectively; and as such, neither had any observed exceedances. The lift-to-drag ratio was reassuringly within the historical solution space (i.e., < 3.18) despite being allowed to extrapolate if necessary up to 3.5 in value.

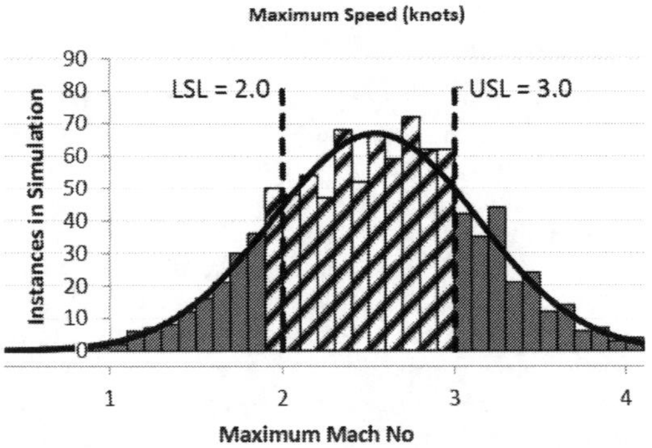

Figure 5. Maximum Speed and Mach No achieved relative to the limits set (hatched inside limits, grey outside)

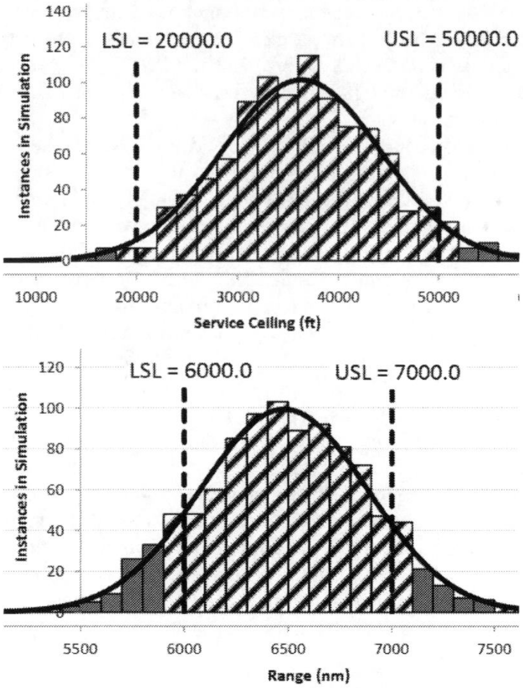

Figure 6. Service Ceiling and Range achieved relative to the limits set (hatched inside limits, grey outside).

7 Sensitivity Analysis for Robust Design

One of the benefits of robust design techniques in DFSS is sensitivity analysis, where the sensitivity of probabilistic relationships to input variations can be determined so as to inform subsequent design iterations where these design margins or tolerances are likely to be most critical [7, pp. 269-274]. Such sensitivity analysis in the optimization of aircraft conceptual designs is similar to recent dissertation research [15] that had critical outputs of range, landing distance and specific range (nautical mile per pound). For the supersonic aircraft regression model in this study there are 32 combinations of sensitivity to consider, based on eight inputs and four outputs. The major concern for the next sizing iteration would be the effect of a shift in any input mean and this is shown in Table 4.

Clearly the greatest sensitivity to input changes is the Service Ceiling followed by the Range, while the most critical of inputs is the wingspan and then *MTOW*, wing area and wing loading. The next iteration of the conceptual design can use this information to prioritize tracking of these key design measures.

Table 4. Sensitivity Analysis results showing effect of absolute change in each output from a +0.5 sigma and a -0.5 sigma shift in the mean of the input.

Inputs	0.5 sigma shift	Outputs Maximum Speed (knots)	Maximum Mach No	Service Ceiling (ft)	Range (nm)
Empty Weight (lb)	500 lb	1.34%	3.12%	32.00%	9.50%
MTOW (lb)	500 lb	5.71%	11.75%	27.99%	13.58%
Length (ft)	1.5 ft	4.26%	7.82%	13.56%	10.56%
Wingspan (ft)	1.5 ft	8.09%	18.13%	73.81%	54.59%
Height (ft)	0.5 ft	2.67%	4.55%	61.18%	7.36%
Wing area (sq ft)	4.5 sq ft	1.33%	2.75%	38.53%	18.35%
Wing loading (lb / sq ft)	0.3 lb/sqft	1.89%	3.14%	40.75%	18.65%
Lift to drag ratio	0.1	1.38%	4.23%	7.11%	1.52%

Legend
significant (>20%) shift in quality for design goals
moderate (10-20%) shift in quality for design goals
insignificant (<10%) shift in quality for design goals

8 Study Limitations and Further Design

There are noteworthy limitations in this Study, typical of early conceptual sizing of a platform based on historical data. For example, all relationships suffered from poor multi-collinearity. Primarily, there is a limited data set of prior designs to inform the historical analysis, while that set of aircraft designs are each somewhat unique and therefore, to some extent, not a readily comparable family. Also these designs are increasingly dated since supersonic transport design has commercially stagnated. Yet, the only alternative is to design without reference to previous attempts. A true validation check of the model is not possible at this time due to the lack of new supersonic aircraft designs being ready for flight testing.

The optimization was based on estimates of the range and travel timings for a future aircraft network without any passenger survey or traffic density analysis and predictive business modelling. Also, the EVA was based on estimated input variations that would need confirming with a team of design specialists before the next conceptual design iteration. Finally, this research has not considered regulatory challenges such as sonic boom issues over land. Ideally the subsequent design iteration should seek a further multidisciplinary optimization that incorporates commercial cost relationships, aircraft aeroacoustic modelling and passenger preference relationships, as well as contemporary propulsive, structural and aerodynamic design factors that go beyond simple weight fractions. These next iterations of the design would expand eventually to a high-dimensional, computationally-expensive and '*black box*' problem [12]. In pursuing increased fidelity of optimization, the authors favour following the DFSS strategies of screening and focusing on the most sensitive factors, as well as examining the merit of breaking the global range and endurance sectors (Figure 3) into mission modules for the optimization, as was outlined recently for design of an unmanned aerial vehicle [16, pp. 199-202]. There is potential to use some of the new adaptable parametric sizing methods proposed in dissertation research [17] and to continue conceptual design optimization with uncertainties [18] — wherever possible still using commercially available software [13].

9 Conclusions

This Study has analyzed historical supersonic transport aircraft designs to determine probabilistic relationships between key design input factors and the responses of maximum speed, maximum Mach Number, service ceiling and range. This study optimizes with a method between the two extremes of the NASA optimization techniques [4] and the Raymer textbook method [5]. Only the equation for service ceiling was statistically weak in significance and had to be persisted with for the completeness of the solution space. The analysis has then used standard software packages and DFSS techniques, which are available to post-graduate students, in order to optimize around an initial concept size using robust design techniques. A simple consideration of a possible intercontinental airport network set a realistic target range of 6,000 nm and travel times of about one-third of current values, based on a speed of 1,500 knots and a Mach Number of 3.0. Aircraft length was used to approximate a number of passengers of between 150 and 200.

The approximate solution is a supersonic aircraft with an empty weight of 224,000 lbs, a *MTOW* of 2.3 times that at 519,000 lbs, a length and height of 200 ft and 42.5 ft respectively, a wingspan of 120 ft, a wing area of 6,570 sq ft and a wing loading and lift-to-drag ratio of 56.6 lb/sq ft and 2.7 respectively. The analysis showed this conception to be reasonably robust to input variations at this design solution, with the most sensitive output to input changes being service ceiling, especially to changes in wingspan. This work is the first supersonic transport aircraft sizing to use commercially-available Excel add-on software and standard DFSS analysis techniques, notably for the sensitivity analyses to guide the next design iteration.

The authors are interested in furthering the use of the DFSS methods used here and in working with an enhanced aircraft design team on the next conceptual design iteration for a commercially viable supersonic transport aircraft. In particular, to examine how the sensitivity results best influence the next design iteration. There is potential to use some of the new adaptable parametric sizing methods and to continue conceptual design optimization with uncertainties — wherever possible still using commercially available software. Given the age of the designs used in this analysis, coupled with advances in technology and design skills, the recommended supersonic transport aircraft design should be much more assured than its predecessors.

References

1. A. Giordani, *Could an airliner that flies anywhere in under three hours be in service by 2030?*, BBC newspaper [online] (15th Sept. 2015)
2. J. Anderson and R. Passman, X-15: *The World's fastest rocket plane and the pilots who ushered in the Space Age* (Zenith Press, 2014).
3. J. *Sobieszczanski*-Sobieski and R. T. Haftka, J. Struct. Opt. **14**, 1-23 (1997).
4. S. N. *Patnaik*, J. D. Guptill, D. A. Hopkins and T. M. Lavelle, J. of Aircraft **35**, 839-850 (1998).
5. D. P. Raymer, *Aircraft design: a conceptual approach* 5th ed. (AIAA, 2012).
6. H. Xue, H. Khawaja and M. Moatamedi, AIP Conf. Proc. **1637** 1202-1210, (2014).
7. L. A. Reagan and M. J. Kiemele, *Design for six sigma: the tool guide for practitioners* (Air Academy Associates, 2008).
8. P. N. Koch, R. J. Yang and L. Gu, J. Struct. Multidisc. Opt. **26**, 235-248 (2004).
9. I. Ferreira, J. A. Cabral, P. Saraiva and M. C. Oliveira, J. Struct. Multidisc. Opt. **49**, 501-521 (2014).
10. K. F. Joiner, M. Kiemele and M. McAuliffe, ITEA J. **37**, 141-152 (2016).
11. J. D. Anderson, *Introduction to Flight* 8th ed. (McGraw-Hill, 2016).
12. S. Shan and G. G. Wang, J. Struct. Multidisc. Opt. **41**, 219-241 (2010).
13. J. Watson, *A high-fidelity approach to conceptual design,* PhD dissertation ProQuest Dissertations Publishing ISBN: 9781339847092 (2016).
14 A. Lenz, *Simple uncertainty propagation for early design phase aircraft sizing*, PhD dissertation ProQuest Dissertations Publishing (2010).
15. M. J. Daskilewicz, B. J. German, T. T. Takahashni, S. Donovan and A. Shajanian, J. Struct. Multidisc. Opt. **44**, 831-846 (2011).
16. P. K. Lewis, V. R. Murray and C. Mattson, J. Struct. Multidisc. Opt. **43**, 191-204 (2011).
17. G. Coleman, *Aircraft conceptual design - an adaptable parametric sizing methodology* PhD dissertation ProQuest Dissertations Publishing ISBN: 9781124055336 (2010).
18. D. Neufeld, Multidisciplinary *aircraft conceptual design optimization considering fidelity uncertainties*, PhD dissertation, ProQuest Dissertations Publishing, ISBN: 9780494714034 (2010)

Design and Validation of Deployable and Lockable Mechanism for Gossamer Spacecraft

SUN Liang[1], Ji Ming [1], Li Haizhi[1], WANG Jin[1], KANG Changxi[1] , YAN Bin [1], YAO Lei [1], KOU Zongyu[1]

[1] *Lanzhou Institute of Physics, Lanzhou, 730000, China*

Abstract. This passage introduces the researches of Deployable and Lockable mechanism for a new Gossamer Spacecraft built by space apply which can be locked and deployed by its own. This mechanism integrates deploying and locking. Its advantages are simple and reliable. It is based on a serious project "Platform Technology of Gossamer Spacecraft" from Lanzhou Institute of Physics. This project manufactured a model machine, finished the ground experiment, passed the project check. This passage introduces work theory particularly, builds the dynamics models, analyzes experiment results factors and presents measurement methods. Results show that the mechanism designed can meet the project requirement; it can build the foundation for the super-large Gossamer Spacecraft of China and improve the degree of the Aerospace technology application.

1 INTRODUCTION

As a new type of spacecraft platform, gossamer spacecraft has attracted wide attention at home and abroad. Pressure of sunlight is used to be a driving force to carry out space flight without fuel, and carry more space loads to achieve the goal of faster, cheaper and better. For on-orbit deep space exploration, gossamer spacecraft platform has a huge opportunity.

There are two basic types of gossamer spacecraft platform, one is non-skeleton, which uses rotating centrifugal expansion, and the other one is skeleton, which uses inflatable expansion. The non-skeleton is widely used in foreign countries, like the NASA's Jet Propulsion Laboratory and Langley Research and Development Center have developed a 20 meter demonstration prototype[1,2] and the European Space Agency has developed a biased reflector antenna, which is a support structure for inflatable telescope hood and a parachute of inflatable expansion. Inflatable research is studied in the United States and Russia[3,4]. At present, China Academy of Space Technology has developed a skeleton and inflatable 25 meter ground demonstration prototype and smooth inflatable work[5].

The ground demonstration prototype which is developed by Lanzhou Institute of Physics is mainly composed by a deployable and lockable mechanism and a gossamer sails body. The aim of the deployable and lockable mechanism is to store gossamer sails body as much as possible in a limited space, to maintain a certain shape, and not to affect the inflatable process of gossamer sails. The deployable and lockable mechanisms can provide effective restraint and protection to the support arms and sails which are folded during the launch phase. Before the operation of the spacecraft, the mechanism is deployed to the designated position and locked to maintain the working state of the locking plane, waiting for the support arms and sails smooth to be deployed.

2 Mechanism composition and design

2.1 Mechanism Research status

The deployable and lockable mechanism is earliest used in the spacecraft mechanism organizations, and is most widely used in the spacecraft mechanism organizations. The deployable and lockable mechanism can be divided into three main types according to the principle of operation: strip semi-rigid, locking jaw type and connecting rod[6].

The typical application of the strip semi-rigid deployable and lockable mechanism is to become a belt separation device which is used adapter connect and separate in satellite and rocket. This technology has become a mature technology, so China and other major space countries have mastered the technology.

The locking jaw type deployable and lockable mechanism is a more mature of lockable and released form in the world. This locking and releasing type is used in solar arrays and antennas. The typical examples are solar arrays lockable and released mechanism of the Intelsat-v satellite and the Astro-AeroSpace company has developed a lockable and released mechanism of AstroMesh flexible antenna. China has used this mechanism in the solar arrays of FY-1 satellite and antennas of ShenZhou spacecraft, as also, there is no relevant application in large and heavy weight object up to now.

The most typical applications of connecting rod deployable and lockable mechanism are the solar arrays and some of the connection separation bolts, and now China and other major countries have mastered this technology, and it is widely used to achieve the connection and separation between solar arrays and cabin in star ships and other spacecrafts.

2.2 Mechanism composition

Taking re-use and energy saving into account, hinge torsion-spring mechanism can be used in this project. Deploy function after releasing will be driven by preloaded torsion springs which have been installed in each hinge of product. According to light weight and small size constraints, the rotary hinge locking device can achieve a high locking force and locking stiffness, can easily carry out the ground unlock action to reset and test experiments, and also can achieve a higher locking repeatability.

In this paper, the deployable and lockable mechanism is used to lock and deploy membrane sail body of the spacecraft, before reaching the track. Its role is to maintain the lockable state, protect and restrain the sail body. After reaching the track, its role is to unlock, deploy and automatic lock. When it reaches the predetermined position angle, it ensures the opening board to be flat, provides a platform for the sail to inflate and deploy. The three-dimensional schematic diagram of the mechanism is shown in Fig1.

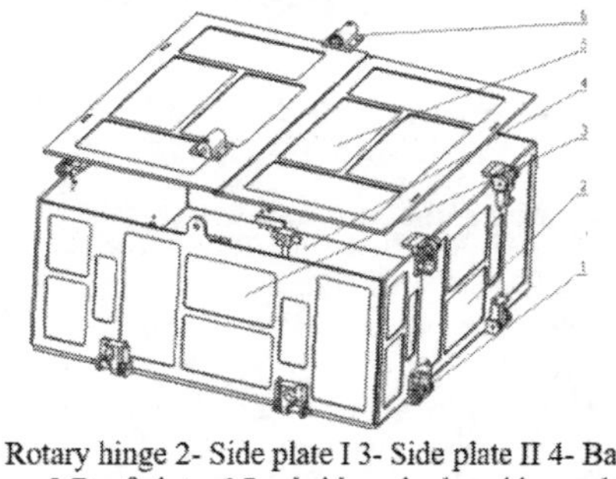

1- Rotary hinge 2- Side plate I 3- Side plate II 4- Base plate 5-Roof plate 6-Lockable and releasable mechanism

Figure.1 Deployable and Lockable Mechanism Model

The base plate serves as a platform for mounting other components. Side plate I has two plates, parallel to each other, side plate II has two plates, parallel to each other too, roof plate has two plates, located in the same plane. Side plate I and side plate II are fixedly attached to the base plate by means of a rotary hinges, roof plate is fixed attached to the side plate I by means of a rotary hinges. The lockable and releasable mechanism is mounted on the roof plate by screws, locking function is finished by compression spring pretension force and limit hole of side plate II, the pressure generated by the intake valve is used as the power source for the release of the mechanism. The torsion spring of rotary hinge is used as the power source for the deploying of the mechanism, and the compression spring is used as the power source for the locking when it reached the scheduled position.

2.3 Deployable and lockable mechanism

The deploying and locking functions are integrated in the rotary hinge assembly. The rotary hinge assembly consists of male hinges, female hinges, rotary shafts, torsion springs, compression springs and locking pole. The three-dimensional schematic diagram of the rotary hinge is shown in Fig2a.

The torsion spring, rotary shaft, male hinge and female hinge complete the deploying function of the mechanism, torsion spring as energy storage component, the initial state is the state of force, the potential energy which is stored will transform into the kinetic energy to rotate the male hinge and the female hinge, the size of the output force can be changed by adjust the torsion spring installation position, the deploying function of mechanism is achieved. The compression spring, locking pole and slide rail are fully engaged in the locking function, the initial state of compression spring is in the working state, and the potential energy stored in the compression spring is converted into the kinetic energy to finish forward movement of the lock pole, when the locking pole is inserted into the locking hole, the locking function of mechanism is achieved. The schematic diagram of the locking function is shown in Fig2b.

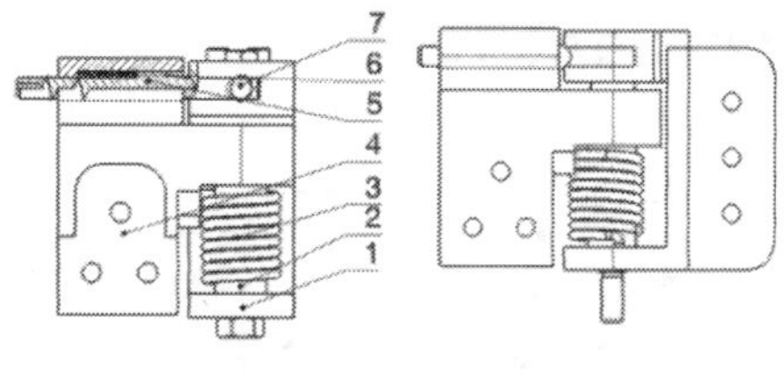

(a)Initial State (b) Lockable state

1- Male hinge 2- Rotary shaft 3- Torsion spring 4- Female hinge 5-Locking pole 6- Compression spring 7- Slide rail

Figure.2 Rotary Hinge Assembly Model

2.4 Lockable and releasable mechanism

The lockable and releasable mechanism is mounted on the roof plate. It consists of two pneumatic valves which can provide locking and release function for the roof plate of gossamer spacecraft.

The pneumatic valve form is used in lockable and releasable mechanism. It mainly contains piston shaft, cylinder, compression spring, end cap and O-shaped rubber ring etc. The three-dimensional schematic diagram of the rotary hinge is shown in Fig3. The working principle is as follows: before the spacecraft works, the roof plate shall is kept in the closed state. At this time, the motive force of the piston rod movement is provided by the deformation of the compression spring. When the piston rod is moved leftward into the limit hole, the pressing function is realized. After the work of the spacecraft, the intake valve through the hole of cylinder aerate into a certain pressure of the gas, pushing the piston rod to be moved by the opposite direction, to promote the compression spring compress and deformation, the piston rod will separate from the limit hole of the side plate, to achieve the release function of spacecraft. The schematic diagram of the locking function is shown in Fig4.

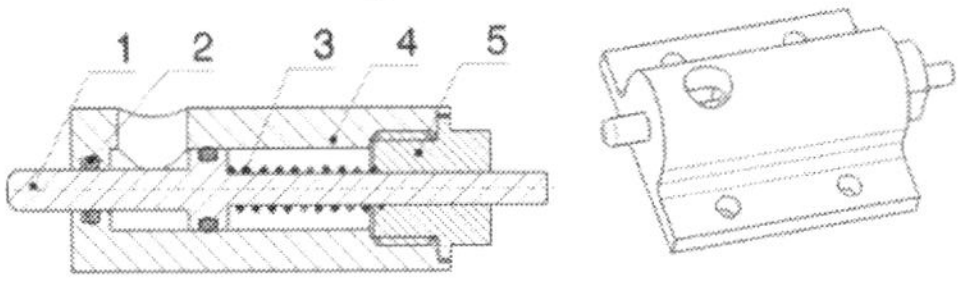

1- Piston shaft 2- O-shaped rubber ring 3- Compression spring 4-Cylinder 5End cap

Figure.3 Lockable and Releasable Mechanism Model

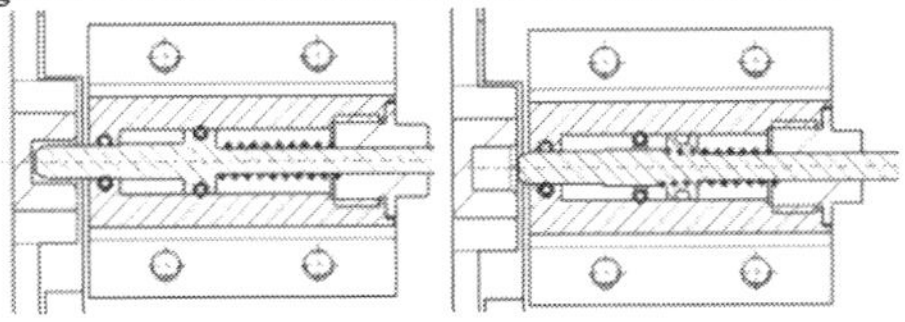

(a)Locking State (b) Releasing State

Figure.4 Lockable and Releasable Mechanism Schematic Diagram

3 Dynamics models

The dynamically modeling for the deployment process of deployable and lockable mechanism is built, and the deployment process will be simplified to that a hinge rotates around the axis. All the driving forces of the mechanism are derived from the torsion springs in the hinges and the drive torque of the torsion spring is related to the rotate angle. This article uses the deployable analysis method of the trusses, $3n$ Cartesian coordinates are used as unknown vectors, and the relationship between unknown vectors and the angle of the torsion spring will be established. The kinetic equation of the torsion spring drive is deduced. Based on the first Lagrangian equation applied to the rigid body, the torsion spring drive has the elastic deformation energy[7]. The kinetic equation is:

$$\sum_{i=1}^{3n}\left(\frac{d}{dt}\frac{\delta T}{\delta \dot{x}_i}-\frac{\partial T}{\partial x_i}-\frac{\partial V}{\partial x_i}\right)\Delta x_i=(\Delta X)^T\left(\frac{d}{dt}\frac{\delta T}{\delta \dot{X}_i}-\frac{\partial T}{\partial X_i}-\frac{\partial V}{\partial X_i}\right)=0 \quad (1)$$

In the equation, V is the kinetic energy of the system, T is the elastic deformation energy of the system, and X is the generalized coordinate array of the nodes.

In the process of deploying, the torsion spring drive force is also changing at times with the angle of change; the torque has the following formula equation:

$$M=K(\alpha_0-\alpha) \quad (2)$$

In the equation, M is the driving torque; K is the stiffness coefficient of torsion spring, is the initial twist angle of the torsion spring, and is the twist angle of torsion spring which is changing in the movement. The elastic deformation of torsion spring can be:

$$V=\tfrac{1}{2}K(\alpha_0-\alpha)^2 \quad (3)$$

Bring the elastic deformation formula equation into the equation (1); you can deduce the generalized driving force of torsion spring driving:

$$F=\frac{\partial V}{\partial x}=K(\alpha_0-\alpha\frac{\partial \alpha}{\partial x}) \quad (4)$$

After calculating the driving force of the torsion spring, all the torsion spring driving forces in the mechanism are assembled according to the finite element method, forming a generalized driving force array. And the generalized driving force array is different at the different time of the movement, which is changed with the change of the different node positions. At now, the dynamic equation of the torsion spring drive structure is:

$$\sum_{i=1}^{3n}\left(\frac{d}{dt}\frac{\delta T}{\delta \dot{x}_i}-\frac{\partial T}{\partial x_i}-\frac{\partial V}{\partial x_i}\right)\Delta x_i=(\Delta X)^T\left(\frac{d}{dt}\frac{\delta T}{\delta \dot{x}_i}-\frac{\partial T}{\partial x_i}-Q(x,t)\right)=0 \quad (5)$$

4 Ground validation and factors

Through above design process and dynamic models, as well as research and analysis a large number of domestic and foreign literatures, we have manufactured a demonstration prototype, conducted a ground experiment.

4.1 Ground demonstration validation

The aluminum alloy 2A12 is selected as the main material of the deployable and lockable mechanism, the stainless steel 1Cr18Ni9Ti is used as the main material of the rotating hinge assembly and the lockable and releasable mechanism, and the wire 65Mn serves as the material of the compression spring and the torsion spring. The physical dimension of the demonstration prototype is: 550mm×550mm×250mm, its weighs about 9.5Kg. The real product of the deployable and lockable mechanism shown as Fig5:

(a) Initial State of Product (b) Locking State of Product

Figure.5 Product of Deployable and Lockable Mechanism

4.2 Factors discussion

From the results of the ground demonstration experiment, it can be found that the main factors affect the acceleration of the deployment process and the time of the deployment process: the friction of mechanism and spring stiffness of torsion spring.

(1) the impact of mechanical friction

In the ground experiment, the parts of the mechanism did not carry out lubrication treatment, astringent phenomenon had appeared in the deploy process, the deploying starts slowly, about 5 seconds, through the measurement and calculation, the friction between the side plate I and the base plate is about 2.6 Newton, The friction between the side plate II and the base plate is about 2.8 Newton.

After the grease was applied to the parts of the mechanism, the astringent phenomenon disappeared, the velocity was relatively increased, deploying time is about 3 seconds, the acceleration is also relatively large, but the smoothness of the deployment process was affected, the impact force is also large when it works, side plates appeared jitter phenomenon.

(2) the impact spring stiffness of torsion spring

In the ground experiment, astringent phenomenon appeared in the deployment process of roof plate, on one hand it is the impact of friction, and on the other hand it is due to the spring stiffness of the torsion spring. After measurement and analysis, the resistance moment of one single roof plate is about 100 Newton millimeter, the torsion spring stiffness coefficient K initial selected is 1.5 times margin, the torsion spring stiffness coefficient K re-winding is 2 times margin, the astringent phenomenon disappeared in the deployment process, the velocity was relatively increased, deploying time is about 3 seconds.

It can be seen that the friction of the mechanism and the stiffness of the torsion spring are the main influencing factors of the acceleration of the deployment process and the deployment time of mechanism. According to the requirements such as deployment time, deployment acceleration and the impact force, it is recommended to apply the lubricating or grease to the mechanism to reduce the friction, the torsion spring stiffness coefficient margin recommended 1.8-2.0.

5 Conclusions

1) For the new deployable and lockable mechanism, design is clever, set deployment and locking function as a whole, takes up a small size, its weight is light, simple and reliable.
2) The torsion spring-type device is used; convert potential energy into the kinetic energy by the use of energy conversion principle, function is achieved, without additional energy, which can be repeated several times in the ground experiment.
3) Manufacture a demonstration prototype, conduct a ground experiment, the deployment and locking of the mechanism are fast and reliable, and analyze the factors which affect the development process, and put forward the countermeasures and suggestions.

References

1. John Huang, Houfei Fang, Richard Lovick, Michael Lou. The Development of Large Flat in floatable Antenna for Deep-Space Communications [C]. AIAA, 2004.
2. Robert R. Romanofsky. Inflatable Membrane Reflector and Shape-Memory Polymer Antenna Developed for Space and Ground Communications Applications [C]. NASA, 2006.
3. Veldman S L, Ver meeren C AJr.Inflatable structures in aerospace engineering-an overview [D] . Composite Material sand Structures Faculty of Aerospace Engineering , Delft university of technology , 2002.
4. Rohwelle D J. @qualification of the inflatable sunshield in space (ISIS) mast [C] . Proceeding of the 36t h Aerospace Mechanisms Symposium. Glenn Research Center, 2002.
5. David Cadogan, Craig Scheir, Anshu Dixit, and Jody Ware, Emily Cooper, Dr. Peter Kopf. Intelligent Flexible Materials for Deployable Space Structures (InFlex). AIAA 2006-1897.
6. Kriss J.Kennedy. Inflatable Habbitats Technology Development. NASA20050182969.
7. Juan R. Cruz, J. Stephen Lingard.Aerodynamic Decelerators for Planetary Exploration: Past, Present, and Future NASA20060028185.

Multi-aircraft Rerouting Method Under Dangerous Weather

Yukun Zheng , Ying Wang, Yun Sun, Yao Qi, and Ning Wang

College of Equipment Management and Unmanned Aerial Vehicle Engineering, Air Force Engineering University, 32018 Xi'an, CHINA

Abstract. In view of the problems that single-aircraft rerouting method existing in civil aviation considered little about the associated effects of multi-aircraft and route network characteristics. A multi-aircraft rerouting method under dangerous weather was proposed. Based on the characteristics of network failures during the occurrence of dangerous weather, the performance indicators and related constraints of temporary routes are analyzed to establish a single- aircraft rerouting model. Then, a multi-aircraft rerouting model based on layout of diversion points was proposed in combination with the performance indicators of the rerouted route network, and solved by NSGA-III algorithm. Finally, a simulation analysis was conducted using the regional flight route network in part of the Beijing Flight Information District as an example and the results show that multi-aircraft rerouting method can effectively avoid flight conflicts and improve the comprehensive characteristics of the network.

1 INTRODUCTION

The structure of China's airspace is complex. There are a large number of "no-fly zones", "restricted zones" and "danger zones" around the route. The sudden dangerous weather may block the route, which has a major impact on the flight safety of civil aviation[1]. In order to solve such problems, it is necessary to reroute the original route.

At present, domestic and foreign scholars have conducted preliminary studies on the problem of rerouting under dangerous weather. Yang Shuangshuang[2] adopted a geometric method to reroute under static dangerous weather. Wang Fei and Wang Lili[3-5] aim at minimizing the cost of rerouting, and established a rerouting model based on genetic algorithm, which improves the accuracy of rerouting planning. Meng Linghang[6] considered the dynamic characteristics of dangerous weather and designed a heuristic algorithm for the specific weather conditions. Li Xiong[7] proposed a multi-target rerouting model, but it was not universal. Luitpold Babe[8] and Wang Shijin[9] introduced the route terminal area into rerouting plan and established a three-dimensional model that circumvents the "three zones," but the model didn't consider the impact of sudden dangerous weather.

The current researches are mostly single-aircraft rerouting under dangerous weather without considering the relevance of rerouted routes and the integrity of route network which increase the possibility of flight conflicts. Based on single-aircraft rerouting, this paper proposes a multi-aircraft rerouting model based on the layout of diversion points, and uses a NSGA-III algorithm to solve the multi-object rerouting model.

2 SINGLE-AIRCRAFT REROUTING

2.1 Rerouting Environment Construction

In order to express the structural characteristics of the aeronautical domain, use the grid method to characterize the rerouting environment[9]. The target airspace is represented as a mesh structure composed of several grids, as shown in Figure1 , and its specific steps are:

(1) Create a grid coordinate network according to the connection between origin and destination of rerouting.

(2) Divide the grid coordinate network into $(G_x+2)\times G_y$ square grid. The side length of grid is m', and the label is n_i.

(3) Assign a specific state value for each grid to indicate the security, and use center point state to represent the grid state.

$$C(n_i)=\begin{cases}1 & n_i \in Z_U \cup Z_W \\ 0 & n_i \in Z_{free}\end{cases} \tag{1}$$

In the formula, $C(n_i)$ is state assignment function, Z_U represents the "three zones" airspace, Z_W represents the dangerous weather scope, and Z_{free} represents free airspace.

(4) Display dangerous weather scope and the "three zone" airspace in the grid network, and the schematic diagram is as follows:

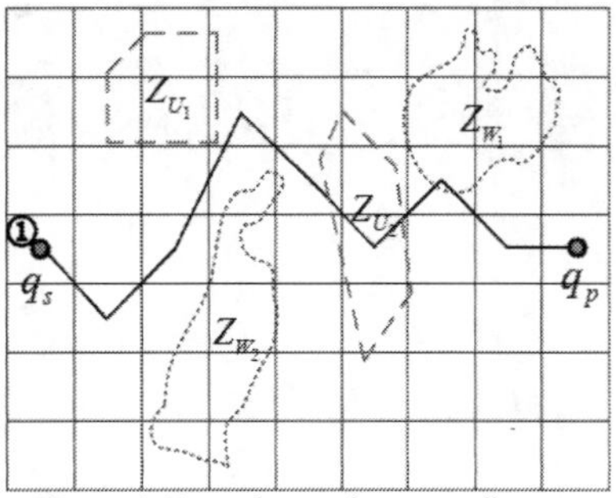

Figure 1. Rerouting environment diagram

In the figure, q_s is original, q_p is destination, and ① indicates the air route.

2.2 Objective Functions and Constraints

The traditional single-aircraft rerouting uses the shortest route distance as the planning objective, which only considers the economic effects while ignoring its safety. Therefore, this paper adopts three objective functions: shortest route distance, minimum route dispersion degree and minimum route slope change amount to enhance the safety, adaptability and stability of the rerouting method. Then introduce a penalty function to ensure the rerouted route meets the obstacle avoidance constraints and range constraints, which are expressed as follows:

$$\min D = d(q_s, q_1) + d(q_{G_x}, q_p) + \sum_{i=1}^{G_x-1} d(q_i, q_{i+1}) + \Phi \quad (2)$$

$$\min S = \left[(y_s - \bar{y})^2 + (y_p - \bar{y})^2 + \sum_{i}^{G_x} (y_i - \bar{y})^2 \right] \times \left[1/(G_x + 2) \right] + \Phi \quad (3)$$

$$\min K = \left| k_{q_1 q_2} - k_{s q_1} \right| + \sum_{i=2}^{G_x-1} \left| k_{q_i q_{i+1}} - k_{q_{i-1} q_i} \right| + \left| k_{q_{G_x} p} - k_{q_{G_x-1} q_{G_x}} \right| + \Phi \quad (4)$$

$$\Phi = \lambda \cdot \left(C(d'_{sq_1}) + \sum_{i=1}^{G_x} \sum C(d'_{q_i q_{i+1}}) + C(d'_{q_{G_x} p}) \right) \quad (5)$$

Among them, q_i represents a grid the rerouted route passed, D represents the total distance, $d(q_i, q_{i+1})$ represents the distance of $q_i q_{i+1}$. S represents dispersion degree of the route, y_i is the ordinate of the grid. K represents slope change of the route, $k_{q_i q_{i+1}}$ is the slope of $q_i q_{i+1}$. Φ represents the penalty function, and λ is a penalty factor, preferably 1000[7], $d'_{q_i q_{i+1}}$ represents the passing grid collection, $\sum_{i=1}^{G_x} C(d'_{q_i q_{i+1}})$ represents state values of rerouted route. At the same time, in order to ensure the rerouted route meets actual needs, the corresponding constraints are proposed:

$$(sq_1 \cup q_i q_j \cup q_{G_x} p) \notin (Z_T \cup Z_W) \; i, j \in [1, G_x] \quad (6)$$

$$q_i \in Z \quad i \in [1, G_x] \quad (7)$$

$$m' \geq 10\text{km}^{[8]} \quad (8)$$

Equations (6), (7), and (8) represent obstacle avoidance constraints, optimization range constraints, and grid length constraints, respectively.

2.3 Algorithm coding

Because the binary code will limit the rerouting angle, when using a multi-objective algorithm to solve the model, the objective function can be compiled by using integer coding. Set the chromosome length to G_x, so the gene value range is $[1, G_y]$. The route of chromosome 434654544 is shown as ① in figure 1.

3 MULTI-AIRCRAFT REROUTING

3.1 Model Assumptions

The rerouted route is also referred as temporary route. When a large-scale dangerous weather occurs in a regional route network and affects the operation of aircraft, a temporary route network for multi-aircraft rerouting may be formed on the basis of single-aircraft rerouting, and a optimization model based on diversion point layout to optimizes the network to reduce flight conflicts and improve the security can be adopted. To simulate the actual operating environment, make the following assumptions:

(1) Aircraft in the network are all located at the same height level.

(2) The connection relationship and flight flow between the waypoints remain unchanged, and the "three zones" and dangerous weather areas cannot be crossed.

(3) The aircraft cruise speed is 800km/h, the specified minimum radar control interval is 20km, and the optimal range of waypoints is ±80 km.

(4) The cost of network optimization is not considered.

3.2 Structure of Temporary Route Network

The regional temporary route network consists of waypoints and route segments. The waypoints include airport, intersections points, and diversion points. Route segments are connected by waypoints. In order to reflect the properties of the temporary route network, it can be expressed as $N(P, D', f)$, and its meaning is as follows:

(1) $P = \{P_1, \cdots, P_n, \cdots P_{n+m}, \cdots P_{n+m+k}\}$ represents the set of waypoints, the former n is airport, the middle m is intersection points, and the latter k is diversion point.

(2) D' is total distance of the temporary routes which can be expressed as:

$$D_i' = \sum d_{ij} \quad (9)$$

In the formula, d_{ij} means the distance of segment P_iP_j, $\sum d_{ij}$ is the sum of segment distance values that make up the network.

(3) f is the network flow parameters, including f_{ij} represents the segment flow of P_iP_j and f_i represents the waypoint flow of P_i.

3.3 Objective Functions and Constraints

In order to improve economy of the temporary route network while ensuring its security, the minimum operating cost, minimum flight collision coefficient, and minimum route angle change are used as the objective functions of the network optimization and can be expressed as follows:

$$\min T = \sum_{l=1}^{h} f_l D_l' + \Phi' \quad (10)$$

$$\min c = \sum_{i=n+1}^{m+n} \sum_{j=1}^{m+n+k} \sum_{k=j+1}^{m+n+k} \left[\left(2 f_{ji} f_{ki} / v\right) \cdot L_{\min} \cdot \arccos\left(\alpha_{jik}/2\right) / \left(f_{ji} + f_{ki}\right) \right] + \Phi' \quad (11)$$

$$\min \theta = \sum \Delta\theta_l + \Phi' \quad (12)$$

In these formulas, T is operating cost, f_l is flow of the temporary route l, h is the number of temporary routes, c is the network flight collision coefficient which is ralated to the angle and flow rate of air route segment, v is aircraft cruise speed, $L_{\min}$ is minimum control distance, α_{jik} is the angle between segment P_jP_i and P_kP_i. θ is angle change of temporary routes in the network, and $\Delta\theta_l$ is the angle change of l. Φ' is a penalty function, expressed as follows:

$$\Phi' = \lambda \cdot \sum C(l) \quad (13)$$

In the formula, $\sum C(l)$ represents the sum of the status values of all the temporary routes.

Combined with the actual needs of network optimization, the following constraints are proposed:

$$l \cap \left(Z_W \cup Z_U\right) = \varnothing \quad (14)$$

$$x_i^{\min} \le x_i \le x_i^{\max}, \; y_i^{\min} \le y_i \le y_i^{\max} \quad (15)$$

Among them, Equations (14) and (15) represent obstacle avoidance constraints and optimization scope constraints respectively. (x_i, y_i) is the coordinate of P_i, $x_i^{\min}$, $x_i^{\max}$ are the abscissa optimization range of P_i, and $y_i^{\min}$, $y_i^{\max}$ are the ordinate optimization range of P_i.

3.4 Algorithm coding

The structure of the temporary route network mainly depends on the layout of the diversion points, and a real-coded method can be used to construct chromosomes to represent their location information, for example: $x_{11}x_{12} \cdots x_{\alpha 1}x_{\alpha 2} \cdots x_{k1}x_{k2}$. Where k is the number of diversion points, $2k$ is the length of the chromosome, and $x_{\alpha 1}$, $x_{\alpha 2}$ are the horizontal and vertical coordinates of point P_α.

4 REROUTING ALGORITHM DESIGN BASED ON NSGA-III

4.1 Basic ideas

Common multi-objective optimization algorithms include multi-objective weighted genetic algorithm, target vector method, etc. Mostly, the multi-objective problem is transformed into a single-objective problem through transformational ideas. The NSGA-III algorithm can obtain the pareto optimal solution of the NP complete problem[10]. This non-inferior optimal solution set is suitable for different decision-making environments and enhances the completeness, robustness and comprehensiveness of the decision-making scheme. On the basis of inheriting the NSGA-II non-dominated sorting operation, the method of solving the reference distance by the association operation replaces the original congestion operator, avoids the premature convergence of the algorithm and improves the population diversity, when the number of objective functions is greater than 3, the planning effect is obviously better than the traditional multi-objective planning algorithm.

4.2 Operators

Combined with the actual needs of network optimization, the following operator can be set:

(1) Selecting operator: Use the roulette selection method and the objective function is set as a fitness function. The higher the fitness, the greater the probability of individual selection.

(2) Crossover operator: Use a method of partial match crossover, the cross points and matching segments of two parent individuals are determined to achieve chromosome rearrangement.

(3) Mutation operator: Adopt the basic mutation operation to randomly select a gene on the chromosome and change its value.

(4) Removing operator: Perform a delete operation on the chromosome whose coordinates of the waypoint is out of range, and regenerate the chromosome.

4.3 Algorithm Steps

The basic steps of the NSGA-III algorithm are similar to the traditional genetic algorithm[10], but there are innovations in selecting mechanism, as shown in Figure 2.

https://doi.org/10.1051/matecconf/201819805003

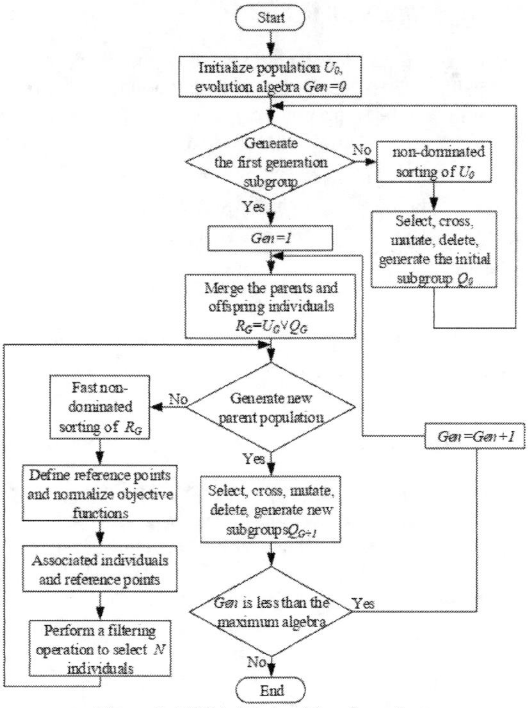

Figure 2. NSGA-III algorithm flow chart

5 SIMULATION ANALYSIS

Beijing flight information region is complex and the "three districts" are densely distributed. When dangerous weather occurs, it is easy to cause obstructions. Therefore, the routes of G212, B251, W41, H8, and B458 in this region are used as examples for simulation analysis. Its route information is shown in Table 1.

Table 1. Route information

origin-destination	air route	flow sor/h
Taiyuan-Gubeikou	G212	8.4
Taiyuan-Dawangzhuang	B251	9.9
Dawangzhuang-Gubeikou	W41	7.7
Pingyi - Shijiazhuang	H8	9.2
Tianzhen - Shijiazhuang	B458	8.1

Among them, the straight distance from Taiyuan-Gubeikou on the G212 route is 506.61km. For convenience, the entire rerouting area is set to 22×10 grid coordinate network and $m'=24.14$ km. Assume that the aircraft move from left to right or from top to bottom, when scattered thunderstorms occur in the area, carry out single-aircraft rerouting of 5 routes and collect them together, the network shown in Figure 3 is available.

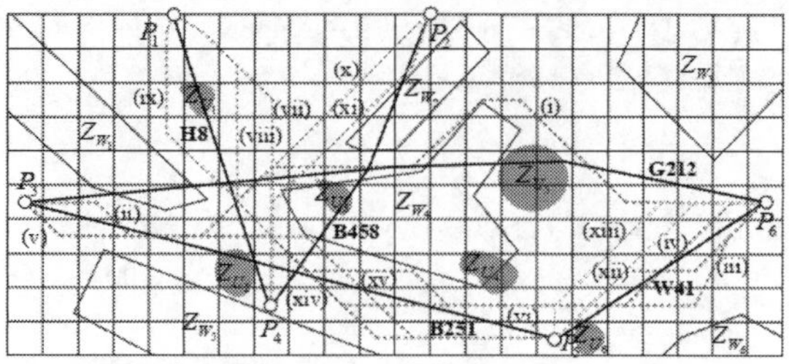

Figure 3. Schematic summary of single-route diversion routes

In the figure, the black solid line is the original route, and the red, purple, yellow, green, and blue dotted lines represent the reroute routes of G212, B251, W41, H8, and B458, respectively. Take optimal individual components of each rerouted route to form initial network of multi-aircraft rerouting, as shown in Figure 4.

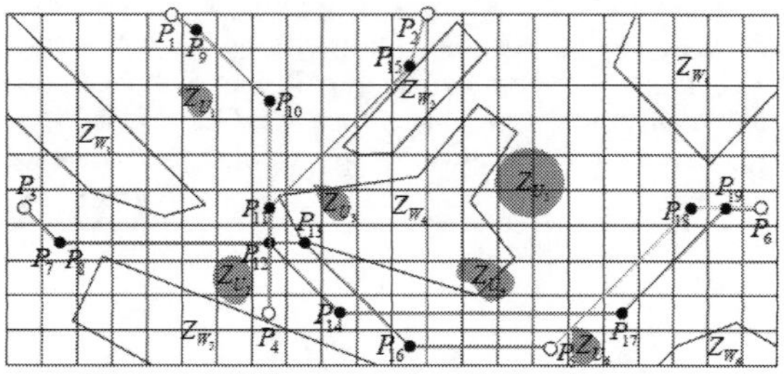

Figure 4. Initial network of multi-aircraft rerouting

In the figure, $P_i\left(i=1,\cdots6\right)$ is airport point or border point, and $P_i\left(i=7,\cdots19\right)$ is diversion point. It is easy to see the initial network composed of single-aircraft rerouting based on grid integer coding is complex and complicated, many routes overlap, and the distance is large. Therefore, it is necessary to carry out multi-aircraft rerouting of the initial network based on the existing diversion points. Set the number of populations to 100, the evolutionary generation to 800, the mutation rate to 0.1, and the hybridization rate to 0.15. Use NSGA-II, NSGA-II and particle swarm optimization algorithm to solve the multi-aircraft rerouting model, the network is shown in Figure 5.

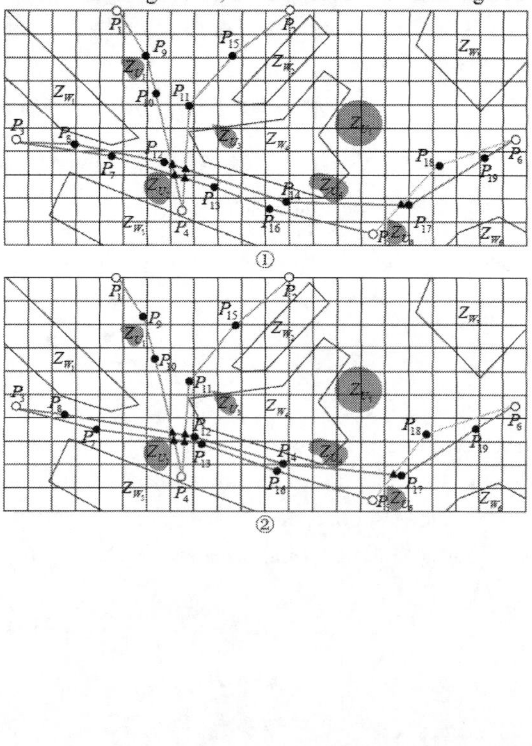

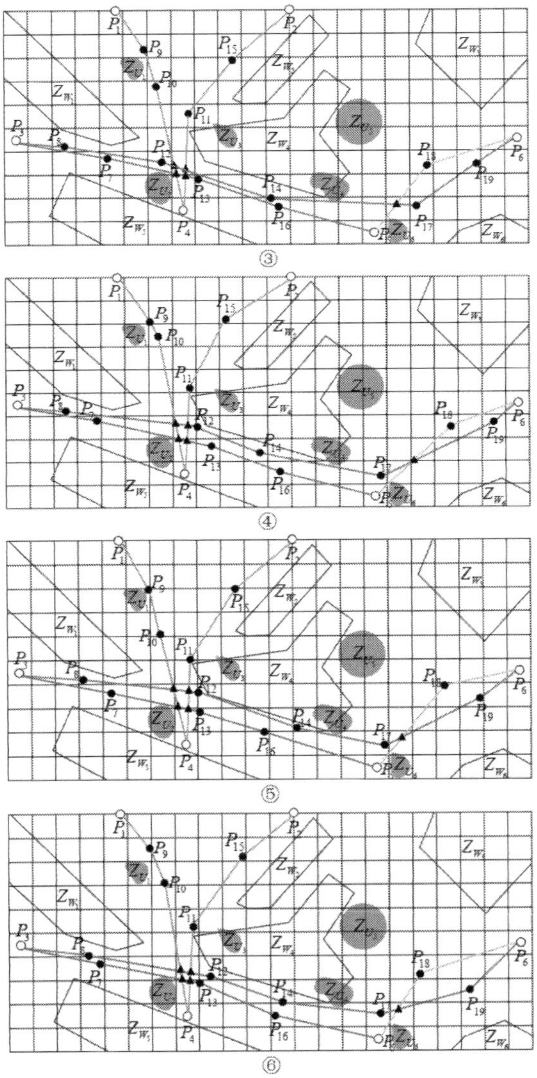

Figure 5. Multi-aircraft rerouting optimization network

Among them, ①-③ are the networks obtained by NSGA-III algorithm, ④ and ⑤ are the networks obtained by NSGA-II algorithm, and ⑥ is the network obtained by particle swarm algorithm. After judgment, all networks meet obstacle avoidance constraints and optimization scope constraints. The specific characteristic values are shown in Table 2.

Table 2. Characteristic value of networks

No.	operating cost	flight conflict coefficient	angle change/ degree
①	13596	1.01	180.1
②	13601	0.95	179.1
③	13640	1.03	178.0
④	13636	1.04	198.5
⑤	13626	1.06	191.0
⑥	13656	0.98	212.5
⑦	14326	1.58	533.7

In the table, ⑦ represents the initial network. By comparison, all the optimized networks are better than the initial network. Among them, the indicators of the networks ①、②and ③ are relatively balanced, and their optimization effect is the best. Some indicators of networks ④ and ⑤ are good, but they are slightly worse in overall integrity. The flight conflict coefficient of network ⑥ is smaller, but its other indicators rank lower. To further highlight the differences, the indicators of each optimized network are divided by the indicators of network ⑦ and make a contrast, as shown in Figure 5.

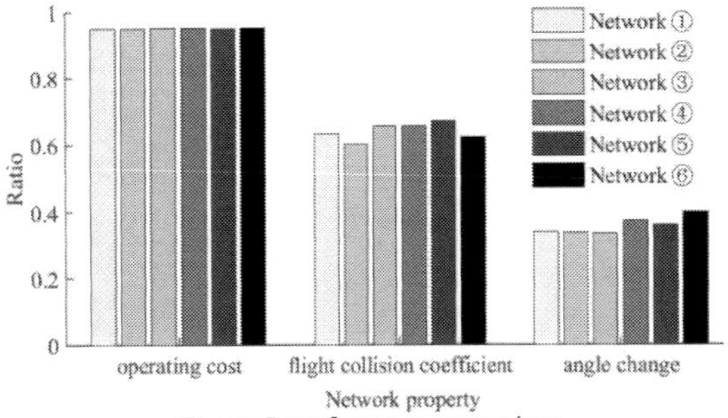

Figure 5. performance comparison

It is easy to see that network ② has the best improvement effect. Its operating cost is reduced by 5%, the flight conflict coefficient is reduced by 39.9%, and the angle change amount is reduced by 66.4%. Therefore, the NSGA-III algorithm can effectively solve the problem of multi-aircraft rerouting based on layout of diversion points and improve the safety and integrity of the temporary route network. What's more, a single operation can provide a variety of attributes with different results which can promote planning personnel to change the route design under different constraints and mission objectives.

6 Conclusion

This paper establishes a multi-aircraft rerouting model to study the route network under dangerous weather, which solves the problems of single-aircraft rerouting effectively and improves the safety and integrity of the temporary route network. Finally, the simulation analysis proves the validity of NSGA-III algorithm in solving multi-objective optimization problem, It can help to reduce the operating cost, flight conflict coefficient and angle change amount effectively which provides a new way for airspace rerouting planning.

References

1. W Yan, Y Wang, Y F Meng. Journal of air force engineering university, 18, 6: 20-16 (2017)
2. S S Yang, Y Tian, L L Wan. Mathmetical in practice and theory, **42,** 13: 87-93 (2012)
3. F Wang, H Y Wang. TSEIT, **14**, 5: 154-160 (2014)
4. L L Wang, X Wang, B Peng. Flight Dynamics, **35**, 2: 84-87 (2017)

 https://doi.org/10.1051/matecconf/201819805003

5. L L Wang, J Zhou. J. Wuhan. Univ. Technol, **36**, 5: 86-90 (2014)
6. L H Meng, X H Xu, S M Li, Z X Geng. J. Southwest.Jiaotong. Univ, **47**, 4: 686-691 (2012)
7. X Li, X H Xu, Y F Zhao, D X Wei. Act. Aer. ET. Astr. S, Acta. Aeronaut. Astronaut. Sin, **30**, 12: 2342-2347 (2009)
8. 8Babe Luitpold. J. Intell. Robot. Syst, **71**: 255-269 (2013)
9. S J Wang, Q Y Li, X Cao, H Y Li. Trans. N. Univ. Aeronaut. Astronaut, **33**, 4: 469-478 (2016)
10. D Kalyanmoy, J Himanshu. IEEE. T. Evolut. Comput, **18**, 4: 577-601. (2014)

Dynamic analysis of wing deployment of high altitude balloon launched small size folding-wing UAV①

Hangyue Zhang[1,2], Yanchu Yang[1,2]

[1]University of Chinese Academy of Sciences, Beijing 100190, China
[2]Academy of Opto-Electronics, Chinese Academy of Science, Beijing 100094, China

Abstract. The batch launch of small folding-wing UAVs by high altitude balloon at high altitude is a new technology in recent years. The wing deployment process at 20km altitude is analyzed, and mathematical model of the wing delivery is established. Using the multibody dynamics analysis software ADAMS simulate the wing deployment process in order to verify the correctness of the model established. Then, the impact of the wing with final velocity on the fuselage is analyzed, and an elastic locking mechanism is designed to reduce the impact. The analysis process provides a certain reference for the task design of balloon-UAVs and the folding mechanism analysis.

1 Introduction

The near space of $20km$-$100km$ is between the traditional aeronautical area and the space flight area [1]. At present, the high-altitude balloon is a relatively mature near space vehicle. The release and launch aircraft by high altitude balloon is a technological means to develop and utilize near space resources and high altitude launch of small UAVs(Unmanned Aerial Vehicles) can make full use of the advantages of height. The mission profile of the UAVs launch by high altitude balloon is shown in Figure 1.

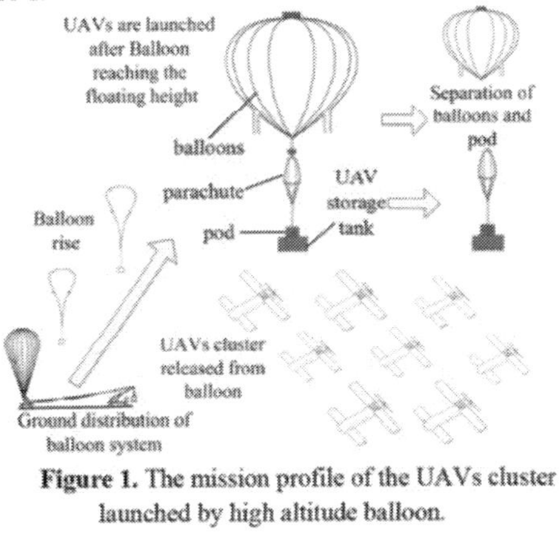

Figure 1. The mission profile of the UAVs cluster launched by high altitude balloon.

Folding wing UAV is the application of folding wing technology in UAV design process. Through the folding wings changes the dimension of the UAV, improving its storage, transportation, launch, recovery and aero-dynamic performance[2], which is beneficial to saving space and increasing the number of UAVs carried in one time. Such as the "Coyote" UAV (air launched, Fig. 2) and "vector hawk" UAV (submarine launched, figure 3), wing and tail both adopt a foldable design.

Figure 2. The "Coyote" UAV.

Figure 3. The "vector hawk" UAV.

The research on folding wing UAV is mostly focused on the study of aerodynamic influence of the front and rear wing layout, or the aeroelastic flutter analysis during the folding and expanding process[3-6] .The study of dynamic analysis and return vibration of the wing process is less [7].The wing expansion performance is related to whether the UAV can fly normally after the launch. In this regard, many scholars have studied the process of wing expansion. Qin Zhang[7]design the folding mechanism with spring as the power source, and the dynamic analysis and simulation are carried out. Wenqiang Zhen[8] also designed the axial and spreading locking mechanism. For analysis method, most of researchers establish mathematical models of mechanism and combined with simulation software. Yushan Zhao[9] set up the expansion process of the numerical solution of Lagrange equation. In addition, Jian Ni[10] carried out the qualitative analysis of the reliability of the unfolding mechanism based on the FEMA principle. Bincai Qin[11] used the sequence of two quadratic programming (SQP)

*master student; research field:Technology of releasing and launching aircraft by aerostat

①→Supported by the National Defense Innovation Program of Chinese Academy of Science(CXJJ-16M206).

method on folding and unfolding mechanism optimization. Considering that there is a certain terminal velocity after the deployment of the wing, the impact on the fuselage structure and the vibration of the wing will occur, which will affect the reliability of the wing deployment and the locking mechanism. In addition, the deployment of folding wing UAV at the stratospheric height, combined with the thin air of the stratosphere, and the smaller dynamic pressure of the wing during the deployment, is conducive to the rapid and reliable deployment of the wing.

2 The establishment of the mathematical model and simulation

UAV folding and unfolding mechanism is designed to carry out the wing reliably and reduce the impact on the fuselage. The UAV has a tandem wing layout. The area of the front and rear wings is roughly the same to increase the lift area. The airfoils select the *Ritz* airfoil, chord length is $8cm$, single span $L_A = 30cm$. The UAV folding and unfolding sketches are shown in Figure 4, 5. The wing is powered by the torsion spring, and the wing is locked through the limit mechanism after the wing is fully unfolded. The wing angle, angular velocity, angular acceleration with time, and the end velocity of wing can be obtained. The final velocity dropped to 0 at the moment of collision with the fuselage, showing the impact on the structure of the body. Because the wing is a long straight beam with a ratio of more than 1, it will bring return vibration and propose a certain design requirement for the structure of the wing.

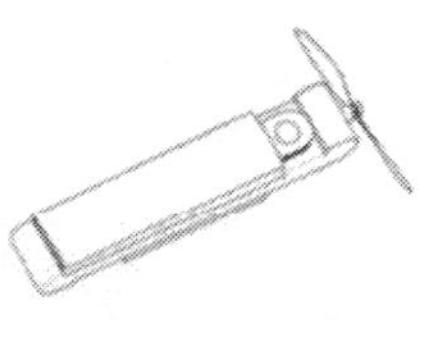

Figure 4. The unfolding shape of UAV.

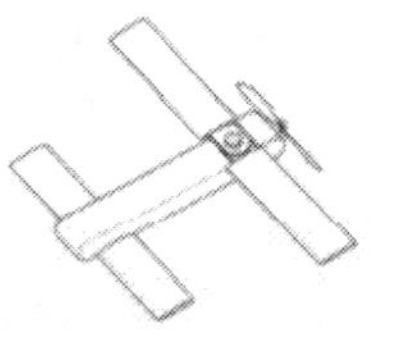

Figure 5. The folding shape of UAV.

The mathematical model shown in Figure 6 is established for the wing expansion process. θ is the unfolding angle ($0 \le \theta \le 90^\circ$), small rotation angle $\Delta\theta$ with tiny time Δt of the wing. M_φ is a torsion spring driving torque. F_L, F_D is the aerodynamic lift and drag. M_D is air drag torque. M_μ is friction torque for rotating wing. J is the moment of inertia around the shaft of the wing.

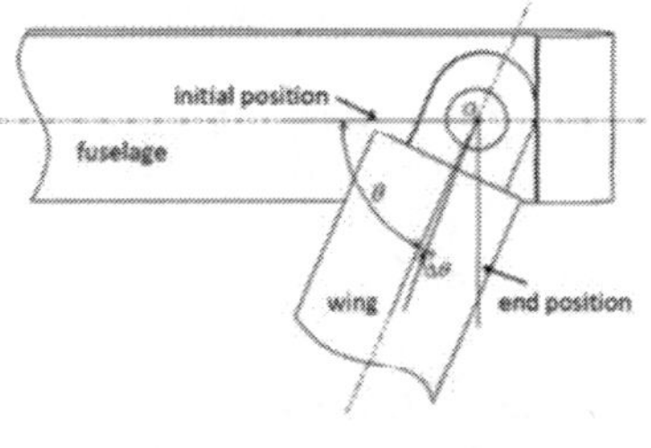

Figure 6. The mathematical model of wing expansion.

The motion equations of the wing are established as follows:

$$M_\varphi - M_D - M_\mu = J \cdot \ddot{\theta} \tag{1}$$

The initial condition $\theta_{(0)} = 0, \dot{\theta}_{(0)} = 0$, and set up SIMULINK module to get numerical solution. As figure 7, firstly, getting the relationship between acceleration with the angle of expansion, the angular acceleration at a certain time can be interpolated, and the angular velocity and angular displacement can get by integrating. And It is solved by ode45 function with the time chosen as the independent variable.

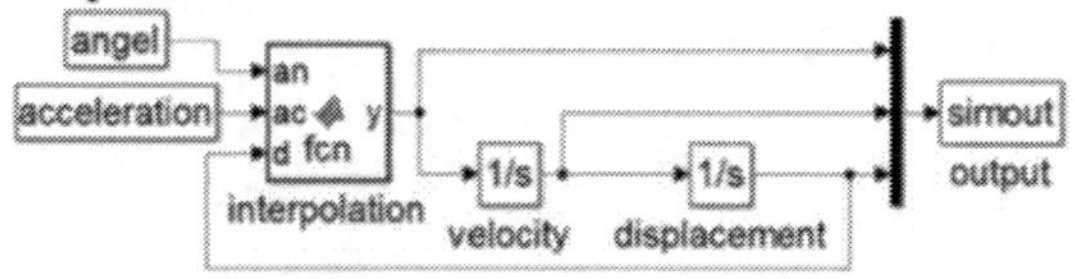

Figure 7. The SIMULINK module for numerical solution.

The aerodynamic drag and torque of the wing have different values in the unfolding process due to the different unfolding angles of the wing. The computational fluid dynamics (CFD) analysis is carried out on the wing with 4° angle of attack under $20km$ altitude and the environment flow velocity is $10m/s$. Using block and bonding technology and O-shaped grid to generate structured body fitted grid, and more than 10 grid nodes are arranged at the bottom of laminar flow. The steady state calculation is done, and the Simple algorithm is applied to get the aerodynamic parameters (lift coefficient $\overline{CL}$ and drag coefficient $\overline{CD}$) of the wing at different expansion angles, and the lift with drag of the wing are calculated. According to formula (2) the aerodynamic drag moment of wing can be derived.

$$M_D(\theta) = 0.5\rho v^2 \cdot S \cdot \overline{CD}(\theta) \cdot L_A \cdot \sin(\theta)/2 \tag{2}$$

Fitting out $\overline{CL} \sim \theta$, $\overline{CD} \sim \theta$ and $M_D \sim \theta$ curves as shown in Figure 8, 9.

The relationship between torsional spring torque and unfolding angle is expressed in formula (3).

$$M_\varphi(\theta) = \frac{Ed^4}{3667Dn}\theta \tag{3}$$

The E is elastic modulus of the twisted spring material. d is the twisted spring line diameter. D is the middle diameter of the torsion spring and the n is the number of twisted reed rings .

The wing is subjected to ring frictional resistance moment at the joint of a wing shaft and it can be calculated according to formula (4).

$$M_\mu(\theta) = 2\frac{2\mu F_L(\theta)(R_1^2 + R_1R_2 + R_2^2)}{3(R_1 + R_2)} \tag{4}$$

μ is the coefficient of friction. R_1, R_2 is the large radius and small circle radius of the ring.

The formula (2-4) is brought into the formula (1) and the formula (5) is obtained, and the results shown in Figure 9 are obtained.

$$J\ddot{\theta} = \frac{Ed^4}{3667Dn}\theta - 2\frac{2\mu F_L(\theta)(R_1^2 + R_1R_2 + R_2^2)}{3(R_1 + R_2)} - M_D(\theta) \quad (5)$$

Now, the angle of the wing, angular velocity, angular acceleration variety with time, and the end velocity of the wing can be obtained as figure 10 displays.

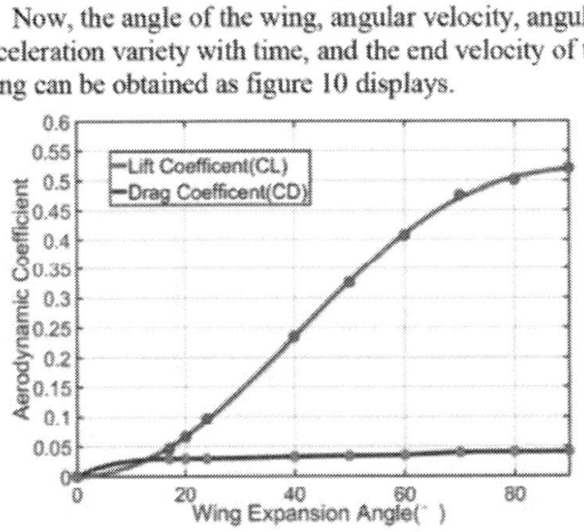

Figure 8. Curve of wing aerodynamic force with angle.

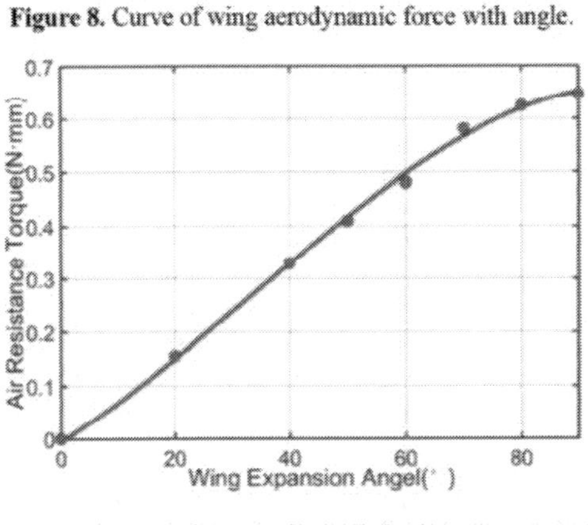

Figure 9. Curve of Wing aerodynamic drag torque with angle.

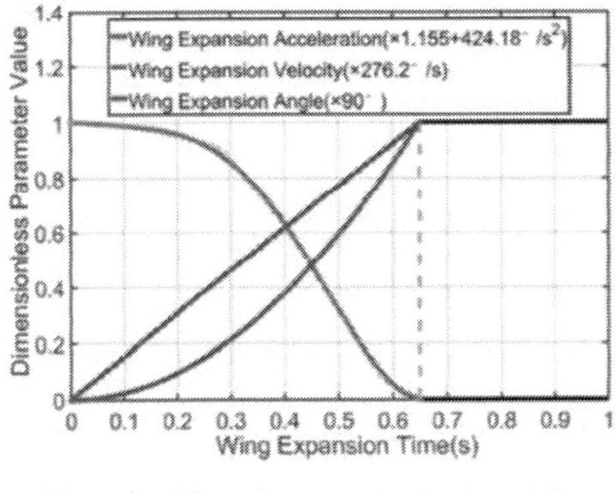

Figure 10. Kinematic curves obtained by solving mathematical models

From the result of the solution, it can be concluded that the wing process is completed in 0.65 seconds, the acceleration change little and angular velocity almost increase linearly. Finally, wing rotation is stopped at 90°.

The dynamics analysis software ADAMS is used for simulation, and the physical models of fuselage, wing, torsion spring and rotating shaft are established, giving the user defined driving force and driving torque of the wing center. The circular friction is realized by defining the indirect contact force of the component, and the nonlinear equivalent spring damping model based on the collision function algorithm is used to calculate. Contact force is expressed as

$$F = K\delta^e + STEP(\delta,0,0,d_{max},c_{max})\cdot\dot{\delta} \quad (7)$$

The F is the normal contact force, K is contact stiffness, permeability δ, strength index e and the step function $STEP$, d_{max} and c_{max} is the maximum penetration depth and the maximum damping value[l]. The simulation results show that the motion curve of the wing process is in agreement with the results of the mathematical model, and the correctness of the mathematical model is verified.

3 Impact Dynamic analysis and design of locking mechanism

The wing has a certain terminal velocity (1.5m/s) at the moment of fully unfolding, and it produces a shock to the fuselage, which is a transient dynamic problem. The direct explicit integration method of finite element analysis software ABAQUS is used to analyze the impact process, and the response curve of the impact force to the fuselage structure is obtained, as shown in Figure 11.

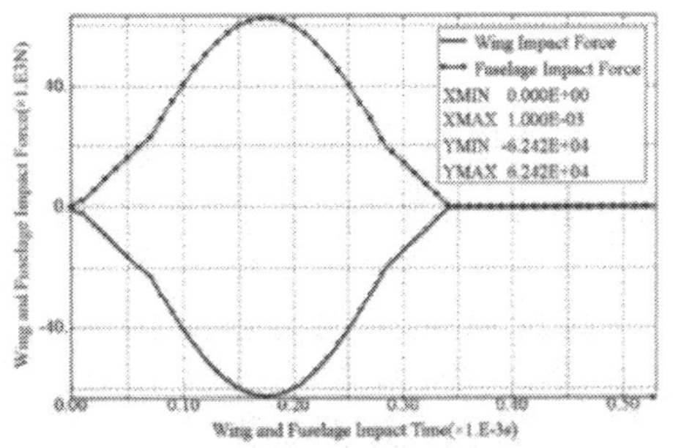

Figure 11. Instantaneous collision force curve of the wing and the fuselage.

From the analysis results, it can be found that the contact force of the collision process changes greatly, and the structure of the wing and the fuselage is certain to be destructive.

In order to reduce the hard impact of the wing on the fuselage, reference documentation [7] proposed the use of rubber cushion to shock absorption and give the simulation analysis. Considering the high throwing height and low temperature, the rubber property will become hardened and crisp, which is not suitable for high-altitude cushioning device. Inspired by the cushioning and damping characteristics of the elastic structure, and using friction to adsorb the kinetic energy, the locking mechanism shown in Figure 12 is designed. The locking mechanism has the elastic deformation of the structure under the impulse of the wing and ensures that the wing is always fixed during flight. Considering ABAQUS has powerful ability to solve nonlinear contact problems, it is used to analyze the locking mechanism dynamics either.

The dynamic analysis of the locking mechanism is carried out under the impact of the wing's terminal velocity with the obtained kinetic energy. The kinetic energy and internal energy variation curve of the elastic

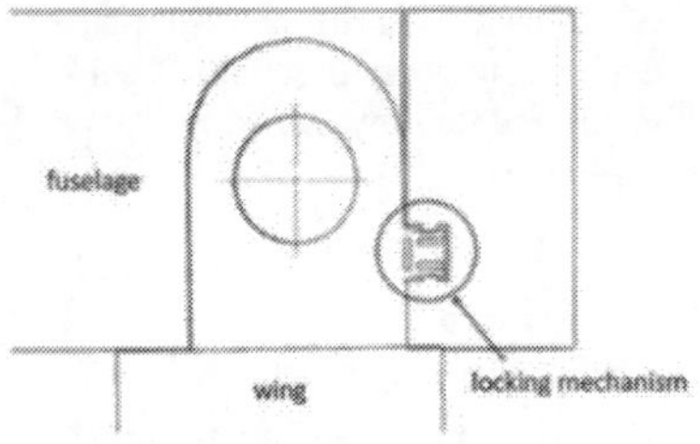

Figure 12. Elastic locking mechanism.

mechanism is shown in figures 13 and 14. It can be seen that the kinetic energy of the wing is completely transformed into the internal energy of the collision, and the internal energy is mainly shown as the strain energy without elastic locking mechanism. And it can be drawn the wing is bounced back. Nevertheless, the kinetic energy of the wing is mainly transformed into friction dissipation with elastic locking mechanism, and the wing is locked tightly with kinetic energy reduced to 0, which reduces the structural damage to the wing and the fuselage.

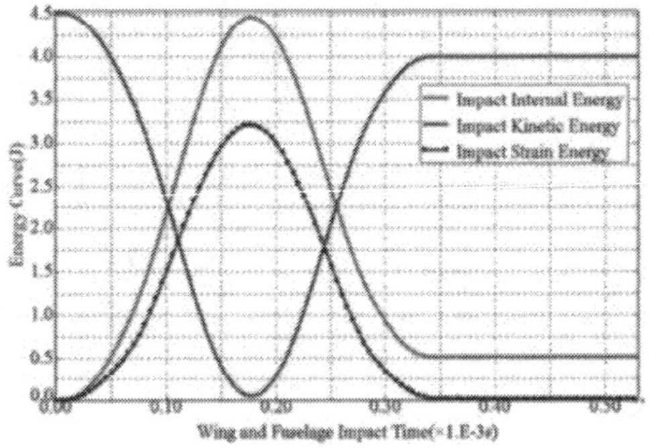

Figure 13. Energy change curves without locking mechanism.

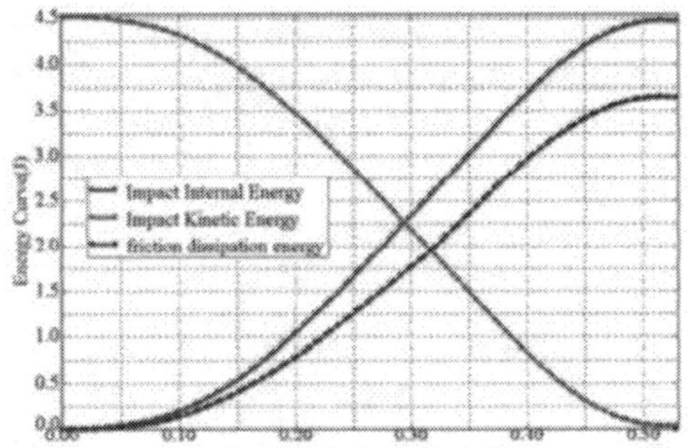

Figure 14. Energy change curves with locking mechanism.

4 Summary

The folding wing UAV is carried by a high altitude balloon to the stratosphere, and the wings unfold independently after the launch. The following several aspects have been analyzed.

①Combined with the design of the deployment mechanism, a mathematical model is established for the deployment process of the wing. The aerodynamic force and moment are calculated by the CFD tool. But only analyzed one wing at the moment, the tandem wing UAV has front and rear wings, should consider its interaction on aerodynamic analysis.

②The impact on the fuselage is analyzed at the end of the wing deployment speed. The impact will result in flutter of the wing and coupling with the structural dynamic behavior of the wing long straight beam. The impact on the structure is still needed to be further analyzed.

③In order to reduce the impact on the fuselage car, the locking mechanism is designed, and the elastic parts are added. Through simulation analysis, it can be concluded that the locking mechanism has a good effect on the reliable deployment and impact reduction of the wing.

The analysis results provide some reference for the design of the same kind of folding unfolding mechanism and the high altitude launch technology of UAV.

References

1. Gang Guo, Maoxing Shen. Present situation and development trend of military application in near space. Information Online. **2**,16-21(2012).
2. Bo Liao, Changsheng Yuan, Yongze Li. The development status and key technology of the folding wing unmanned aerial vehicle. The Mechanical Design. **29**,4(2012).
3. Chuanmei Bao, Changliang Liu, Ye Sun. Launching technology and development of UAV. Unmanned Aerial Vehicle. **2**,56-60 (2012).
4. G.Q.Zhang, S.C.M.Yu. Unsteady Arodynamics of a Morphing Tandem-Wing Unmanned Aerial Vehicle. Journal of Aircraft. **49**,5 (2012).
5. D.H.Lee, Terrence A.Weisshaar.Aeroelastic Studies on a Folding Wing Configuration. 46th AIAA Strucutres, Structural Dynamics & Materials Conference.(2005).
6. Peter J.Attar, Deman Tang. Nonlinear Aeroelastic Study for Folding Wing Structures. AIAA Journal. **48**,10(2010).
7. Qin Zhang, Hong Nie, Ming Zhang. Dynamic analysis of unmanned aerial vehicle folding wing expansion. Mechanical Design and Manufacturing Engineering. **44**,2(2015).
8. Wenqiang Zhen, Yongqiang Ji. Dynamic simulation of folding wing deployment and locking process analysis. Information Technology. **2**,62-65 (2016).
9. Yushan Zhao, Xudong Yu, Caixia Ma. The simulation of the folding wing expansion process. Missile, Arrow and Guidance Journal. **2**,19-23(1997).
10. Jian Ni, Kai Lu, Duo Zhang. Reliability analysis of the launching mechanism of missile folding wing. Shanghai Spaceflight. **3**,33-36 (2000).
11. Bincai Qin, Hua Li. The optimization design of small folding wing. Journal of Missiles and Guidance agencies. **29** ,4(2009).
12. Weidong Guo, Shouzhong Li, Lu Ma. ADAMS2013 application example tutorial. Beihang University press. (2008).

Design and Implementation a Measurement System for Aircraft Cabin Heat Comfort Test

Yan Jiani
ShangHai Aircraft Design and Research Institute,China

Abstract. A measure system is designed for aircraft cabin thermal comfort test, which can measure temperature, wind speed, humidity, pressure, flux, position parameters. The test results can be used to calibrate the model of temperature field simulation and cabin flow field simulation. The system synchronization accuracy is achieved 1mS. The measure system uses distributed configuration which comprises many measure cells. Test datum in the cabin can be set parameters and controlled re are transmitted to computer via internet. The measure equipment motely via internet, so that the measurement interfere affected by human activities can be reduced and the measurement accuracy can be improved. The measure system can communicate with instrument installation, making the cabin airflow field test automatism come true.

1 Background

With the development of civil aircraft design, in addition to meeting the safety and economic requirements for the design of civil aircraft, cabin comfort design is becoming more and more important. Due to the complexity of aircraft flight conditions, aircraft cabin comfort is often affected by many factors, including physical, physiological and psychological aspects. One of the important controllable factors in comfort factors is thermal comfort, including chamber ventilation, wind speed, pressure, temperature, humidity, radiation, etc [1-4]. Figure 1 is a schematic diagram of the cabin environment test platform. The platform uses the test samples produced during the development phase of the aircraft. Through the high and low temperature environment simulation test system and the outside environment simulation facilities such as the sunshine simulation system, a close-to-real aircraft cabin environment is created to develop cabins thermal comfort test study.

Cabin thermal comfort test, a large number of temperature, wind speed, humidity, pressure, flow, location, illumination and other parameters need to be tested. In this paper, a set of distributed test systems that can be monitored remotely is designed to meet the multi-parameter test requirements for cabin thermal comfort tests. In addition to the multi-parameter synchronization test of each measuring point in the cabin, the test system can also interact with the instrument installation platform to complete the automatic collection of cabin flow field data.

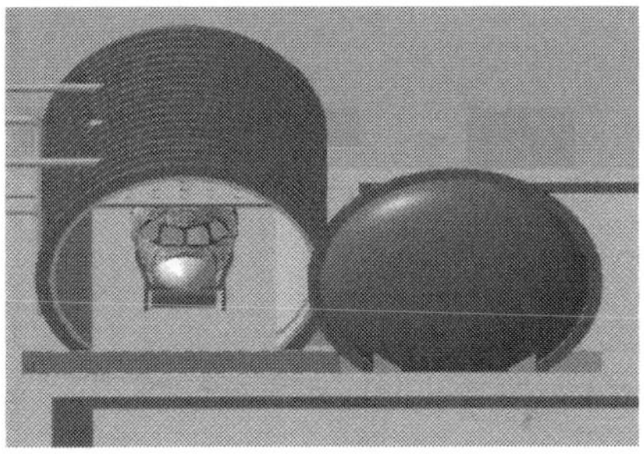

Figure 1 Schematic Diagram of Cabin Environment Test Platform

2 Test system architecture

The number of sensors for cabin thermal comfort test reaches several hundred. The test sensors are distributed outside the cabins of the cockpit and passenger cabin, cabin skin surface, the interior of the cabin structure, the surface of the cabin interior trim panels, and the vicinity of the seats .The distribution of the sensors is very scattered. Therefore, the test system adopts a distributed architecture, that is, the test terminal is composed of multiple test units, placed near the test sensor nearby, and the data collected by the test unit is transmitted to the measurement and control room through Ethernet [5-6]. This architecture requires only a few network cables to pass through the cabin wall and the bulkhead of the high- and low-temperature cabins to reach the monitoring and control room, which reduces the difficulty of wiring the sensors, shortens the transmission distance of the analog signals from the sensors, and reduces the external interference to the analog signals. At the same time, the test personnel can carry out parameter setting and remote control of the in-cabin test equipment through the network and remotely monitor the measurement process at any time, which will effectively reduce the interference of the tester's activities to the cabin environment and improve the measurement accuracy [7]. Figure 2 is a schematic diagram of the test system architecture.

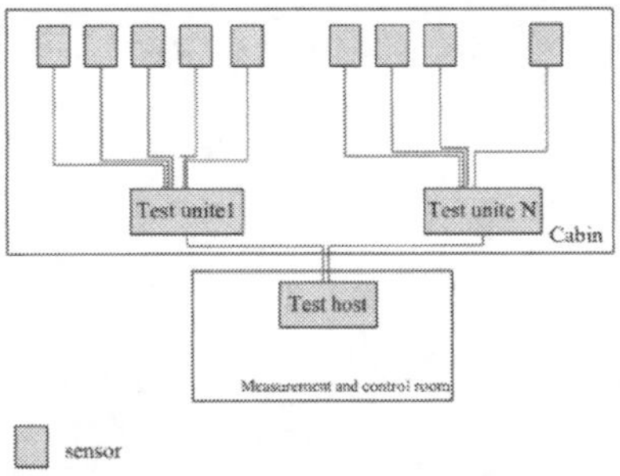

Figure 2 Test System Architecture

3 Key technical analysis and implementation of test system

3.1 Test System Synchronization

The cabin thermal comfort test system adopts a distributed architecture. The system consists of a plurality of test units. Each test unit has its own clock. It is necessary to adjust and maintain the physical or logical clocks of each test unit through some mechanism to ensure that all collected data can be analyzed and processed on the same time axis. Therefore, how to adjust and maintain the physical or logical clock of each test unit is the key to system design.

The full name of the IE1588 protocol is the "precision clock synchronization protocol standard for network measurement and control systems", abbreviated as PTP (Precision Time Protocol). This protocol is the hotspot of clock synchronization research in distributed test systems. It has the advantages of simple implementation, small occupied network bandwidth, and low requirement for system resources, thus becoming the most promising LAN clock synchronization protocol. The PTP protocol can achieve a subtle level of synchronization accuracy.

PTP defines four types of multicast packets: Sync, Follow-up, Delay-Req, and Delay-Resp. The order of interaction of these messages and the information received are related to the current state of the clock.

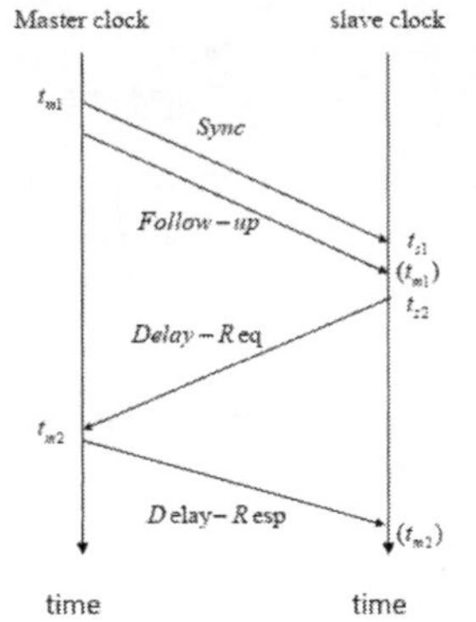

Figure 3 Clock synchronization process

Figure 3 is a schematic diagram of the clock synchronization process. To describe the clock synchronization problem of the test system, we illustrate the clock synchronization process by taking the synchronization process of a master clock and a slave clock as examples.

a) The master clock periodically sends synchronization messages (Sync) to the slave clock at defined intervals. The time stamp of this message is the expected transmission time of the master clock, and the master clock records the actual transmission time tm1; At the receiving end, the text is recorded from the clock at the receiving time ts1.

b) The master clock sends a follow-up message to the slave clock. The message contains the exact transmission time tm1 of the previous synchronization message. The network delay time is Delay and the master and slave clock offsets are Offset:

$$ts1 = tm1 + Offset + Delay \tag{1}$$

c) Send the Delay-Request message from the clock to the master clock (the delay request message interval is set independently). After the message is sent, the slave clock records the exact transmission time, ts2, at the master clock. End record accurate reception time tm2.

d) The master clock returns a delay-response message (Delay-Response) to the slave clock. This message contains the exact reception time tm2 of the previous delay request message. The slave clock uses ts2 and the transmission time tm2. You can get:

$$tm2 = ts2 - Offset + Delay \tag{2}$$

From equations (1) and (2) can be calculated:

$$Delay = [(ts1 - tm1) + (tm2 - ts2)]/2 \tag{3}$$

$$Offset = [(ts1 - tm1) + (tm2 - ts2)]/2 \tag{4}$$

If the current slave clock time is tsOld, then the new slave clock time tsNew that should be synchronized to now can be calculated:

$$tsNew = tsOld - Offset - Delay \tag{5}$$

According to this, in the next synchronization cycle, synchronization between the slave clock and the master clock can be achieved. The accuracy requirement of the chamber thermal comfort test system is 1mS. The PTP clock synchronization accuracy meets the requirements. Besides the hardware support of the equipment itself, there is no need to add additional synchronization cables and the networking is convenient.

3.2 Flow field automation test

In the cabin thermal comfort test, the flow field test inside the chamber needs to be completed. When testing and arranging measuring points in the cabin flow field, no less than three measuring points around the human body (head, waist, ankle, etc.) shall be arranged, and the other measuring points shall be arranged in a uniform manner. The layout of measuring points on a single cross section of the cabin As shown in Figure 4.

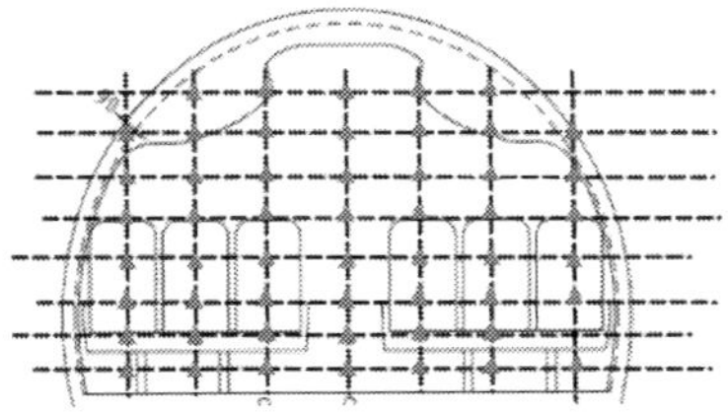

Fig. 4 Schematic diagram of layout of measuring points in cabin flow field (heading)

The cabin has the characteristics of complex space and small space and low wind speed at close distances. When the flow field is tested, if the wind speed sensor is arranged at each measuring point, the distribution of the flow field in the cabin will be seriously affected and the test result will lose significance. Therefore, the laboratory specially designed a set of instrument installation platform. The instrument installation platform can be equipped with 7 wind speed measurement devices. The installation platform can automatically control the measurement device to reach the specified measuring point.

Every time the instrument installation platform controls the measurement equipment to move to the target position, it needs to manually control the test system to start or stop the test task. The test personnel need to frequently operate the test system, and the work is very tedious. How the instrument installation platform interacts with the test system to achieve automated data acquisition and processing of the flow field is the key to the design of the test system.

According to the test requirements for automated data collection in the flow field, the instrument installation platform and test system information interaction methods and workflows were designed. The schematic diagram of information interaction between the instrument installation platform and test system is shown in Figure 5. The instrument installation platform controller receives an instruction from the tester to control the measurement instrument to reach the specified measurement point position, and sends a trigger signal after the "set time 1" waits for the flow field in the chamber to stabilize. The test system receives the trigger signal and starts data acquisition. The instrument installation platform sends a trigger signal according to the "set time 2" waiting for the test system to complete data acquisition. The test system receives the trigger signal and ends the data acquisition. Among them, "set time 1" refers to the time required for the stability of the cabin flow field input by the tester in the interface of the instrument installation platform controller; "set time 2" refers to the test personnel's interface in the instrument installation platform controller. The time of the input data acquisition. "Set time 1" and "Set time 2" need to be determined by tests or previous test experience. According to the workflow input in advance, the instrument installation platform issues a trigger signal, and the test system receives the trigger signal to start or stop the collection of tasks, thereby realizing the automatic collection of the flow field in the cabin.

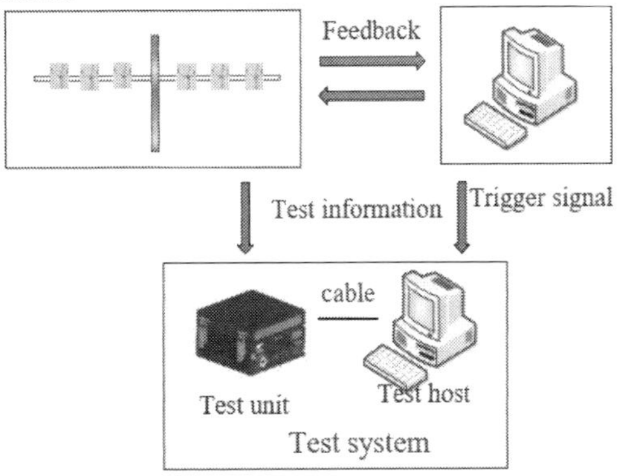

Figure 5 flow field test experimental equipment information exchange diagram

4 Test system software and hardware implementation

4.1 Test System Hardware

The system is mainly composed of test units, switches, operation and display terminals. Since part of the cabin test is required to be performed within the temperature range, the selected test unit must be able to operate normally within the temperature range. The test unit selected in this paper is the VTX company RX series of rugged field data acquisition equipment, the equipment to meet the test accuracy, sampling frequency and other test requirements, at the same time, the test equipment supports PTP protocol, can achieve the test system 1mS synchronization accuracy requirements.

4.2 Test System Software

The cabin thermal comfort test system software has commonly used functions such as system configuration, channel parameter setting, data recording, data display, data export, data playback, etc. At the same time, the system software needs to have certain data when testing the flow field. Processing function. When the cabin flow field test, the test system collects the position sensor signals of the instrument installation platform, and the obtained information includes "time information" and "position information". The test system collects the sensor signals of the measurement point, and the obtained information includes "time information" and "measurement point parameter information". The test system software can match the "position information" and "measurement point parameter information" of the measurement point according to "time information", and the final output of the test text measurement point information includes "time information", "position information", and "measurement point parameter". information". The test personnel can analyze the flow field in the cabin according to the "position information" and "measurement point parameter information" of each measuring point.

5 Conclusion

In this paper, a set of multi-parameter distributed testing system, such as temperature, wind speed, humidity, pressure, flow, illumination and position, is designed based on the test requirements of civil aircraft cabin thermal comfort test. The test system consists of multiple test units. The VTI RX series rugged field data acquisition equipment is used in the test unit. The device can work normally in the temperature range. At the same time, the test equipment supports the PTP protocol and can be tested. System 1mS synchronization accuracy requirements. The test system can interact with the instrument mounting platform to complete the automated collection and processing of cabin flow field data.

References

1. Liu Junjie, Li Jianmin, Li Fei, Shen Yan. Construction of air-conditioned ventilation experiment platform for large passenger aircraft cockpit and experimental research [J]. Journal of Tianjin University (Natural Science and Engineering Technology Edition), 2014.
2. Liu Junjie, Li Bingjun, Jing Jingjing et al. Subjective and objective experimental study of aircraft cabin environment in different seasons [J]. Journal of Tianjin University (Natural Science and Engineering Technology Edition), 2015.
3. Sun Hejiang, Wu Chen. Orthogonal test method for the analysis of air distribution and thermal comfort of large passenger aircraft cockpits [J]. Journal of Tianjin University (Natural Science and Engineering Technology Edition), 2015.
4. Du Guoxing, Sheng Wen, Wang Shengchao. A quasi-LXI bus architecture of remote
5. Zhu Wangchun, Zhang Jingyu, Yang Surong. Network Distributed Virtual Instrument Testing Research [J]. Instrument Technology and Sensors, 2012.
6. Chen Ruiqiong, Liu Ya, Li Xiaohui. Analysis and implementation of key technologies for networked time-frequency test labs [J]. Journal of Electronic Measurement and Instrument, 2014.

Tension modeling of package binding belts in cargo spacecraft

Weifeng Yuan[1], Hailin Dai[1] and Ruizhao Du[1]

[1]*Beijing Institute of Spacecraft Environment Engineering, Beijing, China, 100094*

Abstract. At the site of cargo package binding in cargo spacecraft, the binding belt tension is loaded with a ratchet device. It is determined by experience of operators, therefore it's difficult to quantify the force and the reliability of the binding belt is low. Based on the dimension analysis method, a semi-quantitative model was established between the binding belt tension and the ratchet force of the fastening device. The motion of the binding process was carefully observed and analyzed based on Analysis of Therbligs. On this basis, the standard packing process is developed, which improves the repeatability of loading and packing pre loading. A loading test platform for cargo loading is built with small tension sensor and tension tester, and a large number of tension data and ratchet force data are obtained. Based on the least-squares linear data fitting, four coupling models for the four binding belts were established. The coupling model verification experiments show that the calculated values are in good agreement with the measured values, and the errors are mostly lower than 5%. The model can be applied to the quantification of the binding belt tension.

The cargo spacecraft is one of the main development models of the space laboratory mission and the space station project in China. It is a necessary support vehicle for the third step long term human care space station. When goods are transported to the space station for supplies, the vast majority of goods are loaded through flexible soft cargo bags. The binding belt is usually used in order to prevent the cargo from slipping and flying during the launch and flight of the cargo ship. The current package binding is carried out manually by the operator, and the tightness of the package binding is also determined by the experience of the operator. Therefore, the consistency of the tightness of the package binding is poor, the pre-loading force is difficult to quantify, and the quality and reliability of the binding cannot be guaranteed, and it is difficult to meet the requirement of quantitative loading for preloading force of cargo binding belt.

The binding belt is made of thick cotton ribbons. In addition to the elastic deformation caused by tension, there is also tension relaxation caused by structural slip and fracture. On one hand, it is more different from the usual elastic body and difficult to predict the binding belt tension by theoretical model; On the other hand, the binding belt is a stereotyped product, and no additional sensors are allowed, so the tension of the binding belt cannot be directly measured. At present, most of the research at home and abroad is focused on tensile fracture performance, safety and comfort of automobile and aircraft passenger seat belts, as well as tension measurement of wire rope, damage mechanism, detection and distribution of stress and strain, etc. However, there is almost no research on the binding of bundles, and no similar research experience and models can be used for reference.

The test platform is set up to simulate the assembly site of the cargo package. The ratchet force and the binding belt tension are measured by the tension sensor connected in the ratchet loading device and the end of the binding belt. On this basis, the coupling model of the

 https://doi.org/10.1051/matecconf/201819805006

ratchet force and the binding belt tension is established. In order to improve the repeatability of the coupling model, a set of standard packing process is proposed based on the Analysis of Therbligs. When implementing in the field, operate according to the standard binding process, and predict the binding belt tension according to the ratchet force applied on the ratchet loading device and the coupling model mentioned above. Therefore, the quantitative loading of binding pretension of binding belt can be realized. It has important engineering application value for improving the quality and reliability of package binding.

1 Cargo package, binding belts and the loading device

1.1 Cargo package

In binding test, the package used is a double package (length × width × height 470 mm × 360 mm × 285 mm, maximum wall thickness 10 mm, maximum weight 0. 9 kg, loading capacity 20 kg), as shown in Figure 1.

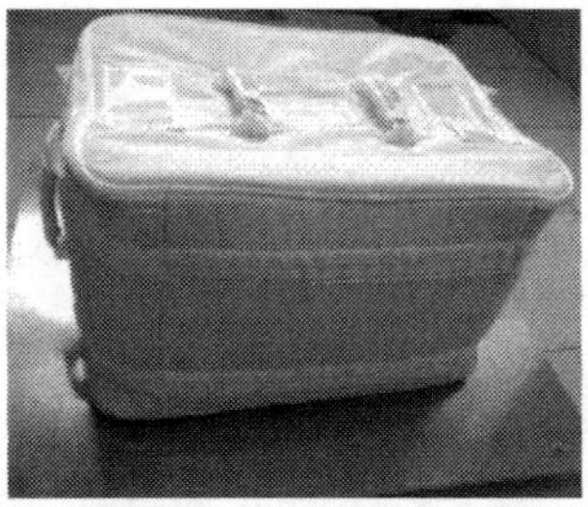

Figure 1. Test package

1.2 Binding belts

The belt is made of straw green thick cotton ribbon, as shown in Figure 2.

Figure 2. Binding Belts and Package Base

According to the actual working condition of the site, 4 binding bands are sequentially bound according to the order of D2, D4, D1 and D3.

1.3 Fillings

Considering that the items in the package are not only hard instruments and equipment, but also soft food and clothing, a flexible paper filler with moderate hardness was selected in the experiment. The filling rate was 95%, leaving an appropriate gap in the interior, and the package could be easily closed.

1.4 Loading device

Because of the simple operation, labor-saving and compact size, the ratchet shown in Figure 3 is used as the fastening and loading device for binding belt.

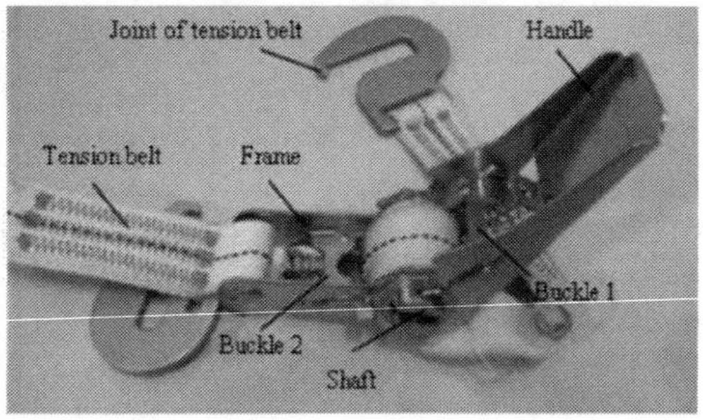

Figure 3. Ratchet fastening and loading device

2 Analysis of semi-quantitative model of binding belt tension

2.1 dimensional analysis

The binding belt is a stereotyped aerospace product, and the external sensors are not allowed to measure the tension directly. Therefore, it is an effective way to measure the ratchet force imposed by the fastening and loading device, and to predict the binding force through the ratchet force and the binding belt tension coupling model, and then to realize the quantitative loading of the binding belt. In the process of package binding, the binding belt tension is not only related to the size of the ratchet force, but also influenced by the loading speed, the binding method, the safety buckle force and so on. The mechanical properties of the binding belt involve deformation and relaxation, which is very complicated. It is difficult to describe the specific mathematical model. For the complex problem with no mathematical model, starting from the basic physics, the semi-quantitative model between the ratcheting force and the binding force can be established by the dimensional analysis, and then the function relation of the two is determined by the experiment [1, 2, 3].

The general formulas are established by selecting the binding belt tensile stress σ (kg / (m ·s^2)), the cross section area S (m^2), the loading time interval V (times/s), the ratchet force F_1 (kg · m/s^2), the safety buckle tension F_2 (kg · m/s^2), the binding belt tension F_3 (kg · m/s^2):

$$f(\sigma, S, v, F_1, F_2, F_3) = 0 \qquad (1)$$

Select quality M, length L and time T as basic dimensions, and select σ, S, V as basic variables. The dimension matrix of the basic variables is

$$\mathbf{A}=\begin{bmatrix} 1 & 0 & 0 \\ -1 & 2 & 0 \\ -2 & 0 & -1 \end{bmatrix} \qquad (2)$$

The value of the determinant is$|\boldsymbol{A}| = \begin{vmatrix} 1 & 0 & 0 \\ -1 & 2 & 0 \\ -2 & 0 & -1 \end{vmatrix} = -2 \neq 0$, so σ, S, v are linear independent, and can be used as basic variables.

According to the π theorem [4, 5, 6], in the six physical variables σ, S, v, F_1, F_2, F_3, formulate the dimensionless function of the remaining 3 variables and basic variables respectively π_1, π_2, π_3:

$$\pi_1 = \frac{F_1}{\sigma^{x_1} S^{y_1} v^{z_1}}$$

$$\pi_2 = \frac{F_1}{\sigma^{x_2} S^{y_2} v^{z_2}}$$

$$\pi_3 = \frac{F_1}{\sigma^{x_3} S^{y_3 1} v^{z_3}} \qquad (3)$$

And: $\pi_3 = f(\pi_1, \pi_2)$ (4)

For π_1: $[\mathrm{MLT}^{-2}] = [\mathrm{ML}^{-1}\mathrm{T}^{-2}]^{x_1}[\mathrm{L}^2]^{y_1}[T^{-1}]^{z_1}$ (5)

According to the principle of dimensional harmony:

$$1=x_1,\ 1=-x_1+2y_1,\ -2=-2x_1-z_1$$

$$x_1=1,\ y_1=1,\ z_1=0$$

$$\pi_1 = \frac{F_1}{\sigma S} \qquad (6)$$

Similarly, apply the principle of dimensionless harmony to π_2 and π_3, it can be obtained:

$$\pi_2 = \frac{F_2}{\sigma S},\quad \pi_3 = \frac{F_3}{\sigma S} \qquad (7)$$

Formula (8) can be obtained after putting π_1 π_2 and π_3 into formula (4):

$$F_3 = f(F_1, F_2) \qquad (8)$$

According to formula(8), it can be reached that the output is the preloading force of the binding belt F_3, and the input is ratchet force F_1 and the safety buckle force F_2,which can be measured through tension sensors. However, it is difficult and inconvenient to fasten the sensor at the end of the binding belt, considering that the measuring device is specially designed to measure the safety buckle force F_2, Therefore, the safety buckle force can be measured by quantitative control of the descending value of ratchet force.

In this way, the coupling model of binding belt tension F_3 and safety buckle force F_1 can be formulated after collecting a large number of data and fitting.

2.2 Analysis of Therbligs

Analysis of Therbligs was founded by the American engineer Frank Bunker Gilbreth. The core idea is to establish standardized operation flow through the analysis and reasonable adjustment of the elements of workers' action, so as to reduce the labor intensity and shorten the operation time, and ultimately, to improve the overall efficiency of the operation flow[7,8,9,10,11,12]. Therefore, applying Analysis of Therbligs to study the binding process of the cargo package, establishing a standardized operation flow, determining key steps key actions and quantifying can enhance the reproducibility of the binding operation of the cargo package, thereby improving the accuracy of the binding belt tension relaxation model.

Analysis of Therbligs is used to analyze and observe the operation of packing binding. Eight basic actions (Transport Empty, Transport Loaded, grasp, assemble, use, Disassemble, Release Load, Inspect) are selected to describe the whole binding process. On this basis, the operation process is divided into six phases: pre-tightening, pre-loading, step loading, applying safety buckle tension, ratchet unloading and removing ratchet, as shown in Figure 4.

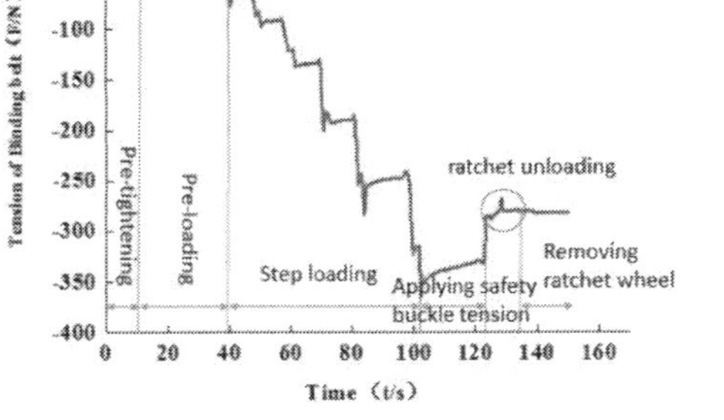

Figure 4. Loading process of binding belts

The six-phase Analysis of Therbligs table is shown in Table 1.

Table 1. Analysis of Therbligs of pre-tightening force of package

basic actions	phases					
	pre - tightening	pre - loading	step loading	applying safety buckle tension	ratchet unloading	removing ratchet
Transport Empty	√	√	√	√	√	√
Transport Loaded	√	√	×	×	×	√
grasp	√	√	√	√	√	√
assemble	√	√	×	×	×	×
use	×	√	√	×	×	×
Disassembl e	×	×	×	×	×	√
Release Load	√	√	√	√	√	√
Inspect	×	√	√	√	×	×

Note: √ indicates the actions required and × indicates not.

Through the analysis of each phase, the standard binding process of six-stage binding belt is formulated:

(1)Pre-tightening: tighten the belt safety buckle and the belt, the key action is to make it fit with the package;

(2)Pre-loading: the ratchet loading device is assembled through safety buckles at two ends of the binding belt, and pre-loading is carried out through the ratchet. The key action is to preload the ratchet force F_{10} to 70 N (the ratchet force is measured by a tension sensor connected in series in the ratchet loading device, as shown in Figure 2), and straighten the belt to make it tightly close to the package;

(3)Step loading: pulling the ratchet wheel to load the binding belt, limiting that only one ratchet tooth can be rotated at a time, then straightening the binding belt to make the binding belt fit with the cargo package; Repeat until the ratchet force reaches the predetermined value.

(4)Applying safety buckle tension: hold the binding belt lead and tighten slowly while observing whether the ratchet force F_{10} falls to the specified range (F_{11}), and then stop.

The optimal value of safety buckle tension obtained through extensive tests is shown in Table 2.

Table 2. The optimal of safety buckle force

Num ber	F_{10}	F_{11}	Number	F_{10}	F_{11}
1	-210	-80	7	-300	-130
2	-230	-90	8	-320	-135
3	-250	-100	9	-340	-140
4	-260	-110	10	-370	-150
5	-280	-120	11	-390	-160
6	-290	-125	12	-400	-165

Note: define the ratchet force before applying safety buckle force F_{10}, the binding belt tension before applying safety buckle force F_{30}; the ratchet force after applying safety buckle force F_{11}, the binding belt tension after applying safety buckle force F_{31}.

(5)Ratchet unloading: press the ratchet loading device buckle to unload the ratchet force.

(6)Remove the ratchet: remove the ratchet hook installed in the belt buckle and remove the ratchet loading device.

3 Data analysis and modeling

3.1 Data acquisition

Set up the test platform for package binding. The tension sensor is connected in the ratchet loading device and the end of the binding belt, and the ratchet force and the binding belt tension are recorded synchronously by the tension measuring instrument. The tension sensor selects the F2808 micro tension sensor from Tecsis company, matched with the Tecsis B6480 handheld tension tester and the software, as shown in Figure 5.

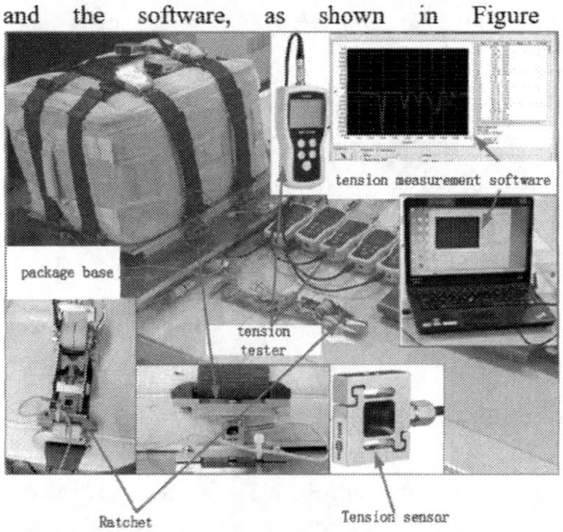

Figure 5. Test platform for package binding

Measurement of ratchet force and binding belt tension is synchronous started in package binding test. A series of data of ratchet force F_{10}, F_{11} and binding belt tension F_{30} and F_{31} can be acquired through the software.

3.2 Steps of coupling modeling

According to the packing process, the coupling analysis of ratchet force and binding belt tension can be divided into 3 steps:

The first step: establishing the coupling model of F_{30} - F_{10};

The second step: establishing the coupling model of F_{31} - F_{30};

The third step: establishing the coupling model of F_{31} - F_{10} by using function transfer relationship.

During the steps, loading according to the standard binding process obtained by the front element analysis and applying safety buckle tension according to Table 2 are the key point to improve the accuracy of ratchet force and binding belt tension coupling model.

3.3 Modeling of binding belt tension

The value of binding belt tension and the ratchet force are in a linear relationship, accordingly a linear data fitting method based on the least square method is used to establish the model[13,14,15].

3.3.1 Establishing the coupling model of F_{30}- F_{10}

Firstly, the ratchet force data F_{10} and the binding tension data F_{30} obtained from the tension measurement software are used for singular value filtering. Then, select F_{10} as the horizontal axis, the corresponding F_{30} as the longitudinal axis. The F_{30} - F_{10} measured and fitting curves for each binding belt are available through liner fitting, as shown in Figure 6.

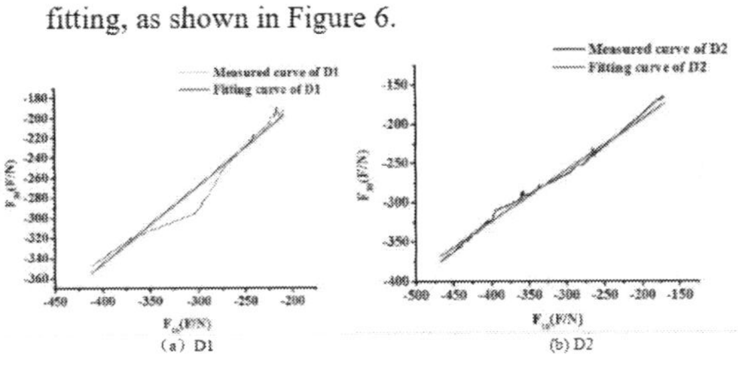

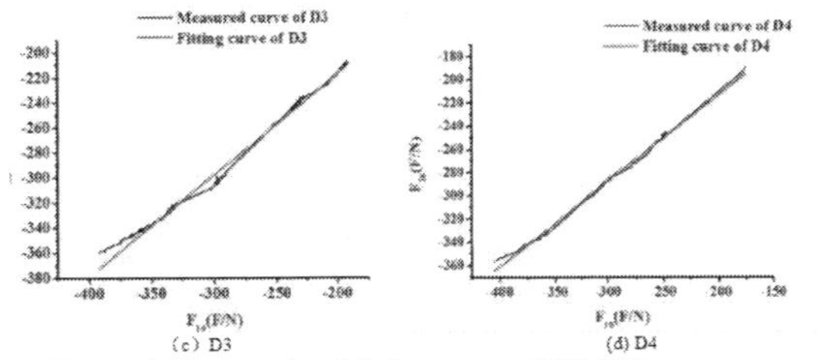

Figure 6. Measured and fitting curve of F30-F10

Corresponding fitting function relations are shown in Table 3:

Table. 3 Coupling formula of F_{30}- F_{10}

Number	linear fitting formula	Correction coefficient
D1	F_{30}=-35+0.77736 F_{10}	0.97328
D2	F_{30}=-63+0.65327 F_{10}	0.99317
D3	F_{30}=-55+0.80969 F_{10}	0.98078
D4	F_{30}=-66+0.73751 F_{10}	0.99721

As can be seen from Table 3, the correction coefficient of the fitting curve of the four binding belts of the package is greater than 0.97, which shows that the fitting degree of F_{30} - F_{10} linear fitting equation is very high and the binding belt tension can be predicted before applying the safety buckle force with a small error.

3.3.2 Establishing the coupling model of F31 – F30

Similarly, the binding belt tension before applying the safety buckle force F_{30} is selected as the horizontal axis, the binding belt tension after applying the safety buckle force F_{31} is selected as the longitudinal axis, the scatter plots and fitting lines of each belt are shown in Figure 7.

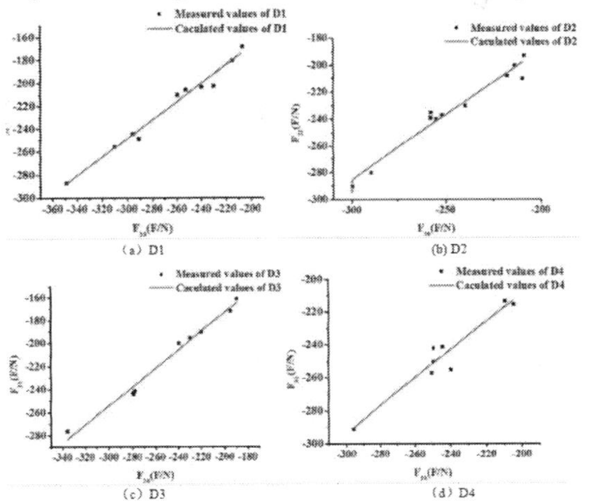

Figure 7. Scatter and fitting curve of F_{31}-F_{30}

Corresponding fitting function relations are shown in Table 4.

Table 4. Coupling formula of F_{31}-F_{30}

Number	linear fitting formula	Correction coefficient
D1	F_{31}=-10.94+0.81118 F_{30}	0.98005
D2	F_{31}=-6.765+0.8051 F_{30}	0.97639
D3	F_{31}=-37.93+0.8529 F_{30}	0.91395
D4	F_{31}=3.978+0.9636 F_{30}	0.95916

Similarly, as can be seen from Table 4, the correction coefficient of the coupling model of F_{31} and F_{30} is above 0.91, which is in a high fitting degree.

3.3.3 Establishing the coupling model of F31-F10

According to the function transfer relation, the F_{30}-F_{10} fitting equation obtained in Table 3 is brought into the F_{31}-F_{30} fitting equation in Table 4.And then, the coupling formula of F_{31}- F_{10} can be obtained, as shown in Table 5.

Table 5. Coupling formula of F_{31}- F_{10}

Number	Coupling formula
D1	F_{31}=-39.33+0.63058 F_{10}
D2	F_{31}=-57.49+0.52595 F_{10}
D3	F_{31}=-84.84+0.69058 F_{10}
D4	F_{31}=-59.62+0.71066 F_{10}

In this way, according to the coupling model of the ratchet force and the binding belt tension established in Table 5, the binding belt tension can be predicted through the ratchet force measured.

4 Verification of coupling model

The measured data of ratchet force under the range of the model are checked and verified by coupling model of F_{10} and F_{31} shown in Table 5. As a result, the scatter plot of calculated F_{31} and measured F_{31} is shown in Figure 8.

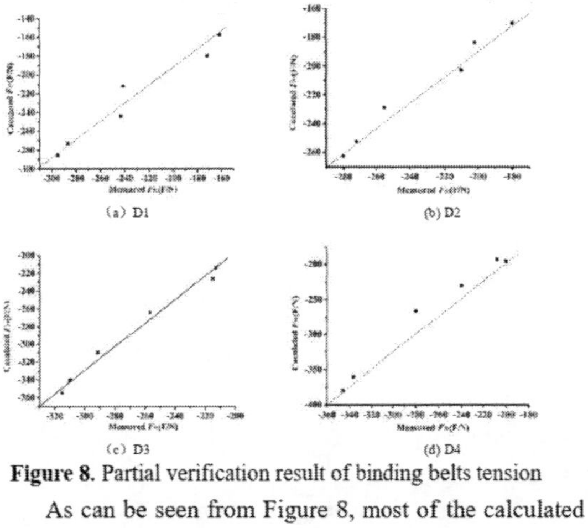

Figure 8. Partial verification result of binding belts tension

As can be seen from Figure 8, most of the calculated values are in good agreement with the measured values, and the errors are mostly less than 5%. The coupling model of F_{10} and F_{31} can be applied to the quantitative loading in the actual working condition.

5 Conclusions

In this paper, the loading of the cargo package binding in cargo spaceship is analyzed deeply. The coupling model of the binding belt tension and ratchet force is established, and the quantitative loading is realized.

1) Based on dimensional analysis, the basic dimensions and basic variables are selected, and a semi-quantitative model between ratchet force and binding belt tension is established.

2) Based on the Analysis of Therbligs, the binding process of cargo package is divided into six phases: pre-tightening, pre-loading, step-loading, applying safety buckle tension, ratchet unloading and ratchet removing. A standard binding process has been established to enhance the repeatability of the binding operation of cargo packages, so as to improve the accuracy of the coupling model of the ratchet force and binding belt tension.

3) The binding test platform is built, which can acquire a lot of ratchet force data and binding belt tension data in real-time; Based on linear regression of least square method, the ratchet force and binding belt tension coupling model of 4 belts were established.

4) The coupling model was checked and verified. The calculated values and measured values were in good agreement, and the errors were mostly less than 5%. Consequently, the coupling model can be applied to the quantitative loading of the cargo package binding in cargo spaceship.

References

1. Y HU. Discussion about Dimension Analysis and its Application [J].JCJEVC, (2006), **23(2)**:52-54.
2. Y Y SHAO. Comprehension and Application for Dimensional Analysis and Rule π [J].JDIT, (2010), **17(3)**:106-109.
3. X H XU, J Z HUANG. Application of dimensional analysis in mechanics [J].JMDJU, (2008), **17(1)**:122-124.
4. B H SUN. Dimensional Analysis and Applications [J].PE, (2016), **26(6)**:11-20.
5. A G XIE, H NIE. Study on Application Theory of

the π Theorem for Dimension Analysis [J].JASIIST, (2007), **20(5)**:4-9.
6. Y S DAI, L YANG. Applications of dimensional analysis and the rule πin chemical engineering-process improvement of teaching in the course of *Principle of Chemical Engineering* [J].JNIT, (2010),**9(1)**:11-16.
7. L ZHANG, M CAO,H G PENG, S L ZHOU. Application of therbligs analysis in assembly line balancing [J].JZJUSE, (2010), **22(2)**:89-94.
8. T GU, B H ZHOU. Application of action analysis method in improvement of cylinder block line balance [J].MM, (2009), **47**:54-57.
9. C WU, J HONG, X Y ZHU. Analysis and Improvement on the Operation Process of Turbocharger Assembly Line Based on Therbligs Analysis[J].VE, (2015),**20**:223-225.
10. X T HU, J ZHAO. The Research of Activity Analysis Design Method Based on the Theory of Motion Analysis [J]. DIMEP, (2011), **24(1)**:59-61.
11. Y Z ZHU, H F YUAN, L SUN. Analysis by the Color Culture in Huian Women's Costume [J].JCHC, (2015), **17(3)**:84-90.
12. R Slama, H Wannous, M Daoudi. 3D human motion analysis framework for shape similarity and retrieval [J].IVC, (2014), **32**:131-154.
13. Q LIU, Partial Least-squares Regressive Analysis and Modeling for Tool Wear [J].JBHU, (2000), **26(4)**:457-460.
14. X X SONG. A General Data Fitting Method Based on MATLAB [J].JSXDTU, (2014), **30(4)**:1-4.
15. L Q ZOU. Least - Square - Multiplication Principle and its Simple Application [J].STI, (2010), **23**:282-283

A Nonlinear Second-order Spacecraft Attitude Tracking Control Model for Control System Stabilization

Xiaoyi Wang[1]
[1]*Center for Control Theory and Guidance Technology, Harbin Institute of Technology, Harbin, China*

Abstract. A control model for the direct parameter approach for spacecraft attitude tracking is presented in this paper. First of all, the spacecraft attitude tracking control model is built up by the error equation of the second-order nonlinear quaternion-based attitude system. A problem of control system stabilization is raised based on the control model. Compared with other control models, the second-order can offer the advantages of no- approximation and clear control states. The basic spacecraft control model has to focus on to the two variables which are angular rate and attitude quaternion, however, the new attitude control problem is only with respect to one variable which is the spacecraft attitude quaternion. Therefore, the second-order model is simpler and clear than basic first-order model.
Keywords: spacecraft attitude tracking; nonlinear system; quaternion representation; second-order control model

1 Introduction

As a significant part of spacecraft navigation control, spacecraft attitude control has been a hot problem for several years. Many different methods are applied on the problem of spacecraft attitude control as in [1-6] and [9-11]. For instance, Bang designed a new sliding model control method which can deal with flexible spacecraft attitude maneuver problem in [5]. Singh utilized adaptive output feedback control method to track the flexible spacecraft attitude in [6]. Additionally, Guan raised a direct parametric approach to stabilize the nonlinear second-order spacecraft attitude model, and the controller can turn the original model into a stable linear constant system certain degrees of freedom in [1].

According to the research background of spacecraft attitude control, it is suggested that spacecraft attitude tracking needs further research and has more profound applications. Compared with Euler representation, quaternion representation can describe large-angle attitude tracking without singular points which should appear when the changing of Euler angle is bigger than 90°. Moreover, this paper raises a second-order system based on the spacecraft attitude dynamical and kinematical differential equations. Although most of methods would like to deal with the spacecraft attitude control problem by first-order models, this paper takes advantage of the second-order model. Furthermore, the second-order control model is with no-approximation, which means it can precisely describe the process of spacecraft attitude tracking. Moreover, the second-order model is only with respect one control variable which is simpler than first-order model with two control variables. In addition, the second-order model has broad application prospects, for instance, Duan offered a direct parametric approach which can turn a nonlinear second-order system into a stable linear constant system by the parametric controller as in [1-3], and [7-8].

2 Basic spacecraft attitude tracking problem

According to the attitude dynamics and kinematics of a rigid spacecraft in the inertial system, the error attitude dynamics of a rigid spacecraft relative to a mobile tracking target in the inertial system is given by

$$\begin{aligned} J\dot{\omega}_e &= -\omega \times J\omega \\ &\quad + J\left(\omega \times C(q_e)\omega_r - C(q_e)\dot{\omega}_r\right), \\ &\quad + u + d \end{aligned} \tag{1}$$

where

$$J = diag\left(J_x, J_y, J_z\right) \tag{2}$$

is the rotating inertia matrix of the spacecraft,

ω is the angular rate vector, ω_e is the error angular rate vector relative to the mobile tracking target, ω_r is the angular rate vector for target, u is the input control torque vector, and d is the disturbance torque vector. Besides, ω, ω_e, ω_r, u and d belong to $\mathbf{R}^{3\times1}$.In addition, the $\mathbf{C}$ (q_e) is the transition matrix and its expression is

$$C(q_e) = \left[C_1(q_e) \quad C_2(q_e) \quad C_3(q_e)\right], \tag{3}$$

where

$$C_1(q_e)=\begin{bmatrix} q_{e1}^2-q_{e2}^2-q_{e3}^2+q_{e0}^2 \\ 2(q_{e1}q_{e2}-q_{e3}q_{e0}) \\ 2(q_{e1}q_{e3}+q_{e2}q_{e0}) \end{bmatrix}, \tag{4}$$

$$C_2(q_e)=\begin{bmatrix} 2(q_{e1}q_{e2}+q_{e3}q_{e0}) \\ -q_{e1}^2+q_{e2}^2-q_{e3}^2+q_{e0}^2 \\ 2(q_{e2}q_{e3}-q_{e1}q_{e0}) \end{bmatrix}, \tag{5}$$

$$C_3(q_e)=\begin{bmatrix} 2(q_{e1}q_{e3}-q_{e2}q_{e0}) \\ 2(q_{e2}q_{e3}+q_{e1}q_{e0}) \\ -q_{e1}^2-q_{e2}^2+q_{e3}^2+q_{e0}^2 \end{bmatrix}, \tag{6}$$

and

$$q_e=\begin{bmatrix} q_{e0} & q_{e1} & q_{e2} & q_{e3} \end{bmatrix}^T . \tag{7}$$

The ω_e can be expressed by

$$\omega_e=\omega-C(e)\omega_r . \tag{8}$$

Then the kinematics of a rigid spacecraft in the inertial system based on error quaternion is given by

$$\begin{bmatrix} \dot q_{e0} \\ \dot q_{e1} \\ \dot q_{e2} \\ \dot q_{e3} \end{bmatrix}=\frac{1}{2}\begin{bmatrix} -q_{e1} & -q_{e2} & -q_{e3} \\ q_{e0} & -q_{e3} & q_{e2} \\ q_{e3} & q_{e0} & -q_{e1} \\ -q_{e2} & q_{e1} & q_{e0} \end{bmatrix}\begin{bmatrix} \omega_{ex} \\ \omega_{ey} \\ \omega_{ez} \end{bmatrix}, \tag{9}$$

where ω_e is the error angular rate vector relative to the mobile tracking target, and q_e is the error quaternion relative to the target.

To solve the attitude problem, a suitable controller u is given by

$$u=f(\omega_e, q_{e0}, q_{e1}, q_{e2}, q_{e3}), \tag{10}$$

which can make the close-loop spacecraft attitude system stable.

When the system is stable, which means

$$\begin{cases} \lim_{t\to\infty}\omega_e=0, \lim_{t\to\infty}q_{e0}=1, \lim_{t\to\infty}q_{e1}=0 \\ \lim_{t\to\infty}q_{e2}=0, \lim_{t\to\infty}q_{e3}=0 \end{cases}, \tag{11}$$

the spacecraft will successfully achieves the attitude tracking goal.

3 Spacecraft attitude tracking problem based on the second-order control problem

In this section, a second-order spacecraft attitude control model is built by attitude dynamical equations and kinematical equations. Additionally, that model is described by quaternion and without any approximation. The problem of quaternion-based spacecraft attitude tracking is raised based on the attitude control model. Generally, the error quaternion q_e has the constraint which is

$$q_{e0}^2+q_{e1}^2+q_{e2}^2+q_{e3}^2=1 . \tag{12}$$

Based on the (12) a simple variable e can be expressed as

$$e=\begin{bmatrix} e_1 & e_2 & e_3 \end{bmatrix}^T=\begin{bmatrix} q_{e1} & q_{e2} & q_{e3} \end{bmatrix}^T . \tag{13}$$

Additionally, the q_{e0} can be denoted as

$$q_{e0}=e_0=\sqrt{1-(e_1^2+e_2^2+e_3^2)} . \tag{14}$$

With the expression of e, the (5) can be written as

$$\dot e=\frac{1}{2}T(e)\omega_e=\frac{1}{2}\begin{bmatrix} e_0 & -e_3 & e_2 \\ e_3 & e_0 & -e_1 \\ -e_2 & e_1 & e_0 \end{bmatrix}\omega_e . \tag{15}$$

Taking derivative of the equation (12) with respect to time t, it yields

$$\ddot e=\frac{1}{2}\frac{dT(e)}{dt}\omega_e+\frac{1}{2}T(e)\dot\omega_e . \tag{16}$$

The expression for ω_e can be derived from the (12), it yields

$$\begin{aligned}\omega_e&=2(I_3-ee^T)^{-1}T^T(e)\dot e \\ &=2\Phi(e)T^T(e)\dot e\end{aligned}, \tag{17}$$

where

$$\Phi(e)=(I_3-ee^T)^{-1}=\begin{bmatrix} \Phi_1(e) \\ \Phi_2(e) \\ \Phi_3(e) \end{bmatrix} . \tag{18}$$

Thus the variables of the ω_e can be expressed as

$$\begin{cases} \omega_{ex}=2\Phi_1(e)T^T(e)\dot e \\ \omega_{ey}=2\Phi_2(e)T^T(e)\dot e \\ \omega_{ez}=2\Phi_3(e)T^T(e)\dot e \end{cases} . \tag{19}$$

The next step is to summarize the second-order spacecraft control model through the (1) and (16). First of all, the expression for $\dot\omega_e$ can be derived by (1) as

$$\begin{aligned}\dot\omega_e&=J^{-1}[\omega_e\times(J\omega_e)] \\ &-J^{-1}\{\omega_e\times[JC(e)\omega_r]\} \\ &-J^{-1}\{[C(e)\omega_r]\times[J\omega_e]\} \\ &-J^{-1}\{[C(e)\omega_r]\times[JC(e)\omega_r]\} \\ &+[\omega_e\times(C(e)\omega_r)] \\ &-C(e)\dot\omega_r+J^{-1}(u+d)\end{aligned} . \tag{20}$$

When these expressions are denoted as,

$$\Gamma(J,e,\dot e)=\begin{bmatrix} 0 & 0 & \Gamma_c \\ \Gamma_a & 0 & 0 \\ 0 & \Gamma_b & 0 \end{bmatrix}, \tag{21}$$

$$\begin{cases} \Gamma_a=\dfrac{J_z-J_x}{J_y}\Phi_3(e)T^T(e)\dot e \\ \Gamma_b=\dfrac{J_x-J_y}{J_z}\Phi_1(e)T^T(e)\dot e , \\ \Gamma_c=\dfrac{J_y-J_z}{J_x}\Phi_2(e)T^T(e)\dot e \end{cases} \tag{22}$$

$$G=\begin{bmatrix} G_1 \\ G_2 \\ G_3 \end{bmatrix}=\begin{bmatrix} [JC(e)\omega_r]_x \\ [JC(e)\omega_r]_y \\ [JC(e)\omega_r]_z \end{bmatrix}, \tag{23}$$

$$H=\begin{bmatrix}H_1\\H_2\\H_3\end{bmatrix}=\begin{bmatrix}\left[C(e)\omega_r\right]_x\\\left[C(e)\omega_r\right]_y\\\left[C(e)\omega_r\right]_z\end{bmatrix}. \quad (24)$$

and the model can be written as

$$\ddot{e}=\begin{pmatrix}\frac{dT}{dt}\Phi(e)+2T\Gamma(J,e,\dot{e})\Phi(e)\\-T\begin{bmatrix}H_2\Phi_3-H_3\Phi_2\\H_3\Phi_1-H_1\Phi_3\\H_1\Phi_2-H_2\Phi_1\end{bmatrix}\\+J^{-1}T\begin{bmatrix}H_2J_z\Phi_3-H_3J_y\Phi_2\\H_3J_x\Phi_1-H_1J_z\Phi_3\\H_1J_y\Phi_2-H_2J_x\Phi_1\end{bmatrix}\\+J^{-1}T\begin{bmatrix}G_3\Phi_2-G_2\Phi_3\\G_1\Phi_3-G_3\Phi_1\\G_2\Phi_1-G_1\Phi_2\end{bmatrix}\end{pmatrix}T^T\dot{e} -\frac{1}{2}T\left\{J^{-1}\left[H\times(JH)\right]+C(e)\dot{\omega}_r\right\}+\frac{1}{2}TJ^{-1}(u+d). \quad (25)$$

In addition, the spacecraft attitude tracking model (25) can be written in the second-order form as

$$A_2(\theta,e,\dot{e})\ddot{e}+A_1(\theta,e,\dot{e})\dot{e}+A_0(\theta,e,\dot{e})e+\xi(\theta,e,\dot{e},t)=B(\theta,e,\dot{e})u, \quad (26)$$

where

$$A_2(\theta,e,\dot{e})=I_3, \quad (27)$$

$$A_1(\theta,e,\dot{e})=-\begin{pmatrix}\frac{dT}{dt}\Phi(e)+2T\Gamma(J,e,\dot{e})\Phi(e)\\-T\begin{bmatrix}H_2\Phi_3-H_3\Phi_2\\H_3\Phi_1-H_1\Phi_3\\H_1\Phi_2-H_2\Phi_1\end{bmatrix}\\+J^{-1}T\begin{bmatrix}H_2J_z\Phi_3-H_3J_y\Phi_2\\H_3J_x\Phi_1-H_1J_z\Phi_3\\H_1J_y\Phi_2-H_2J_x\Phi_1\end{bmatrix}\\+J^{-1}T\begin{bmatrix}G_3\Phi_2-G_2\Phi_3\\G_1\Phi_3-G_3\Phi_1\\G_2\Phi_1-G_1\Phi_2\end{bmatrix}\end{pmatrix}T^T, \quad (28)$$

$$A_0(\theta,e,\dot{e})=0_3, \quad (29)$$

$$B(\theta,e,\dot{e})=\frac{1}{2}T(e)J^{-1}, \quad (30)$$

$$\xi(\theta,e,\dot{e},t)=\frac{1}{2}T\begin{Bmatrix}J^{-1}\left[H\times(JH)\right]\\+C(e)\dot{\omega}_r\\-J^{-1}d\end{Bmatrix}. \quad (31)$$

To solve the second-order spacecraft attitude tracking problem defined by (25), a suitable controller u is given by

$$u=f(e_1,e_2,e_3,t), \quad (32)$$

which can make the close-loop spacecraft attitude system stable.

When the system is stable, which means

$$\begin{cases}\lim_{t\to\infty}e_1=0,\lim_{t\to\infty}e_2=0,\lim_{t\to\infty}e_3=0\\\lim_{t\to\infty}\dot{e}_1=0,\lim_{t\to\infty}\dot{e}_2=0,\lim_{t\to\infty}\dot{e}_3=0\end{cases}, \quad (33)$$

the spacecraft will successfully achieves the attitude tracking goal.

4 Conclusion

A control model is built by the error attitude dynamics and attitude kinematics for spacecraft in this paper. This model is based on the quaternion representation which can be suitable for the spacecraft large attitude angular tracking. Additionally, this model is a nonlinear second-order no-approximate attitude system. As mentioned earlier, the control model can precisely describe the error attitude of spacecraft relative to mobile target in the space. Compared with the first-order model, the second-order model is only with respect one variable which is the spacecraft attitude quaternion. Thus, the second-order model can be considered simpler and clearer than first-order model. Finally, the model is expressed in the general second-order quasi-linear system form, which has profound application prospects on nonlinear control and linear control methods.

Acknowledgement

I am greatly indebted to my supervisor, Professor Guangren Duan, for his valuable instructions and suggestions on my paper and thesis, as well as his beneficial approach provides the theory basis for this paper.

Reference

1. Duan, Guang Ren. Quaternion-based satellite attitude control—A direct parametric approach. *International Conference on Control, Automation & Systems IEEE*, pp. 935-941, (2014)
2. Duan, Guang Ren. Satellite attitude control—A direct parametric approach. *Intelligent Control and Automation IEEE*,,pp. 3989 - 3996, (2014)
3. Duan, Guang Ren. Direct parametric control of fully-actuated second-order nonlinear systems—The normal case. 2014 *China Control Conference (CCC) 33rd Chinese IEEE*, pp. 2406-2413, (2014)
4. Tewari,Ashish. Spacecraft Attitude Control. *Aerospace Technologies Advancements*. pp. 357-390, (2010)
5. Bang, Hyochoong, C. K. Ha, and H. K. Jin. Flexible spacecraft attitude maneuver by application of sliding mode control. *Acta Astronautica*, pp. 841-850, (2005).
6. Singh, Sahjendra N., and R. Zhang. Adaptive output feedback control of spacecraft with flexible appendages by modeling error compensation. *Acta Astronautica* , **54**, pp. 229-243,(2004)
7. Duan, Guang Ren, and B. Zhou. Solution to the second-order Sylvester matrix equation

MVF2+DVF+KV=BW. *IEEE Transactions on Automatic Control*. **51**, pp. 805-809, (2006)
8. Duan, Guang Ren. Generalized Sylvester Equations: Unified Parametric Solutions. *Crc Press*, (2015)
9. Gao, Chunyan, Q. Zhao, and G. Duan. Robust actuator fault diagnosis scheme for satellite attitude control systems. *Journal of the Franklin Institute*, **9**, pp. 2560-2580, (2013)
10. Lu, Kunfeng, and Y. Xia. Adaptive attitude tracking control for rigid spacecraft with finite-time convergence. *Automatica*. **49**, pp. 3591-3599, (2013)
11. Yoon, Hyungjoo, and P. Tsiotras. Spacecraft Adaptive Attitude and Power Tracking with Variable Speed Control Moment Gyroscopes. *Journal of Guidance Control & Dynamics* . **25**, pp. 1081-1090, (2002)

Topological Optimization of Aircraft Frame Structures with the Variable Density Method

Li ying-lei[1], Cao zong-jie[1] and Wang Zi-li[2]

[1] *Aviation University of Air Force, School of Aviation Operations and Services, Changchun, 130022, Jilin, China*

[2] *Aviation University of Air Force, Foreign Training Group, Changchun, 130022, Jilin, China*

Abstract. In this paper, a variable density topological optimization method is derived on the basis of Lagrange function, and the RAMP interpolation model is selected to optimize the frame structure of aircrafts with variable density method. For an example, the ordinary frame structure of the domestic planes is taken to illustrate validity of the presented method. The numerical model of the aircraft frames is obtained. The optimal design analysis of the model structural distribution and the weight loss requirement of the aircraft frame structures are realized. In conclusion, the structural distribution law of the topology optimization is summarized according to the topology optimization structures.

1 Introduction

With the development of aviation, the aircraft with the better comprehensive performances is adapt to the demand of modern wars [1]. Therefore, not only the aerodynamic performance of aircraft structures is necessary [2], the implementation of aviation structural lightweight also has significant practical application value in the structural design [3]. Topology optimization is an optimal method to achieve optimal material layout and lightweight structure when meeting the design constraints [4]. Variable density method, such as the topological optimization methods, is widely used in the topological optimization design of structural continuum [5]. For example, Airbus has applied the topological optimization technique to A380 aircraft on leading edge floor and fuselage door deep floor to achieve a weight reduction of one ton [6]. In the design of the F-35 fighter, the whole structure of the aircrafts was topologically optimized and 25 percent weight reduction of aircraft structures was carried out by Lockheed Martin [7]. In the design of the Boeing 787 aircrafts, a weight reduction of about 25%-45% compared to the corresponding edge floor part of Boeing 777 aircrafts can be achieved with the topological optimization methods [8]

With the topological optimization method based on variable density method, the material distribution of a certain type of domestic aircraft frame structures according to the characteristics about the ordinary frame of the aircrafts is optimally designed in this paper. The best material distribution of the structures is obtained and the weight reduction. to the maximum optimization of the structural materials is realized under the constant loads.

2 Principle of Variable Density Method

The variable density method is also called the pseudo density method [9]. It is assumed that there is a functional relationship between the material density of the structure and several physical parameters of the material. The density of the materials can be obtained between the intervals [0,1]. The pseudo density about the continuous values arbitrarily in [0,1] of the topological optimization structures is obtained, there are three different situations. If the unit is an empty hole, then the pseudo density of the unit is 0. If the unit is an entity, its pseudo density unit should be 1, and if the pseudo density near 0.5, then introducing the power exponent to penalize this intermediate value, and to make the function value incline to the end value 0 or 1 as much as possible. The pseudo density formulation [10] of the design variables can be expressed as: $\rho = X_e \rho_0$ where X_e is the relative density of each unit; ρ_0 is the intrinsic density of each unit; ρ is the density of each unit. When X_e is equal to 1, which unit of the materials will be preserved. When X_e is equal to 0, which unit of the materials will be removed. Because ρ is not a continuous function, a penalty factor p is given in order to let the function called as K_e continue. $K_e = (X_e)^p K_0$ .Where K_0 is the element stiffness of original structures; K_e is the element stiffness of deformation structures. The structural deformation energy is: $C = F^T U = U^T K U$.

where C represents compliance of structures; F is the node load vector and U is the node displacement vector.The mathematical topological optimization condition [18] of aircraft structures can be built as:

$$\begin{cases} Find: \ X = (X_1, \ X_2, \ \cdots X_n) \\ Minimize: \ C = F^T U \\ Subject\ to: \ f = \dfrac{V - V_1}{V_0} \\ 0 < X_{\min} \le X_e \le X_{\max} \\ F = KU \end{cases} \tag{1}$$

where V_0 is the initial volume of material; V is the effective volume of material at various design point when the design variable ρ is less than 1; V_1 is the material volume when the design variable ρ is equal to 1; f is a percentage of material used; $X_{\min}$ and $X_{\max}$ are the lower and upper limit of the unit relative density, respectively.

3 Algorithm topological optimization with variable density method

Usually Lagrange multiplier method is adopted to derive the mathematical model of the frame structure of a plane[10]. When $X_{\max}$=1 and $X_{\min}$=0.01 , Lagrange function in the formula (1) can be changed as:

$$L = C\lambda_0(V - fV_0 - V_1) + \lambda_1^T(KU - F) + \sum_{e=1}^{N}\lambda_2^e(X_{\min} - X_e) + \sum_{e=1}^{N}\lambda_3^e(X_e - X_{\max}) \tag{2}$$

where λ is Lagrange multiplier. Taken the derivation of objective function:

$$\frac{\partial L}{\partial X_e} = \frac{\partial U^T}{\partial X_e}KU + U^T\frac{\partial K}{\partial X_e}U + U^TK\frac{\partial U}{\partial X_e} + \lambda_0 V_e + \lambda_1^T\left(\frac{\partial K}{\partial X_e}U + K\frac{\partial U}{\partial X_e}\right) \tag{3}$$

The structural equilibrium equation of aircraft structures is:

$$F = KU \tag{4}$$

After the derivation of equation(4), we have:

$$\frac{\partial K}{\partial X_e}U + K\frac{\partial U}{\partial X_e} = 0\left(e = 1, 2, 3, \cdots, N\right) \tag{5}$$

According to equation(4) and (5), the equation can be obtained as follows:

$$\frac{\partial L}{\partial X_e} = U^T\frac{\partial K}{\partial X_e}U + \lambda_1^T\frac{\partial K}{\partial X_e} + \frac{\partial U}{\partial X_e}\left(2U^T + \lambda_1^T K\right) + \lambda_0 V_e \tag{6}$$

Since λ_1^T can be an arbitrary value, we can obtain the equation from equation (5).

$$-P\left(X_e\right)^{P-1}U_e^T K_0 U_e + \lambda_0 V_e = 0 \tag{7}$$

Combination of equation (5), multiply X_e^i is

$$PU_e^T K_0 U_e X_e^{i-1} = \lambda_0 V_e X_e^i \tag{8}$$

since

$$U_e^T K U_e X_e^{i-1} = \theta_e \tag{9}$$

From (8), we can obtain:

$$X_e = \left[\frac{P\theta_e}{\lambda_0 V_e}\right]^{\frac{1}{i}}\left(X_{\min} < X_e < 1\right) \tag{10}$$

Combination of equation (9) and (10), where j represents the structural units with constant density, j=1, 2, 3, …, n, and k represents the structural unit when $X_e < 1$. The iterative equation of X_e is :

$$\begin{cases} X_e^{m+1} = \left[\dfrac{\theta_e}{V_e}\right]^{\frac{1}{i}}\left\{\dfrac{fV_0 + \sum_{j=1}^{k} V_j X_{\min}}{\sum_{j=1}^{k} V_j\left[\dfrac{\theta_j}{V_j}\right]^{\frac{1}{i}}}\right\}\left(X_{\min} < X_e < 1\right) \\ X_e^{m+1} = 1\left(X_e \ge 1\right) \\ X_e^{m+1} = X_{\min}\left(X_e \le X_{\min}\right) \end{cases} \tag{11}$$

4 Numerical modelling and analysis of aircraft frame structure

A certain type of frame structures in domestic aircrafts is selected in this paper, as shown in Figure 1. Specific design parameters are shown in Table 1. Stiffener of the fracture is equal to 5mm. The frame structure is made of aluminum alloy (type 7075-T7451). Because of its super high strength, the 7075 aluminum alloy is commonly used as aviation manufacturing material, such as high-strength bearing parts like aircraft wing beam, frame and stiffener. According to GB/T/16475-1996, the material properties of the 7075-T7451 aluminum alloy are shown in Table 2. A certain type of three-dimensional geometric aircraft frame is established, as shown in Figure 2. After the completion of the geometric modeling, the workpiece information is transmitted to the preprocessor of the CAE platform in the form of x_t data, as shown in Figure 3. The physical parameters of the frames include the maximum length, the maximum height and the maximum width. The maximum length of the frames is equal to1800mm. The maximum height of the frames is equal to 1800mm. The maximum width of the frames is equal to 60mm. The volume of the frames is $6.5357\times10^{-3}m^3$and the weight of the frames is 51.305kg.

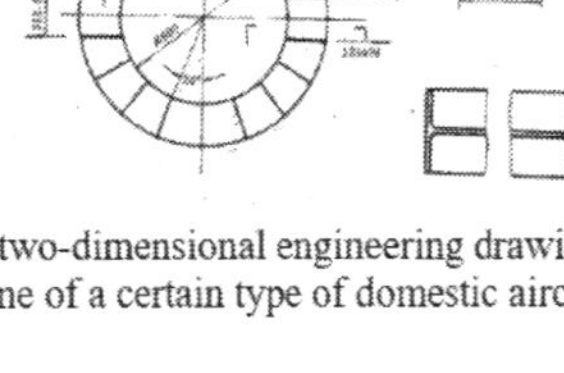

Figure 1. A two-dimensional engineering drawing of an ordinary frame of a certain type of domestic aircraft.

Table 1. The design parameters of aircraft structures alloy

Outer race radius (mm)	Inner race radius (mm)	Frame thick-ness (mm)	Annulus thick-ness (mm)	Stiffener thickness (mm)	Fracture width of the frame (mm)
900	60	2.5	3	3	312.6

Table 2. The material properties of the 7075-T7451 aluminium

Parameter type	Modulus of elasticity (MPa)	Poisson ratio	Strength limit (MPa)	Yield strength(MPa)	Density (g/mm^3)
7075-T7451	7.2×10^4	0.33	510	470	2.83×10^{-3}

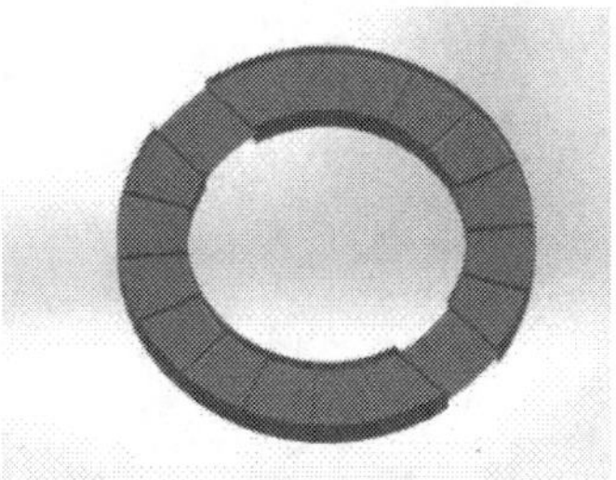

Figure 2. A three-dimensional geometric modeling of an ordinary frame of a certain type of domestic aircraft.

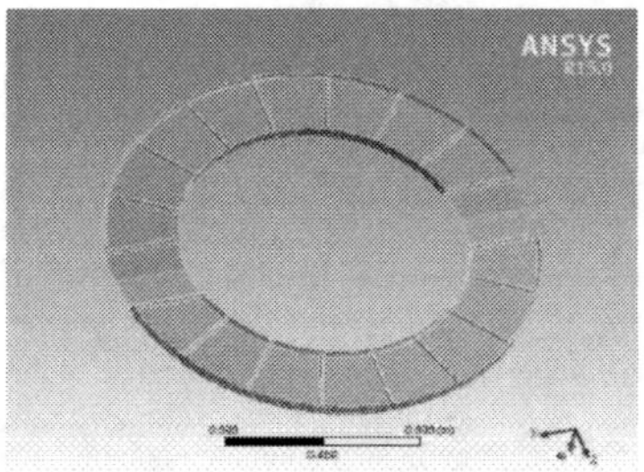

Figure 3. A three-dimensional model of an ordinary frame of a certain type of domestic aircraft.

The partition method selection of frame structure is based on the automatic partition method of entity model and the correlation of grid is set to 0~80. Based on the selected physical field characteristics, correlation controls the grid size and to adjust the global grid. The minimum boundary size of the grid is 1.1mm. As shown in Figure 4, the correlation of the frame structure grid is 80, as shown in Figure 5, the correlation of the frame structure grid is 0.

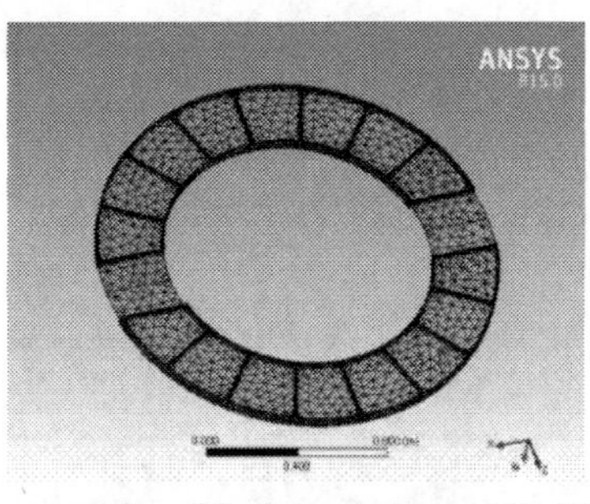

Figure 4. The correlation of the frame structure grid is 80.

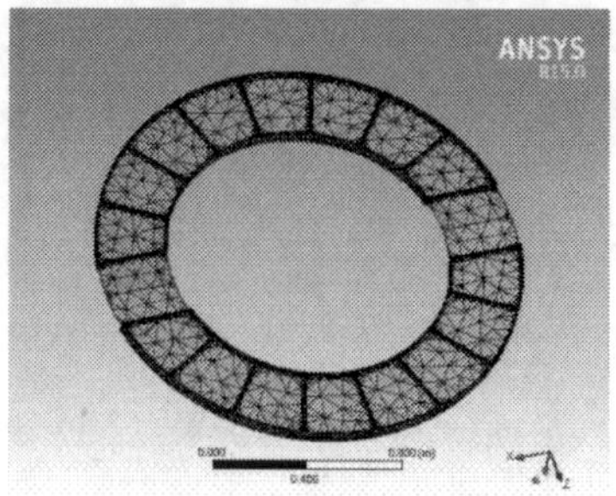

Figure 5. The correlation of the frame structure grid is 0.

In engineering practice, when the grid correlation is 0 or 80, the relative error of the workpiece is generally below 3%. The relative error will be reduced with the increase of the grid correlation. This shows that the numerical simulation results have no dependencies on the grid quality because the number of unit and node is more when the grid correlation is greater, then the operation will be more complex, So subsequent analysis in this paper, we take 0 as the correlation of the grid in order to meet the calculation need.

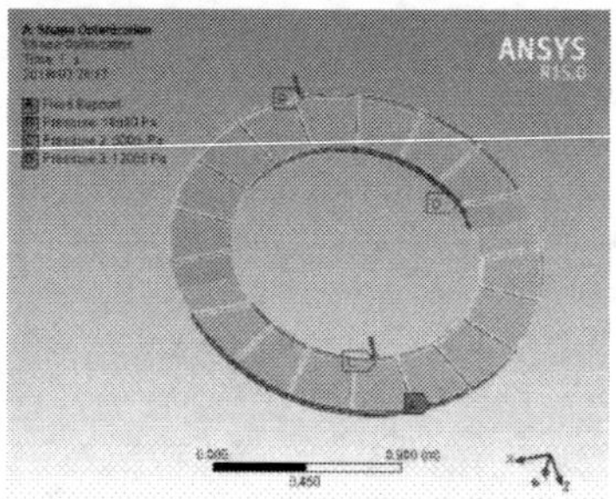

Figure 6. The boundary condition and load of the frame.

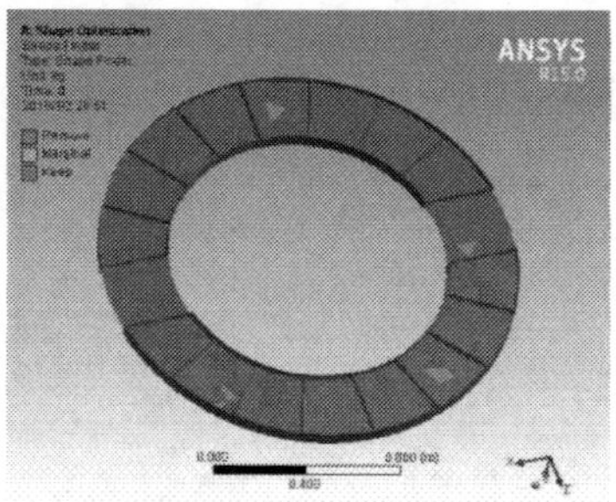

Figure 7. Frame topological optimization nephogram/

The function of the ordinary frame main keeps the shape of the fuselage and provides elastic support to the compressed fuselage skin to prevent them from losing stability. The boundary condition and load of the ordinary frame can be reasonably simplified into the form as shown in Figure 6.

The load distribution will determine the frame optimization space. The percentage of materials is related to the availability of the optimized topological model and the generation of convergence structures. Parameter setting experience of topological optimization in this paper is set the initial value of reducing material percentage to 50%. After completing the above steps, optimized iterative calculation is carried out. Actually, the optimization structure with multiple iterations has better static and dynamic performance of the frames. The topological optimization method is automatic multiple iterations in this paper. The topological optimization of an aircraft frame is obtained, as shown in Figure 7.

In Figure 7, the red part is the removable area of the frame, the gray part is reserved area and the brown part is the edge part. When the partial red structural material is removed, the weight of the whole frame can be significantly reduced. The weight of the frame structure is 51.338kg before optimization, and it is reduced to 31.726kg after optimization, which is reduced by 38.2%. From the analysis results of the frame topological nephogram, it can be concluded that:

(a) The distribution of the material has a great relationship with the load position and load magnitude of the frame. The lower part of the frame structure has a weaker load bearing, so the material is less reserved. And the middle and upper part have a stronger load bearing, so the retention of the material is more concentrated.

(b) The less the percentage of target material is reduced, the less the structure material will be preserved, correspondingly the more difficult the iterative convergence is calculated. But in the design space, most of the unwanted material have been selected certainly in the initial iteration, the subsequent iteration is just used to check the selection process of the boundary material topology.

5 Conclusion

The paper mainly makes research on topological technology based on variable density method. By using the three-dimensional modelling of a certain type of domestic aircraft frame, selecting the Lagrange multiplier method as the solving algorithm to derive the variable density method, choosing RAMP as the interpolation model, topological optimization finite element modelling, optimal parameter setting and optimization analysis are all involved in the paper. At last, the reasonable frame structure is optimized, and the material distribution law of the topological optimization is summarized according to the optimized structure analysis.

References

1 Jin Dongping, Ji bin. Topology optimization of flexible support structure of rear flange of wing [J]. Acta aeronautical journal, 2015,36 (08): 2681-2687.
2 Zhang Ji, Qiu Fusheng. The design of vertical design of aircraft [J]. aircraft based on topology optimization, 2010,30 (03): 31-34.
3 Tong Gang, Liu Tongfei, Deng Yangchen, a wing structure design based on topology optimization. China Science and technology, 2013,10:64-66.
4 Wei Juye. The research of structure topology.

MATEC Web of Conferences **198**, 05009 (2018) https://doi.org/10.1051/matecconf/201819805009
MEAE 2018

Research on Near Space Plasma Vacuum Environmental Simulation Technology

Fu Chunyu[1], Dong Yize[2]

[1]*Beijing Insititute of Spacecraft Environment Engineering, Beijing 100094, China*
[2]*Space Star technology co., LTD, Beijing 100095, China*

Abstract. The operation reliability of spacecraft in orbitis affected by the interaction with space plasma, research on the space enviroment by advanced testing method is important. It is costly to carry out research and analysis with space flight. Therefore, tests using ground vacuum environment simulation system are of significance. This paper proposed the structure design, simulation analysis and numerical calculation methods for the three subsystems of plasma environment simulation system, including vacuum vessel, vacuum acquisition and vacuum measurement and control. The simulation results show that the maximum stress of the vacuum vessel is $113.2MPa$, and the maximum deformation is $0.59mm$, at the same time, vacuum technology index, structural stability and human-machine interaction performance of the system can meet the experimental research requirement.

1 Introduction

The traditional research method of space plasma is that launching a deep space rocket or satellite to obtain the data parameters of space plasma. The space testing instruments are expensive and non-reusable, which is difficult to accurately research many microphysical processes. Because of that the laboratory plasma has the characteristics of easily repetition, accurate diagnosis, and low cost[1-3], it's a new method and a development trend that using laboratory plasma to simulate space physical processes[4-7]. The large proximity space plasma vacuum environment imitation system is an important part of the national major science and technology infrastructure "space environment ground simulation device (SESRI)", it's the main part of the space plasma environment simulation and research system.

This paper describes a large-scale near space plasma vacuum environment simulation technology. It is used to provide the vacuum testing environment for space plasma environment simulator, at the same time, it also provide testing environment for the key components such as antenna groups, target analog components, magnet coils, plasma sources, as well as installed base and interfaces for other units. The system is not only a large vacuum system but also a complex mechatronic system. The most basic requirements are system vacuum specifications, structural stability and high reliability[8]. The system consists of negative pressure vessel system, vacuum vessel system, and control system. The main functional indicators of this system include:

-The effective space of vacuum vessel: diameter is $\Phi 5000mm$ column length is $6000mm$;

-Extreme vacuum degree[9,10]: $\leq 10Pa$ (This indicator will be reached within 6 hours from atmospheric pressure);

-Operating vacuum degree: when the vessel extracted air achieves the extreme vacuum degree, it filled with argon can continuously adjust the vacuum degree. When the air inflow is in the range of $20SLM$ □ $60SLM$, the working pressure can be achieved the scope of $100Pa$ □ $1000Pa$, and the control accuracy should surpass $\pm 5\%$;

-Total leak rate of vacuum system: sum of leakage rate at each leak points are $\leq 1\times 10^{-5} Pa\, L/s$.

2 Vacuum vessel system

2.1 Main structure

The structure of near space plasma environment vacuum vessel is horizontal cylindrical, One side of the vessel is a gate, which uses electric switch. It is designed according to pressure vessel design standards: it uses material is 304 stainless steel, its inner surface is polished, the roughness is no greater than $1.6\mu m$, the length of the straight section is $6000mm$, the inner diameter is $\Phi 5000mm$, its total length is about $8454mm$. Its structure is shown in figure 1.

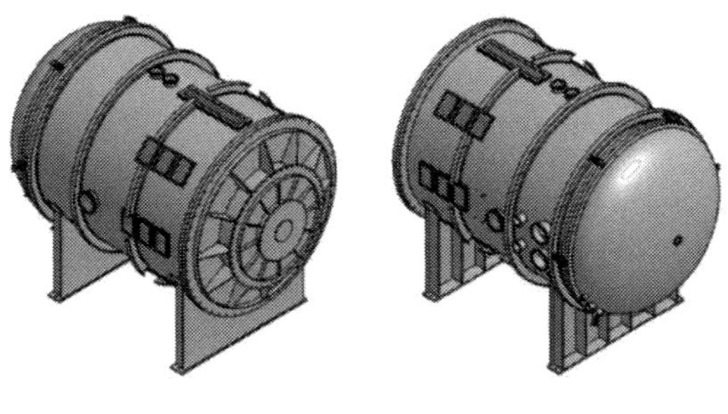

Figure 1. The structure diagram of near space plasma environment vacuum vessel

2.2Finite element analysis of container system structure

Finite element analysis of container system structure is a calculation of intensity proofread and structural stability about vacuum vessel in all kinds of working conditions. It also analyzes the accuracy and verifies the convergence of the calculation results. According to the calculation results, it will give suggestions about improvement and optimization of vacuum vessel's structure.

The 3D model of vacuum vessel is geometric cleaned, and meshed under the physical structure, which is shown in figure 2.

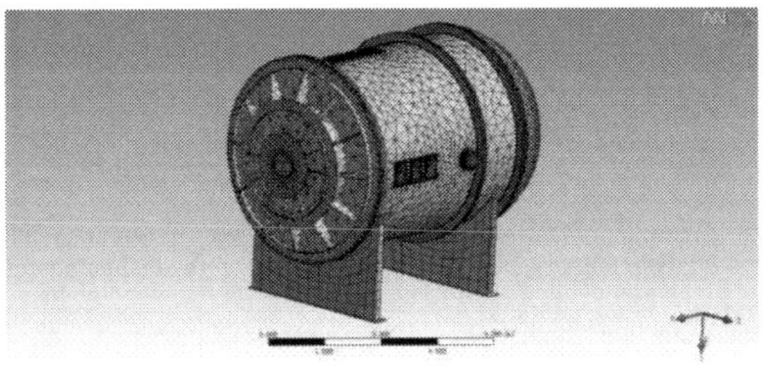

Figure 2. Grid diagram of the near space container

The physical structure is divided into physical units. Added material properties are that the observation window is quartz glass and the rest is stainless steel as required.

The boundary conditions of the model are set as: six DOFs of one saddle are constrained, vertical direction of the other one is constrained. Experimenters submit the analysis of this model, and observe local stress/displacement or distributed stress/ larger displacement area's stress/displacement analysis chart and maximum stress/displacement value.

2.2.1 Calculation results of finite element analysis

Overall stress distribution and the location of the maximum stress is shown in figure 3. The maximum stress is $113.2MPa$. Areas with higher stress are mainly concentrated at the transfer flange of the vacuum vessel flat end cap.

Figure 3. Overall stress distribution and maximum stress position of the vacuum vessel

Overall displacement distribution and maximum displacement position of the vacuum vessel is shown in figure 4. The maximum deformation displacement is $2.85mm$, which appears in the middle of the flat end cap subassembly.

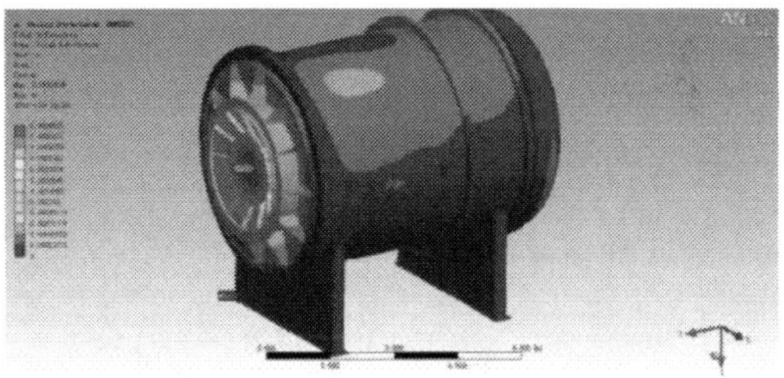

Figure 4. Overall displacement distribution and maximum displacement position of the vacuum vessel

Local deformation extracted from the straight section is shown in figure 5. It can be seen that the structurally deformed area is mainly located on one side of the vessel wall at flange opening, the maximum structural deformation displacement of this area is $0.59mm$.

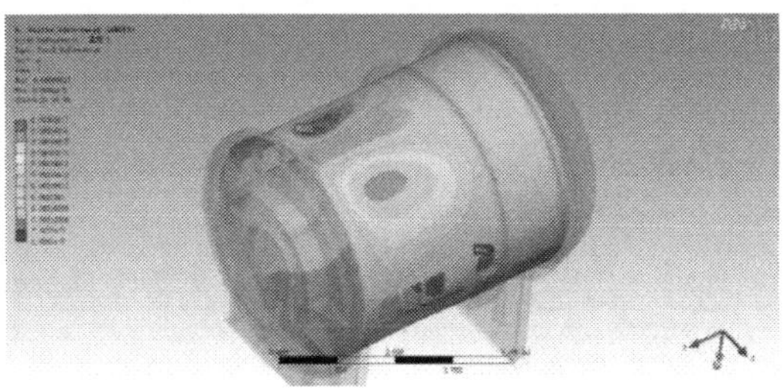

Figure 5. Displacement of the straight section

2.2.2 Results and discussion

(1) The vacuum vessel has a low overall stress level. The maximum stress is $113.2MPa$, which is less than the allowable stress of stainless steel. The structural strength of the vacuum vessel meets requirements.

(2) The local deformation of the straight section shows that the maximum structural deformation displacement is $0.59mm$. Therefore, the offset of the antenna position caused by the vacuum deformation of near space vessel does not exceed $0.59mm$, which meets requirements of the mission.

(3) It can be seen from the finite element calculation results that all of the strength and deformation of the container system meet requirements, and the container will be stable.

3 Vacuum system

3.1 Function and composition

The near space plasma vacuum system is mainly used to provide a vacuum acquired capacity, which can meet the requirements of clean vacuum quality and the device operation. The system schematic is shown in figure 6.

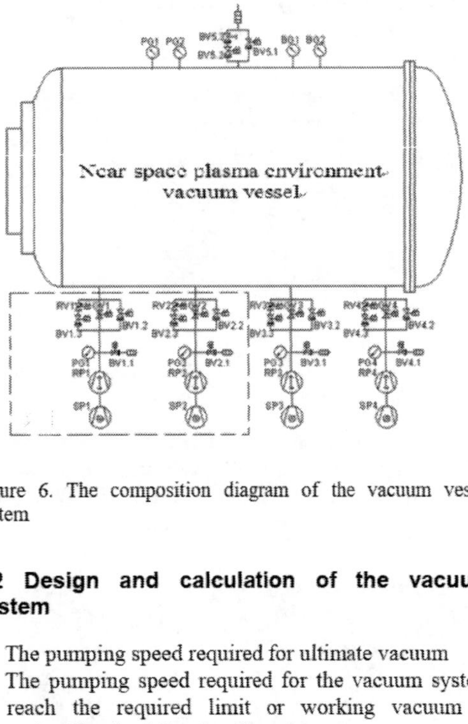

Figure 6. The composition diagram of the vacuum vessel system

3.2 Design and calculation of the vacuum system

(1) The pumping speed required for ultimate vacuum

The pumping speed required for the vacuum system to reach the required limit or working vacuum is determined by the following formula:

$$P_g = P_0 + \frac{Q}{S_{eff}} + \frac{Q_1}{S_{eff}} \quad (1)$$

P_g —The ultimate vacuum that can be achieved by the pod, Pa;

P_0 —The ultimate vacuum of vacuum pump, Pa;

Q_0 —The gas load of the vacuum vessel (no-load) after a certain period of evacuation (including leakage Q_L, material surface leak Qe), $Pa\,L/s$;

Q_1 —The gas load of vacuum vessel;

S_{eff} —Effective pumping speed near the exhaust port of the vacuum vessel, L/s.

In formula (1), the ultimate vacuum of the vacuum vessel is always lower than the ultimate vacuum of a vacuum pumping unit, which is $P_0 \le P_g$. Under certain condition of effective pumping speed, the ultimate vacuum of the vacuum vessel with no-load is proportional to the air leakage and exhalation of the vacuum vessel.

According to the size of the vacuum vessel and related data of other subsystem, this paper calculates the total deflation load released from the surface of the vacuum vessel. The specific data is shown in table 1 & 2.

Table 1. The calculated value of the gas outlet from the vacuum vessel

Outgassing meterial		**Pumping time**			
		1	2	5	10
Inner surface area of the vacuum vessel ($180m^2$)	Outgassing rate $Pa\,L/s\,cm^2$	2.8×10^{-7}	1.73×10^{-7}	9.33×10^{-8}	5.33×10^{-8}
	Gas volume $Pa\,L/s$	0.50	0.31	0.18	0.10

Table 2. The calculated value of the gas outlet inside the vacuum vessel

Deflating object	Outgassing amount in unit time / Pam^3s^{-1}	Quantity
Composite absorber	Outgassing amount of individualmodule 2×10^{-1}	Not less than 400

The leakage rate must be limited to ensure that the air leakage does not affect the ultimate vacuum with no load, The leakage rate of vacuum vessel equipment should be 10 times lower than the outgassing rate. Because the container's flanges and connectors easily lead to leakage, the leakage rate of these components accounts for a large part of the allowable leakage rate. The weld of the container will not be dismantled. According to technical requirements, the total leakage rate of the vacuum system is $\le 1\times10^{-5} Pa\,L/s$.

Therefore, when the vacuum vessel is unloaded (there is no device inside), the internal gas load is determined as $5.0Pa\cdot L/s$, the amount of air release in the vacuum vessel ,which is assembled the microwave darkroom and all supporting components, is about $8.0\times10^4 Pa\,L/s$. Replace the amount of vent gas with the required ultimate vacuum into formula (1), it can be seen that the required pumping speed for the vacuum obtaining system needs to be not less than $8000\,L/s$ (approximately $28800m^3/h$).

(2) Calculating pumping speed required for pumping time

Vacuum system also requires that the vacuum vessel pressure $\leq 10Pa$ in 4 hours (after pumping from atmospheric pressure); vacuum vessel system satisfies the pumping time requirements of vacuum storage compartment. Calculating the pumping speed according to the following formula:

$$Se = 2.3\frac{V}{t}K \times Lg\frac{P_1}{P_2} \quad (2)$$

t ——Pumping time, h;

Se ——Effective pumping speed of dry pump, m^3/h;

V ——Container volume, m^3;

P_1 ——Pressure at the start of pumping, Pa;

P_2 ——Pressure after t time, Pa;

K ——Correction factor.

Above all, to satisfy the vacuum indicator needed for vacuum vessel of near space plasma vacuum environmental simulation subsystem, the minimum pumping speed required for the vacuum system is not less than $8000\,L/s$ (approximately $28800\,m^3/h$).

(3) Review and recalculate of the vacuum system

According to the preliminary layout position of the roughing unit, it can be assumed that the length of the roughing main pipeline is estimated to be within about $10m$. The conductance of rough pipe can be calculated by the following formula for viscous flow and long pipe flow conductance:

$$U = 1.34 \times 10^3 \frac{d^4}{L}\bar{P} \quad (3)$$

U ——Conduction of the main pumping pipe, m^3/s;

d ——The inside diameter of the main pumping pipe, m;

L ——The length of the main pumping pipe, m;

$\bar{P}$ ——Average pressure in the pipeline, Pa.

Review and recalculate the capacity of the selected vacuum unit, according to formula (1), rough pumping unit can pump the vacuum vessel to $\leq 10Pa$, which meet the requirements of ultimate vacuum. Evaluate the extraction time according to formula (2), vacuum system draws container to desired pressure within 1.5 hours, which meet the time requirements for pumping.

4 Control system

4.1 System structure

The control system of the near space plasma vacuum system adopts local control and remote control modes, it is mainly used to control and monitoring the vacuum vessel system and accessory equipment. The structure is shown in figure 7.

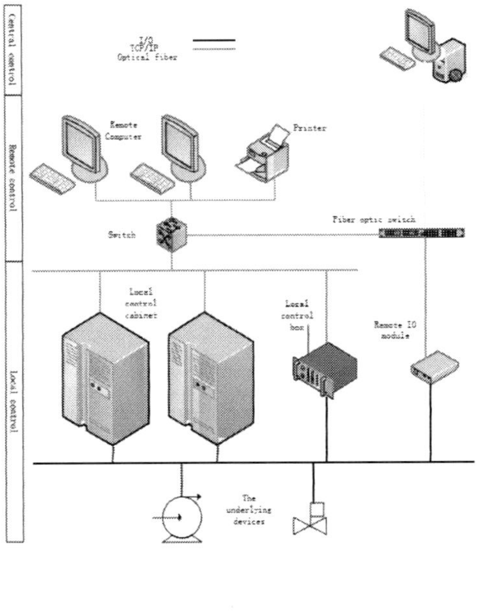

Figure 7. The structure of the control system

The local control system selects the distributed control architecture controlled by S7-300 controller, it is integrated into the local control cabinet. The configuration software can control all controllable equipment and monitor the measurable parameter. The remote control adopts the fusion mode composed by a scheme based on industrial control universal configuration software and a scheme based on universal programming language self-developed software.

System software

The software design of control system includes the design of the PLC bottom layer program and the design of the configuration software. The overall structure of the software is shown in figure 8.

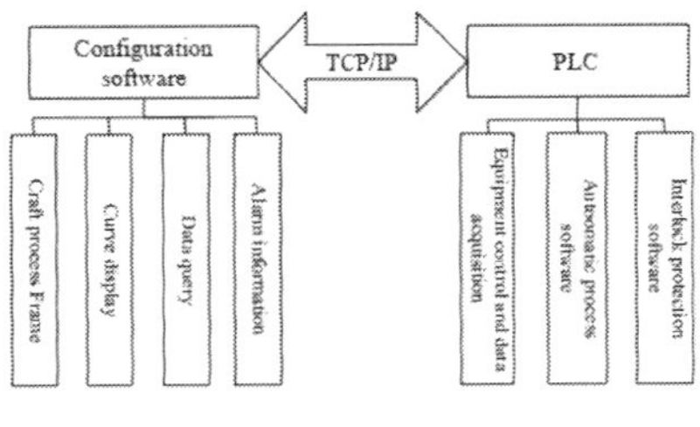

Figure 8. General structure of the software

The PLC bottom program is developed with STEP7, it implements the following man-machine interface:

(1) Craft process frame: According to the vacuum system process design, the craft process frame is established, and the operation status and real-time date of each device can be viewed on screen. In addition, devices,

which have accessible control, can be controlled by buttons on screen. Technological flow chart is shown in figure 9.

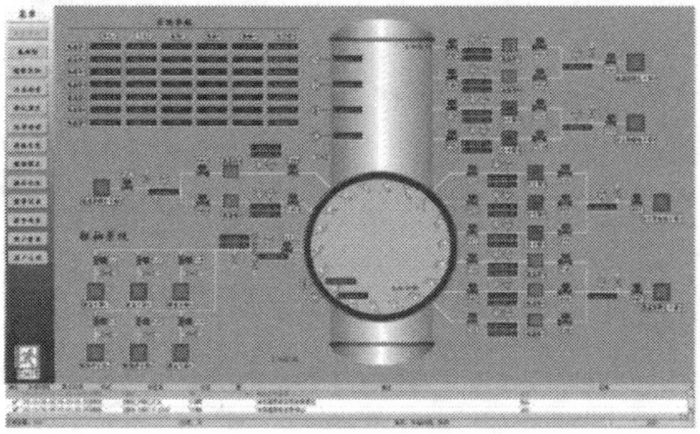

Figure 9. Technological flow chart

(2) Curve display: All the collected real-time date is displayed in the form of curve. Multiple curves can be displayed simultaneously or separately, and history curves can be queried. The schematic of curve display is shown in figure 10.

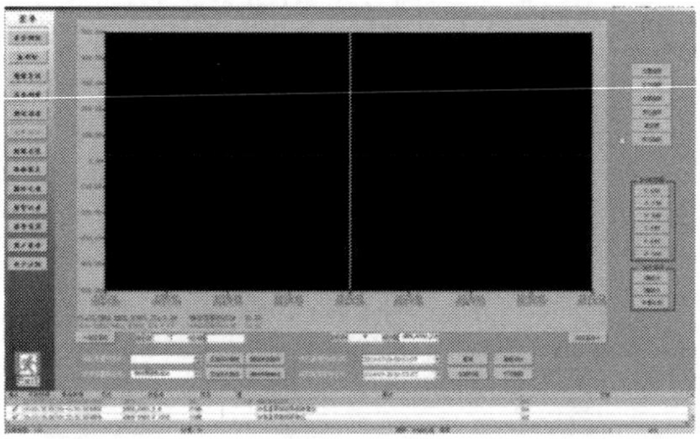

Figure 10. The schematic of curve display

(3) Date query: Establishing an independent data query can query the historical operating state and the historical date of the system, including operation logs and key historical date. The schematic of data query is shown in figure 11.

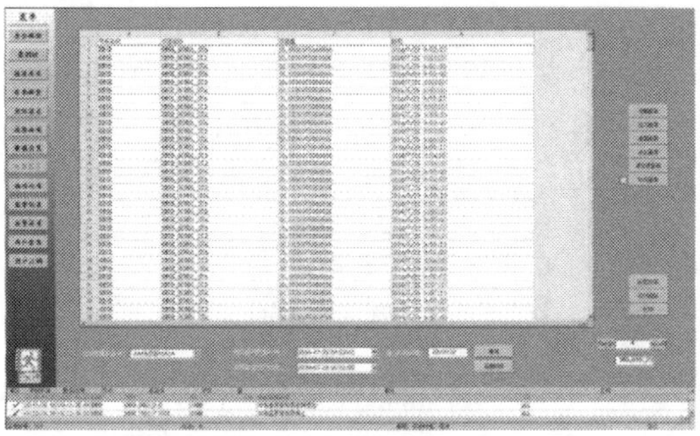

Figure 11. The schematic of data query

(4) Alarm information: Alarm information of the control system include the real-time alarm information and historical alarm information. The current real-time alarm can be manually processed. The schematic of Alarm information is shown in figure 12.

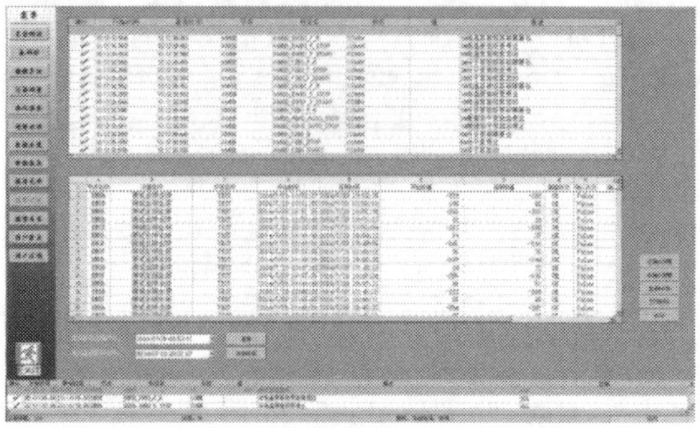

Figure 12. The schematic of alarm information

5 Conclusion

This paper introduces a large-scale near space plasma vacuum environment imitation technique. Through calculation and analysis, this paper provides a design scheme required for vacuum working environment. The negative pressure vessel system, vacuum subsystem, vacuum system is introduced in detail.

The finite element analysis and numerical calculation of the mechanical structure and the vacuum capacity verify the feasibility of the design. The indicators of the system proposed in this paper are qualified. It can effectively reduce the influence of wall effect in the research of space plasma.

References

1. K. Kuriki, H. Kuninaka,. Numerical analysis of interaction of a high-Voltage solar array with ionospheric plasma. *J Spacecr. Rockets*, **24(6)**:512-517(2015)
2. M. Cho, D.E. Hastings. Dielectric Charging Processes and Arcing Rates of high Voltage Solar Array, *J Spacecr. Rockets*, **28(28)**:698-706(1991)
3. H. Thiemann, R.W. Schunk. Computer experiments on arcing processes as observed in ground tests. *J Spacecr. Rockets*,**31(31)**:929-936(2015)
4. H. Thiemann, K. Bogus. High-Voltage solar cell modules in simulated low-earth-orbit plasma. *J Spacecr. Rockets*, **25(4)**:278-285(2015)
5. Upschule B L et al. Arcing of Negatively Biased Solar Cells in Low Earth Orbit. AIAA 92-0578
6. M.R. Carruth et al. Experimental Studies on Spacecraft Arcing. *J Spacecr. Rockets*,**30(3)**:323-327(2015)
7. D.E. Hastings, M. Cho, H. Kuninaka. Arcing rates for high voltage solar arrays: theory, experiment and predictions. *J Spacecr. Rockets*, **29(4)**:538-554(1992)
8. Quin S, Windra G E O. Use of Stress-Strengh Model in Determination of Safety Factor Pressure Vessel Design[J]. *Journal of Pressure Vessel Technology*, **118(1)**:27-32(1996)
9. I. Elishakoff. Essay on Uncertainties in Elastic and Viscoelastic Structures: From A. M. Freudenthal's Crutucusms to Modern Convex Modeling[J]. *Computers & Strultures*, **56(6)**:871-895(1995)
10. Y.M. Zhang, Q.L. Liu. Reliability-Based Design of Automobile Components[J]. Proceedings of the Institution

of Mechanical Engineers Part D, *Journal of Automobile Engineering*, **216(6)**:455-471(2002)
11. B. Huang. Space environment test facility for manned spacecraft[J]. *CSST*, **22(3)**: 1-3(2002)
12. Lee Jin,N.S. Harris,N. Tenwick, et a1. Modern vacuum practice[M]. 3rdedition. BOC Edwards. 201-266(2000)
13. J. Tong,L. Sun,R. Jia,Y. Liu, et al . Experimental Techniques in Simulated Space Plasma Environment on Ground [J]. *Chinese Journal of Vacuum Science and Technology*, **28(3)**:203-207(2008)
14. M. Ma,T. Chen,J. Lei, et al. Design of Large-Sized Space Plasma Simulation System for Low Earth Orbit Environment[J]. *Chinese Journal of Vacuum Science and Technology*, **36(6)**:680-685(2006)
15. Cho M, Shiraishi K. Laboratory Experiments on Mitigation a-gainst Arcing on High Voltage Solar Array in Simulated LEO Plasma Environment. AIAA2002-0629
16. Tribble A C. Low Earth Orbit Plasma Effect on Spacecraft, A-IAA93-0614
17. Hastings D E. A review of plasma interactions with spaceraft in low earth orbit. *J. Geophys. Res*, **100(A8)**:14457-14483(1995)

Finite Element Analysis of Space Science Experimental Equipment Based on HyperMesh and Nastran

Huang Sinian[1,2], Wang Anping[2]

[1]*University of Chinese Academy of Science, Beijing 100049, China*
[2]*Technology and Engineering Center for Space Utilization, Chinese Academy of Science, Beijing 100094, China*

Abstract. In the design process of space science experimental equipment, in order to better grasp the dynamic characteristics of the structure, a large number of finite element analysis is needed. This paper comprehensively use HyperMesh and Nastran to establish a set of finite element analysis process for space science experimental equipment, combine the advantages of each software to perform finite element analysis comprehensively, which greatly improves the accuracy and efficiency of the finite element analysis, the problems of this kind of engineering analysis has certain reference value.

1 Introduction

With the further development of the manned space engineering, the research prospects for the application of space science in China have been presented to our eyes[1]. Space science experimental equipment is an important structure for the installation of payloads in spacecraft capsules and is an important part of space application system. In the design process of space science experimental equipment, a large number of finite element simulation analysis are required, through finite element analysis, the designed space science experimental equipment can be optimized based on the analysis results, shortening the development cycle and saving costs[2,3].

Modal analysis is the most important one in the finite element analysis process of the space science experimental equipment, it often generates violent vibration during the spacecraft launching process, which has a huge impact on the space science experimental equipment in the spacecraft. Therefore, how to calculate the inherent characteristics of the system by determining the reference frequency of the science experimental equipment and to avoid resonance in the emission process is necessary. This paper comprehensively use the different characteristics of Creo, HyperMesh and Nastran to establish a set of finite element analysis procedures, the modal analysis of the space science experimental equipment obtains its natural frequency and mode shape, which has certain reference significance.

2 Comprehensive Application of Nastran and HyperMesh

For a simple geometric model, generally with a single CAE software can complete the analysis process, however, when performing finite element modeling analysis of a complex model, various software is often comprehensively used to complement each other, and the advantages of each software are combined to perform finite element simulation analysis, it can quickly and efficiently solves engineering problems.

2.1 HyperMesh Features

HyperMesh is a finite element pre-processing software developed by Altair Corporation of the United States, it has powerful finite element pre-processing function and can provide users with high-quality and high-efficiency mesh division technology, it can complete pole beam, plate shell and solid mesh division, through the interactive operating environment helps users create various complex finite element models. HyperMesh provides powerful geometric cleanup capabilities that support a variety of automated or manual geometric cleanup and repairs, such as surface stitching, surface repairs and removal of chamfered round holes. HyperMesh supports various mainstream CAD and CAE software currently on the market, it can read different types of geometric model files and can also output finite element model files for different solvers.

HyperMesh is powerful in grid processing, however there are fewer types of materials and units in the finite element analysis, and the solution method is more complex, there is also a certain gap between the finite element analysis results and post-processing[4].

2.2 Nastran Features

Nastran is a series of products of the United States MSC.Software Corporation, has been widely used in engineering simulation analysis of various industries, MSC.Software products cover almost all research institutions, design departments, universities and other institutions of China Aerospace[5]. Nastran has a variety of material libray with a variety of linear and other nonlinear models, which have the advantages of high accuracy and fast convergence in analytical. Figure 1 show a general flow chart for Nastran modeling analysis.

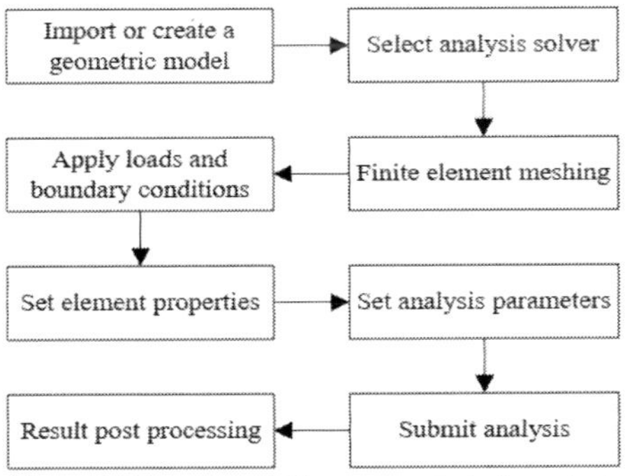

Figure 1. Nastran modelling and analysis flow chart.

Compared with other finite element analysis software, Nastran's meshing capabilities and geometric cleanup capabilities need to be further strengthened.

2.3 Comprehensive use of Creo, HyperMesh and Nastran Analysis process

HyperMesh and Nastran have their own advantages when perform finite element analysis, therefore, it is possible to comprehensively utilize the features of each software, complement each other, establish a set of geometric models using Creo, HyperMesh and Nastran to perform finite element analysis, which greatly improve the efficiency of finite element analysis[6]. The geometric model created by Creo was imported into HyperMesh, and the steps of geometric model cleaning, mesh division, material attribute assignment, unit attribute assignment and boundary load setting in the finite element analysis process were completed in turn, completing the establishment of the finite element model. Then the finite element model was submitted to Nastran for solving calculations, completing the finite element analysis process. As shown in Figure 2, it is a finite element simulation flow chart.

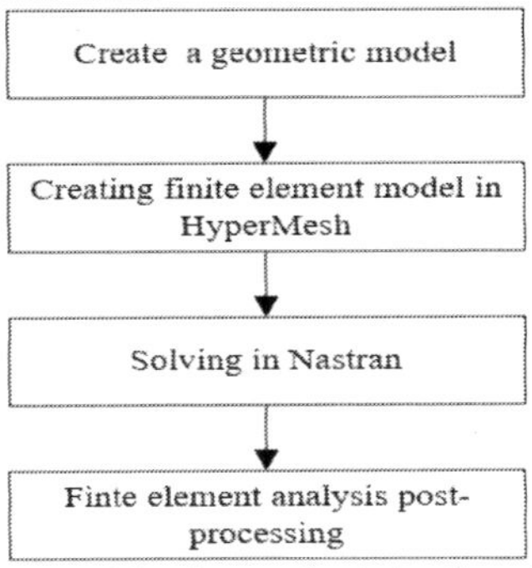

Figure 2. Finite element simulation flow chart.

3 Modal Analysis of Space Science Experimental Equipment Based on HyperMesh and Nastran

3.1 Space Science Experimental Equipment Modal Analysis Theory

The finite element modal of space science experimental equipment in this paper belongs to the multi-freedom undamped free vibration system[7], the free vibration equation of the undamped system is as follows

$$M\ddot{y} + Ky = 0 \quad (1)$$

In the formula, M is the mass matrix, K is the stiffness matrix, $\ddot{y}$ and y are the acceleration vector and the displacement vector respectively, the general solution of (1) cab be expressed as the sum of each special solution, and one set of special solutions is as follows

$$y = A\sin(\omega t + \varphi) \quad (2)$$

This is equivalent to assuming that each mass body performs a simple harmonic motion and that the vibration frequency is the same as the initial phase angle, and differentiated twice for t in (2), the following results are obtained:

$$\ddot{y} = -\omega^2 A\sin(\omega t + \varphi) \quad (3)$$

Substituting (2) and (3) into (1) can be obtained

$$(K - \omega^2 M)A = 0 \quad (4)$$

So the system's characteristic equation is

$$\left|K - \omega^2 M\right| = 0 \quad (5)$$

The solution obtains n positive real roots of ω_i, finding n frequencies ω_i, $(i=1,2,3...,n)$, where the lowest frequency ω_i is the lowest natural frequency, which is the basic frequency of the system, the remaining frequencies are collectively referred to as high-order frequencies.

3.2 Establishment of Finite Element Model of Space Science Experimental Equipment

Space science experimental equipment are generally relatively regular plate structures, which are mainly composed of external frames and internal electronic components and pipelines. In the finite element modeling analysis, the surface is generally treated as a shell element, different thicknesses are assigned to different parts, and reinforcing ribs and mass points are set in the corresponding positions to better simulate the trueness of the structures. In the process of processing and manufacturing, in order to reduce the quality while ensuring the structural strength, the commonly used materials is aluminum alloy.

Taking the science experimental equipment shown in Fig.3 as an example, the geometric model created by Creo was imported into HyperMesh, and choosed the Nastran as the solver template to establish the finite element model. In the finite element meshing, the main unit used is four-node shell element and tetrahedron element, the former mainly simulates the structural PCB structure and the latter is mainly used for simulating the external panel, the connection of each part is implemented by the rigid element RBE2. As shown in Fig.4, the structure finite element model established by HyperMesh has a total of 163,000 finite element nodes.

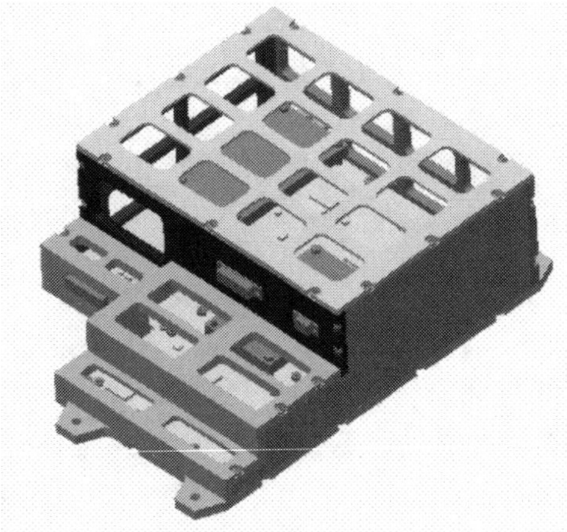

Figure 3. Geometric model of the space science experimental equipment .

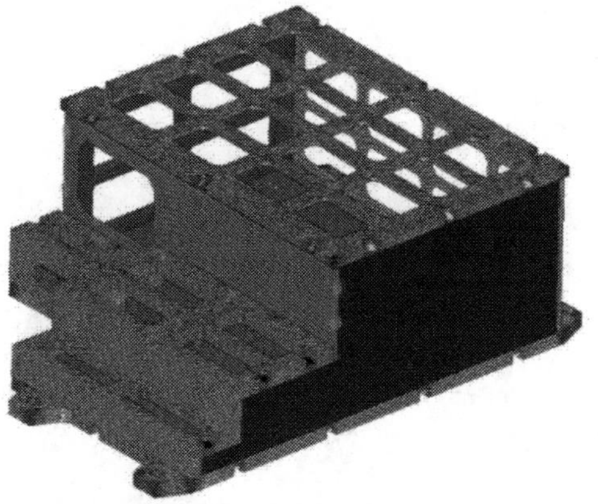

Figure 4. Finite element model of space science experimental equipment.

The experimental equipment is mainly composed of aluminum alloy and internal PCB structure, when defining the material properties, it is necessary to define two kinds of materials at the same time, and set parameters such as desity, Poisson's ratio and elastic modulus of different materials, the material properties of different materials are shown in Table 1.

After the assignment of the matertial properties, the assignment of the unit attributes is performed according to the actual conditions, the thicknesses of the simplified shell units are given different thicknesses, and the corresponding entity attributes are selected for the body units. As the same time, in order to better simulate the real situation of the equipment, different mass points are set on the surface of the equipment. In general, the space science experimental equipment is fixed inside the spacecraft. According to the actual installation form, the four lug positions of the structure are fixed during the analysis. Set the solution parameters in HyperMesh, complete the establishment of the finite element model, and output the finite element model to Nastran for solution calculation.

Table 1. Materials properties.

Materials	Density ($kg \cdot m^{-3}$)	Poisson's ratio	Elastic Modulus
Aluminu m alloy	2700	0.33	7×10^{10}
PCB	1700	0.30	1.4×10^{10}

3.3 Modal Analysis Based on Nastran

The natural frequency and mode shape of the space experimental equipment are determined by the geometry of the structure, material properties and constraints. In the modal analysis of the structure, the Lanczos Method with a fast convergence rate is selected for analysis, in general, during the modal analysis, it is necessary to understand the local modality and the overall modality of the structure, so it is necessary to find the overall structure starting frequency. In the calculation process, the first 20 natural frequencies of the structure are extracted, and the overall starting frequency of the structure is found through analysis. As shown in Table 2, it is the first 8th order modal frequency and mode result of the structure and the 17th order overall start-up of the structure frequency.

Table 2. Modal analysis results.

Order	Frequency (Hz)	Mode description
1	167.7	the vibration of the PCB and internal electronic components
2	170.8	the vibration of the PCB and internal electronic components
3	221.7	the vibration of the PCB and internal electronic components
4	258.2	the vibration of the PCB and internal electronic components
5	272.5	the vibration of the PCB and internal electronic components
6	319.2	the vibration of the PCB and internal electronic components
7	336.5	the vibration of the PCB and internal electronic components
8	347.4	the vibration of the PCB and internal electronic components
17	567.3	the vibration of the PCB and external structure

Select the representative first-order and second-order local vibrations and the 17th-order global vibration, the modal shapes are shown in Figs. 5-7.

Figure 5. First-order mode shape.

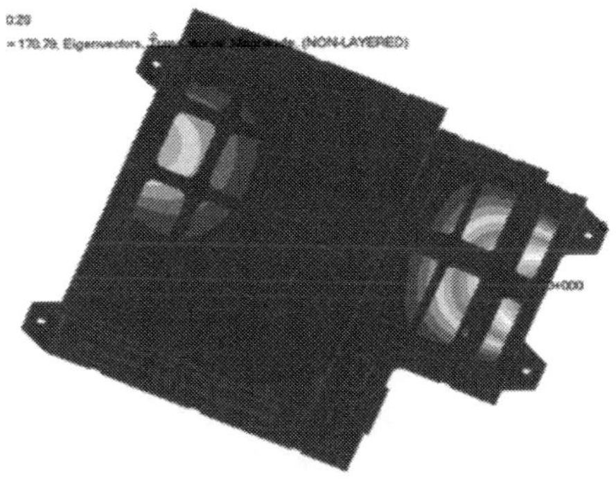

Figure 6. Second-order mode shape.

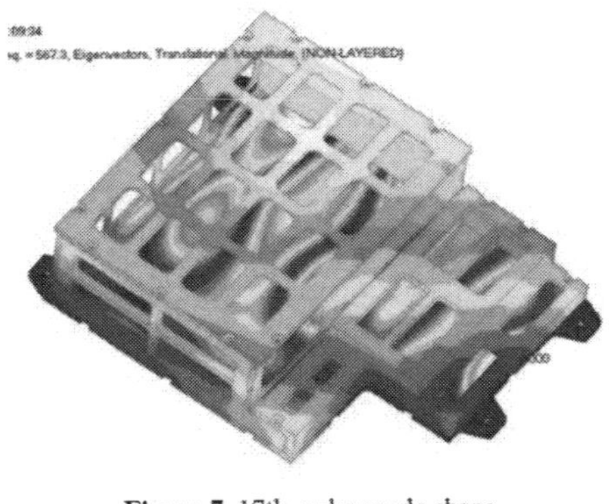

Figure 7. 17th-order mode shape.

From the modal shape diagram and the modal frequency, the basic frequency of the space science experimental equipment is 167.7Hz, which is the vibration of the PCB and internal electronic components, it is a local vibration mode; the first-order overall modal frequency of the structure is 567.3Hz. The results have certain reference value for structural design and optimization.

4 Concluding Remarks

In this paper, taking the typical structure of the space science experimental equipment as an example, a set of procedures for finite element analysis using Creo 3D modeling software, HyperMesh meshing software and finite element solution software Nastran integrated finite element analysis was established, taking full account of the advantages and disadvantages of each software, giving full play to the advantages of each software, the efficiency and accuracy of finite element analysis have been greatly improved. Through the finite element analysis of the space science experimental equipment, the basic frequency and the overall starting frequency of the structure are found, which can be used as a reference in the design process to avoid resonance. The finite element analysis process has certain reference value and application value for engineering applications.

References

1. Wang Anping. Modal Analysis and Model Optimization of Space Science Experimental Equipment[J]. ENGINEERING & TEST, 2017, 57:1-4.
2. Wang Dapeng. Optimal Design of Payload Cabinet Structure[J]. Spacecraft Engineering, 2010, 19(2):34-39.
3. Ding Tao, Peng Wang. Application of HyperMesh Secondary Development in Pre-processing of Bus Finite Element Analysis[J]. Bus Technology and Research, 2015,6:47-49.
4. Jiang Xuewu, Wu Xinyue et al. Comprehensive Application of UG, HyperMesh and MSC Marc in Finite Element Analysis[J]. Computer Aided Engineering, 2007, 16(12) :11-14.
5. Bai Kunxue, Yin Shaoping. Finite Element Analysis for Torpedo Seawater Pipe Based on HyperMesh and Nastran[J]. TORPEDO TECHNOLOGY, 2012, 20(1) :14-18.
6. Yin Mei, Wang Chenglong. Analysis of Brake Disk Modal Optimization Based on HyperMesh and Nastran[J]. Design Research, 2015, 30-33.
7. Faycal Benyahia, Abdulmohsen Albedah. Elliptical and Circular Bonded Composite Repair under Mechanical and Thermal Loading in Aircraft Structures.
8. Clough R W. The finite element method in plane stress analysis[A]. Proceedings of ASCE Conf Electron Computation, Pittsburg, PA, 1960.
9. Courant R. Variational methods for the solution of problems of equilibrium and vibrations[J]. Bulletin of the American Mathematical Society, 1943, 49:1-23.
10. Turner M J, Clough R W, Martion H C,et al. Stiffness and deflection analysis of complex structures[J]. Journal of the Aeronautical Sciences, 1956, 23:805-823.

Analysis of Energy Reception Characteristics of Solar Aircraft in the Tropics of Cancer and Capricorn

Kangwen Sun[1], Jiaqi Dong[1], Xiaobo Sun[2,*], Ziying Zhang[1], Mou Sun[1] and Jian Zhang[3]

[1]*School of Aeronautic Science and Engineering, Beihang University, Beijing 100191, China*
[2]*China Construction Eighth Engineering Division.CORP.LTD, Shanghai 200135, China*
[3]*Forging Factory, North General Power Group Co., Ltd, 037000 Datong, Shanxi, China*

Abstract. This paper mainly aims at the energy reception characteristics of solar aircraft flying in the near space between the Tropics of Cancer and Capricorn. Based on the solar radiation intensity calculation model, the energy reception characteristics including maximum solar radiation intensity, daytime average solar radiation intensity, total solar radiation energy received per day on the unit area., and daytime hours of solar aircraft was calculated varying with date, longitude, and latitude. By analyzing the corresponding calculation results, it can provide some reference value for the design of solar aircraft flying in this area.

1 Introduce

Solar aircraft is a new type of aircraft which can realize High Altitude Long Endurance(HALE) flight [1] and has an irreplaceable advantage compared with conventional aircraft [2]. It can convert solar radiant energy into electrical energy by photovoltaic conversion. Near space generally refers to the airspace 20-100km from the ground, and it has a special strategic position because of the unique height and environmental conditions. Using the stable atmosphere of near space and the inexhaustible supply of solar energy, HALE solar aircraft has a wide application prospect in the tasks of environmental detection, regional communication, border monitoring, disaster warning and monitoring, high resolution, etc., and it is one of the hot and frontier fields of international research [3]. Solar aircraft uses solar energy as the main power source [1]. Therefore, the analysis of solar energy reception characteristics is the basis and necessary prerequisite for the study of such aircraft [4]. Due to the influence of earth-sun distance, time and latitude, the solar radiation received in the near space will change. On account of the existence of the obliquity of the ecliptic, sunlit point is always moving back and forth between the Tropics of Cancer and Capricorn, which causes the solar energy in this area is abundant [5,6].

In view of this, in this paper, the energy reception characteristics of solar aircraft which flies in the near space between the Tropics of Cancer and Capricorn were analysed, with the change of time, latitude and longitude, in order to provide some reference value for the design of this kind of aircraft.

2 Calculation model of solar radiation intensity [7]

Assuming that the influence of cloud, moist, dust, etc. is ignored, the calculation model of solar irradiation intensity is as follows.

2.1 Solar altitude

The solar altitude is the angle between the incident ray and the horizontal plane, and it will vary with time and place. The solar altitude α can be calculated as follows:

$$\sin\alpha = \sin\Phi\sin\delta + \cos\Phi\cos\delta\cos\omega \quad (1)$$

Where: Φ - local latitude, δ - sun declination angle, ω - solar hour angle.

$$\delta = 23.45°\sin[\frac{360°(d_n+284)}{365}] \quad (2)$$

Where: d_n- date number in a year.

$$\omega = (t_t - 12)\cdot 15° \quad (3)$$

$$t_t = t_s \pm \frac{L-L_S}{15} + \frac{e}{60} \quad (4)$$

$$e = 9.87\sin 2B - 7.53\cos B - 1.5\sin B \quad (5)$$

This work was supported by the National Natural Science Foundation of China under Grant No. 51775021, Key Laboratory of Aircraft Advanced Design Technology (Beihang University) Ministry of industry, the fifth youth top talent support program of Beihang University, and the Fundamental Research Funds for the Central Universities.

$$B = \frac{360(d_n - 81)}{364} \quad (6)$$

Where: t_t- true solar time, t_s- local standard time, L - local longitude, L_S- the longitude of the local standard time, e- the equation of time between true solar time and mean solar time.

2.2 Eccentricity correction coefficient associated with earth-sun distance

The sun is in a slightly eccentric position on the earth's elliptical orbit, so the distance between the sun and earth is not constant, and there are perihelion and aphelion. The distance between the sun and earth is usually corrected with the eccentricity correction coefficient ξ_0.

$$\xi_0 = (\frac{r_0}{r})^2 = 1 + 0.033\cos(\frac{2\pi d_n}{365}) \quad (7)$$

Where: r_0 - the average earth-sun distance, r - the earth-sun distance of the observation point.

2.3 Atmospheric transparency coefficient [8]

The atmospheric transparency coefficient is the ratio of transmitted radiation to incident radiation through an air mass. Under the different weather conditions, the atmosphere thickness will be different. Considering the weather factors, the atmospheric transparency coefficient P_{CO} can be calculated as follows:

$$P_{CO} = 0.56 \cdot k(e^{-0.56AMh} + e^{-0.096AMh}) \quad (8)$$

Where: k - weather influence factor, AMh - highly modified air mass.

$$AMh = AM \cdot \eta_h \quad (9)$$

$$AM = \frac{1}{sin\alpha} \quad (10)$$

$$\eta_h = real(1 - \frac{0.0065}{288}H)^{5.256} \quad (11)$$

Where: AM - air mass, η_h - atmospheric height correction factor, real- the value is the real part of the formula, H- height.

2.4 Sunrise and sunset time [9]

Sunrise time t_{sr} is the time when the sun appears on the horizon. Regardless of the curvature of the earth's surface and the refraction of the atmosphere to the light, it can be calculated as follows:

$$t_{sr} = \arccos(\tan\Phi\tan\delta) \cdot \frac{180}{15\pi} \quad (12)$$

The sunset time t_{ss} and the sunrise time t_{sr} are symmetric at 12 noon.

$$t_{ss} = 24 - t_{sr} \quad (13)$$

Daytime hours t_{day} are the difference between the sunset time and the sunrise time.

$$t_{day} = t_{ss} - t_{sr} \quad (14)$$

2.5 Solar radiation intensity

Considering the effects of seasons, earth-sun distances, height, and time, solar radiation intensity I can be calculated as follows:

$$I = I_{sc} \cdot \xi_0 \cdot P_{co} \cdot \sin\alpha \quad (15)$$

Where: I_{sc}- solar constant, refers to the received solar radiation energy at the boundary of the earth's atmosphere surface perpendicular to the light in unit area and unit time. In 1981, the eighth session of the world meteorological organization instrument and observation method committee defined the solar constant as $1367W/m^2$.

2.6 solar energy reception characteristics

The intensity of solar radiation increases from sunrise to noon and then decreases before sunset. It can be reduced to a sine curve, and the highest point is obtained at noon, which is the true solar time $t_t = 12$.

Maximum solar radiation intensity at noon I_{max} can be calculated as follows:

$$I_{max} = I_{sc} \cdot \xi_0 \cdot P_{co} \cdot \sin\alpha_{(t_t=12)} \quad (16)$$

The change function of solar radiation intensity I with time t during the day is:

$$I = I_{max} \cdot \sin\left(\frac{t - t_{sr}}{t_{day}}\right) * \pi \quad (17)$$

Daytime average solar radiation intensity I_{level} can be calculated as follows:

$$I_{level} = \frac{1}{t_{day}} \int_{t_{sr}}^{t_{ss}} I_{max} \cdot \sin\left(\frac{t - t_{sr}}{t_{day}}\right) * \pi\, dt = \frac{2I_{max}}{\pi} \quad (18)$$

The total amount of solar radiation received per unit area E_s in one day is:

$$E_s = I_{level} \cdot t_{day} \quad (19)$$

According to the above calculation model, this paper will analyse the variation of I_{max}, I, I_{level}, E_s and t_{day} with date, longitude and latitude. Relevant constant value parameters are chosen as shown in table 1.

Table 1. Parameters of solar irradiation strength calculation.

Parameter	Value	Unit
L_S	120	°
k	0.85	-
H	20000	m

3 Energy reception characteristics varies with date and latitude

First of all, this paper analyses the changes of energy reception characteristics of the Tropics of Cancer and Capricorn in a year. In the calculation, the latitude north is positive, the south is negative, and the range of change is $\Phi = [-23.5, 23.5]°$; The range of dates is $d_n = [1,365]$; The geographical longitude is $L = 110°$. In addition, the energy reception characteristics of spring equinox $d_n = 80$, summer solstice $d_n = 173$, autumnal equinox $d_n = 266$ and winter solstice $d_n = 356$ are compared and analysed.

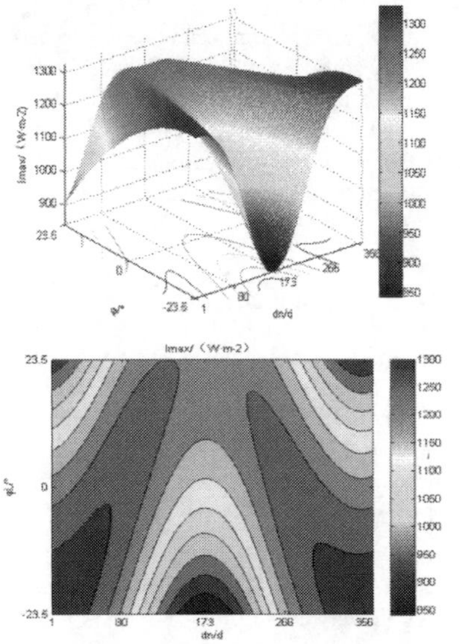

Figure 1. Maximum solar radiation intensity I_{max} varies with date d_n and latitude Φ.

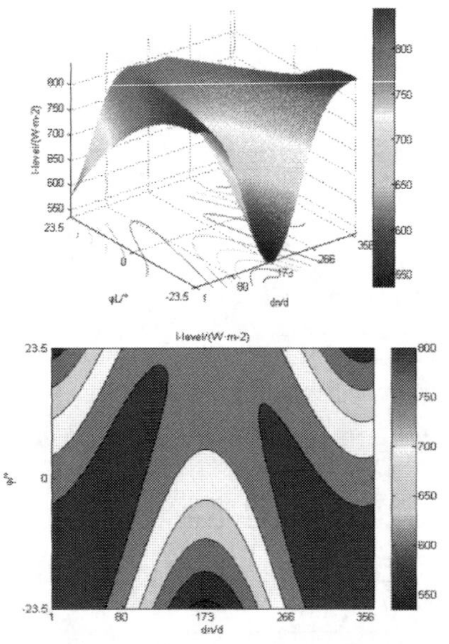

Figure 2. Daytime average solar radiation intensity I_{level} varies with date d_n and latitude Φ.

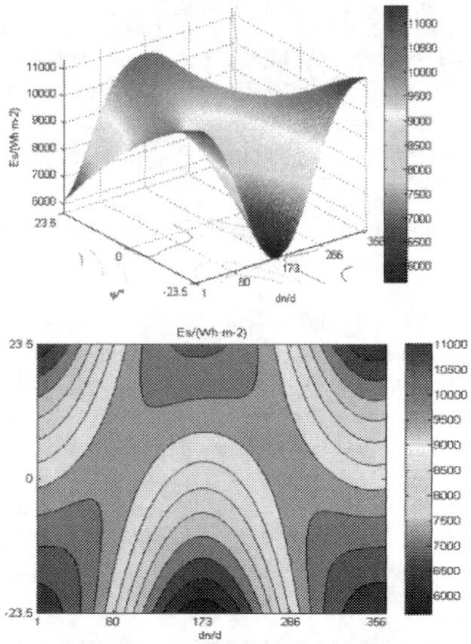

Figure 3. The total amount of solar radiation received per unit area E_s in one day varies with date d_n and latitude Φ.

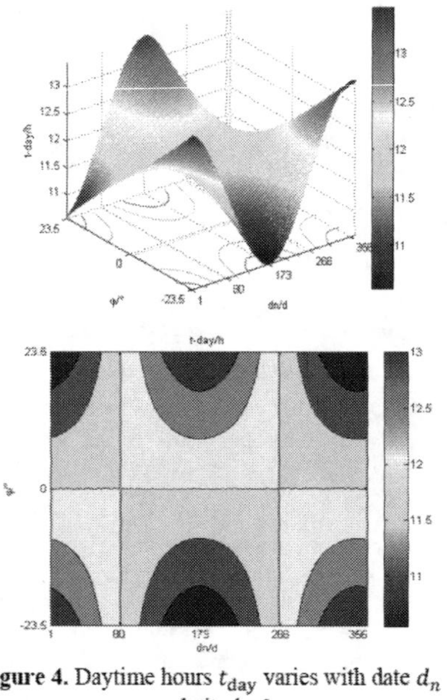

Figure 4. Daytime hours t_{day} varies with date d_n and latitude Φ.

Table 2. The maximum and minimum of I_{max}, I_{level}, E_s and t_{day} and the corresponding Φ and d_n

		minimum				maximum			
		I_{max} (W/m^2)	I_{level} (W/m^2)	E_s (Wh/m^2)	t_{day} (h)	I_{max} (W/m^2)	I_{level} (W/m^2)	E_s (Wh/m^2)	t_{day} (h)
[0,23.5]°	value	898.7	572.13	6036.32	10.55	1294.63	824.18	10633.76	13.45
	Φ/°	23.5	23.5	23.5	23.5	0	0	23.5	23.5
	d_n/d	354	354	354	355	70	70	168	172
[-23.5,0]°	value	842.15	536.13	5656.45	10.55	1325.73	843.99	11346.79	13.45
	Φ/°	-23.5	-23.5	-23.5	-23.5	-23	-23	-23.5	-23.5
	d_n/d	173	173	173	172	365	365	357	355

figure 1 and figure 2 are the curves of $I_{\max}$ and I_{level} with date d_n and latitude Φ. The changes in figure 1 and figure 2 are basically the same. Near the Tropic of Cancer, the double peaks of $I_{\max}$ and I_{level} appear in the vicinity of the spring equinox and the autumnal equinox, and fall in the summer solstice, and the trough value appears near the winter solstice; Near the Tropic of Capricorn, the peaks of $I_{\max}$ and I_{level} appear near the winter solstice, and the trough value appears near the summer solstice, showing a relatively perfect sinusoidal trend.

figure 3 and figure 4 are the curves of E_s and t_{day} with date d_n and latitude Φ. The change trend of figure 3 and figure 4 are basically the same. In the vicinity of the tropic, both E_s and I_{level} show a relatively perfect sinusoidal trend. Near the Tropic of Cancer, the peak value appears near the summer solstice, and the trough value appears near the winter solstice; And it's opposite near the Tropic of Capricorn. In addition, figure 4 shows that the day and night time are equal at the vernal equinox and autumnal equinox between the Tropics of Cancer and Capricorn. At the equator, the time of day and night are the same all year round. This is consistent with the actual situation, and it verifies the reliability and accuracy of the calculation model to a certain extent.

Table 2 shows the maximum and minimum values and the corresponding latitude and date of the energy reception characteristics between the Tropic of Cancer and the equator [0,23.5]° , the equator and the Tropic of Capricorn [-23.5,0]° .

4 Energy reception characteristics varies with longitude and latitude

On the basis of the above analysis, this paper analyses the changes of solar energy reception characteristics between the Tropics of Cancer and Capricorn in a day with longitude and latitude. When calculating, the spring equinox day $d_n = 80$ is chosen as the research object. The latitude range is $\Phi = [-23.5,23.5]^\circ$, and the range of longitude is $L = [105,118]^\circ$.

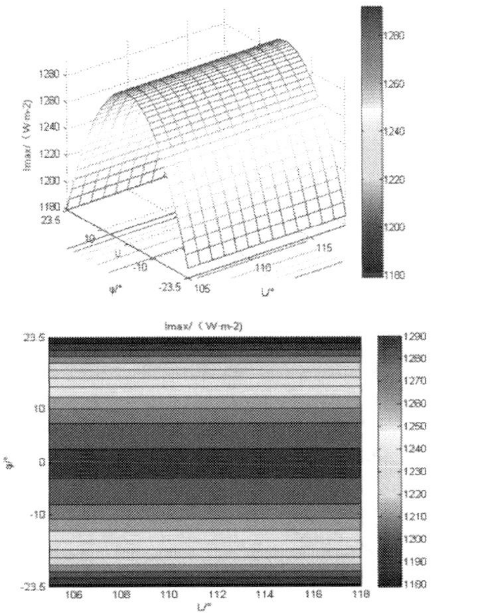

Figure 5. Maximum solar radiation intensity $I_{\max}$ varies with longitude L and latitude Φ.

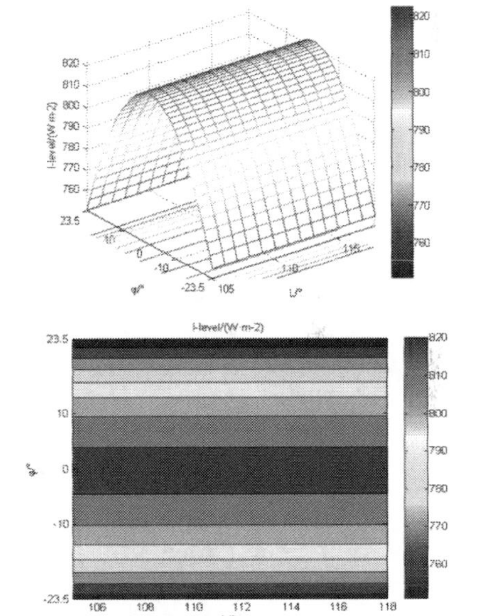

Figure 6. Daytime average solar radiation intensity I_{level} varies with longitude L and latitude Φ.

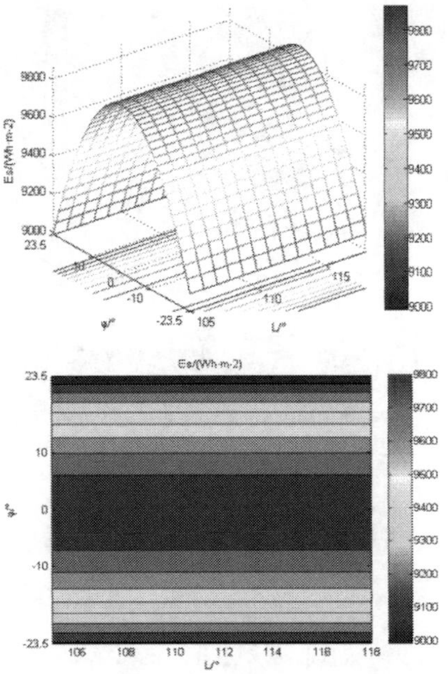

Figure 7. The total amount of solar radiation received per unit area E_s in one day varies with longitude L and latitude Φ.

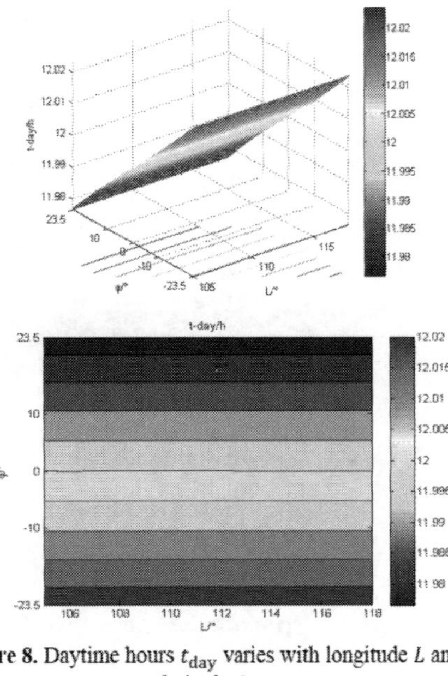

Figure 8. Daytime hours t_{day} varies with longitude L and latitude Φ.

Figure 5-8 shows the variation curves of I_{max}, I_{level}, E_s and t_{day} with longitude L and latitude Φ at vernal equinox. It can be seen that the I_{max}, I_{level}, E_s and t_{day} in a day don't vary with longitude, but changes with latitude. On the vernal equinox day, the I_{max}, I_{level} and E_s are sinusoidal between the Tropics of Cancer and Capricorn, and the maximum values are obtained near the equator. The length of the day is around 12h in this area.

5 Conclusion

In this paper, the energy reception characteristics including I_{max}, I_{level}, E_s and t_{day} of 20000m high altitude area between the Tropics of Cancer and Capricorn are calculated with the change of date d_n and latitude Φ within one year, and with the longitude L and latitude Φ within one day (vernal equinox).

Through analysis, it can be seen that:

1) In a year, the energy reception characteristics including I_{max}, I_{level}, E_s and t_{day} of a certain longitude will vary with the date and latitude;
2) In a day, the energy reception characteristics including I_{max}, I_{level}, E_s and t_{day} only change with latitude, not with longitude.

Solar radiation is abundant in the near space between the Tropics of Cancer and Capricorn. In this paper, the energy reception characteristics of solar aircraft are analysed, and the corresponding calculation results are given. In particular, the maximum and minimum values of energy reception characteristics can provide some reference for the design of solar aircraft for different flying missions in the region.

References

1. H.T. Sang, *D.(Lanzhou Traffic University)* (2016)
2. W.T. Hu, *D.(Tianjin University)* (2013)
3. X.Z. Gao, *D.(National University of Defense Technology)* (2014)
4. X.D. Mei, J.L. Sun, Z.Q. Li, X.F. Chen, J. Xing, H. Xu, L.L. Qie, Y. Lv, M. Li, and L. Li, Spectroscopy and Spectral Analy. **36(03)** , 609-617 (2016)
5. Y. Zhou, China Mod. Edu. Equ. **20**, 44-45 (2013)
6. S. Huang, A Vast View on Publishing, **01**, 38-39 (2015)
7. C.P. Lu, K. Ji, and F. Wang, Tech. Wind, **24**, 62 (2015)
8. J. Yang, Z.Z. Liu, B. Meng, and N. Li, Energy Eng. **01**, 35-38 (2011)
9. L.J. Ding, H.Y. Li, Ningxia Engin. Tech. **14(01)**, 7-9,15 (2015)

Design on Spaceborne Three Orthotropic Thin-Wall Tubular Antennas Deployment Mechanism

Wenjie Fan[1], Lin Wu[1], Hongmin Pen[2]

[1]*Key Laboratory of Microwave Remote Sensing, National Space Science Center, Chinese Academy of Sciences, Beijing 100190, China*
[2] *Tianjin Aerospace Electromechanical Equipment Research Institute, Tianjin 300458, China*

Abstract. The purpose of spaceborne three orthotropic thin-wall tubular antenna is for space science probe which consists of three orthotropic tubular antennas. Three orthotropic thin-wall tubular antennas deployment mechanism is designed which consists of upper plate, lower plate, reel, burn-wire and tension spring. Flattened antenna tube can be wound in a reel which can be hold-down and locked by Dyneema string. Release by giving current to resistor, the string will melt while its temperature reaches its burning point, then the upper plate will be pulled up and the antennas can be deployed by stored strain energy in itself. According to a certain project requirements three orthotropic thin-wall tubular antennas mechanism prototype was developed and tested, antenna tube was broken due to "blossom" phenomenon in deployment process. For avoiding "blossom" during deployment of antennas mechanism, the flange of reel is improved, so antennas can be deployed smoothly. Compared to motor-driven antenna mechanism, this antennas deployment mechanism has advantages of simple structure, lower mass and lower power consumption.

1 Introduction

Deep space exploration has became frontier field competed in space countries in 21st century, while space distribution longest wavelength radio interference imaging can offer the unique observation data for it. Low-frequency radio spectrometer has three meter-class metallic antennas which are orthotropic in space. To achieve exploration objective of space science and meet requirements of the envelope of rocket and limits of mass and power, the exploration antenna should meet requirements of wide frequency, low mass and volume, low power, easy mounting, high and low temperature proof and deployable capacity.

The three orthotropic antennas in this article adopts thin-walled tubular antennas which have simple structure, high deployment reliability and belong to one dimension space deployable mast[1-4]. At present, research on thin-walled tubular antenna mainly focus on root drum model(RDM)—the rotating drum remains at the base of the unit during extension. Usually RDM use a motor for the purpose of retraction; the motor also can be used to control extension velocity, but its mass, volume and power is high and can't be used on micro-satellite. Research on multi-deployment mast mainly concentrate on the deployment masts lie in one plane and they are all wound on one drum deployed by one motor[3-5], so it can't be used in orthotropic antennas which don't lie in one plane.

Three orthotropic antennas deployment mechanism is designed which is comprised of upper plate, lower plate, drum, burning wire and tension spring. The tubular antenna is flattened into a flat tape, then coiled onto a drum for storage, so it contain stored elastic energy. The upper plate, the lower plate and Dyneema string can be used to hold and lock the drum. Give current to resistor for releasing antenna, the burn wire melt when the temperature reach the melting point, then the upper plate is pull up and the antenna deploys with the drum under itself elastic energy. The three orthotropic antennas deployment mechanism not only has lower extension velocity, low shock, but also avoid blossom phenomenon which can destroy antenna tube. The three orthotropic antennas deployment mechanism has low volume, low mass and low power, also complete vibration and thermal test.

2 Design of three orthotropic antennas deployment mechanism

To achieve space science exploration objective, exploration antennas are full polarization and three orthotropic style which are comprised of three pairwise orthogonal antennas. Fig.1 shows antenna deployment mechanism in deployed status. Due to envelope limit of rocket, antennas should have retraction capacity. Slit thin-walled tubular antennas have many cross section which are overlapping single element, overlapping double element, inter-locking double element and lenticular welded double element, and each has its advantages and disadvantages. The process of inter-locking double element is complicated; overlapping double element needs two drums to retract two antennas and has big volume; the lenticular welded double element has high torsional stiffness, but the process isn't good for metal and the fundamental modal of antenna is bending mode. Overlapping single element has simple structure, perfect process, low mass and low volume, so it is chosen for three orthotropic antennas. The cross-section of overlapping single element is shown in Fig. 2. There is a part of overlap in this kind of antenna and this makes antenna has better stiffness.

As is shown in Fig.3, three orthotropic antennas are comprised of three slit thin walled antennas independently and mounted in one base which mounted on a box that printed circuit board is in it. There are three pillars on the box and one axis is between each two pillars. The upper plate is mounted on one axis and rounds on it, but lower plate is fixed on the base. Fig.4 shows magnification of local of antenna deployment mechanism. It can be seen that three antenna are flatten and wound on three drums, however the other side of each antenna is fixed. Teeth were set on the upper plate and drum for gearing and constrain the rotational degree of freedom, and the lower plates are constrained in the axial. The Dyneema strings go through the holes of upper plates and lower plates for pushing down the drums tightly. Thus the 6 dof are constrained for hold down and lock. Give current to the resistors for heating and melting the strings. The upper plates are pull up after the strings are melt, then the antennas deploy by the elastic strain energy. For high reliability, a spring-pull pillar is used for kick off.

Figure 1. Antenna deployment mechanism in deployed status

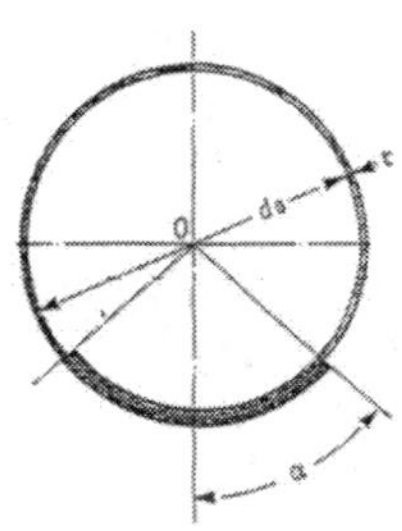

Figure 2. Section of antenna tube

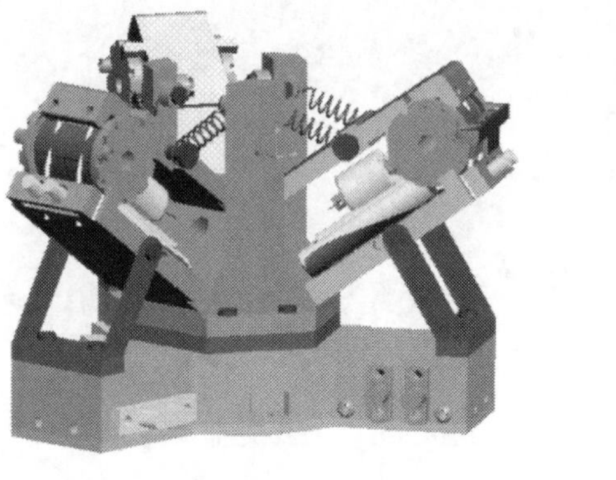

Figure 3. Deployment Mechanism of antenna in stowed status

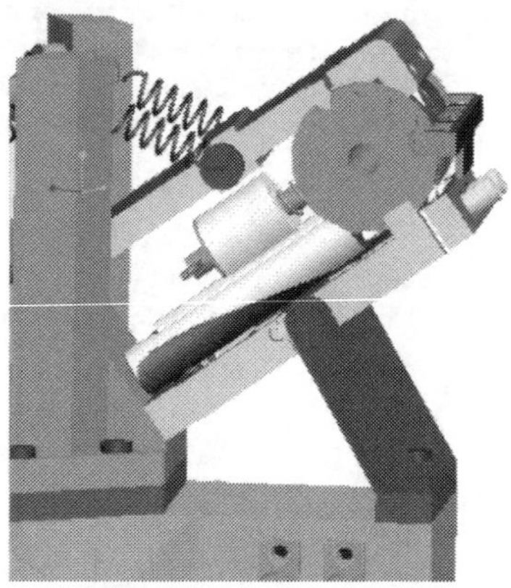

Figure 4. Local magnification of deployment mechanism

Figure 5. Product of mechanism

3 Example for three orthotropic antennas deployment mechanism

3.1 Three orthotropic antennas deployment mechanism

Detection frequency of payload is 1～30 MHz, so the antenna is 1 meter long,0.05mm thick and has 6mm in diameter. The overlap angle is 77.5° and the mass is only 10g for one antenna. The antennas are flatten and wound on the drums. The drums move with deployment of the antennas and can reach higher speed at the end of deployment. The drums are connected in the antennas with strings so that the drums don't become debris in space. Fig. 5 shows the product of three orthotropic tubular antennas deployment mechanism in stowed status. This kind of antenna deployment mechanism has high stow ratio. The envelope is 110 mm×110 mm×75 mm in stowed status and 1740 mm×1740 mm×620 mm in deployed status. The antennas are made of beryllium cooper, as it has not only high yield strength but also perfect electromagnetic properties. The upper plates and lower plates are made of polyimide. Compare to motor-driven concept, it's lighter, smaller and little power.

3.2 Deployment test and structure improvement for drum

For validating performance of antenna deployment mechanism, single antenna deployment test are performed before three orthotropic antennas deploy. Antenna tube was broken during deployment, as is shown in Fig.6. The cause of fracture of antenna tube in the initial deployment phase are analyzed and the drum didn't deploy with antenna tube, so 'blossom' phenomenon happened in antenna tube, as is shown in Fig.6(a). Thus the drum can't deploy with antenna tube, so it bend under gravity, as is shown in Fig.6(b). So avoiding 'blossom' phenomenon is the key element for deployment successfully.

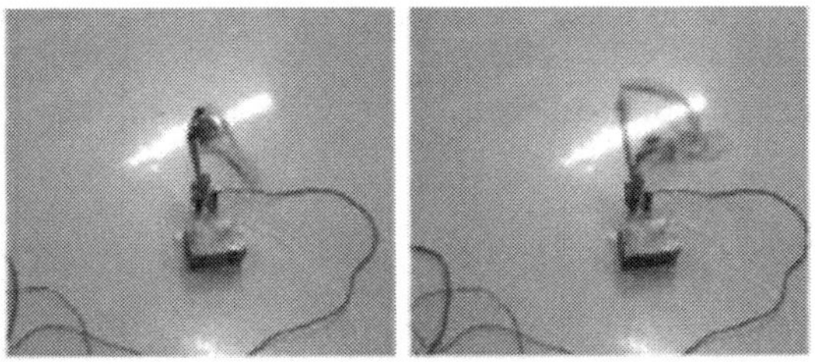

(a)

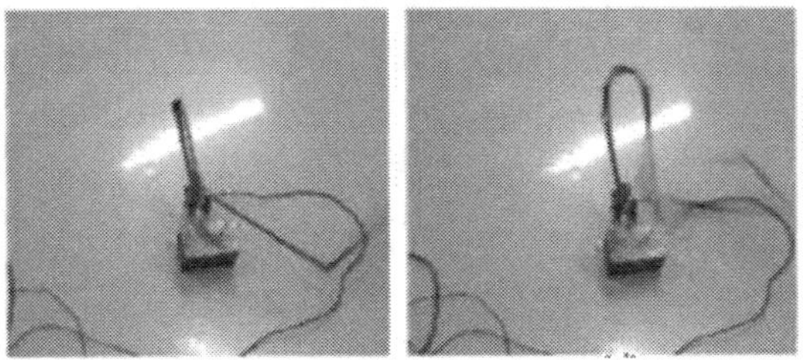

(b)

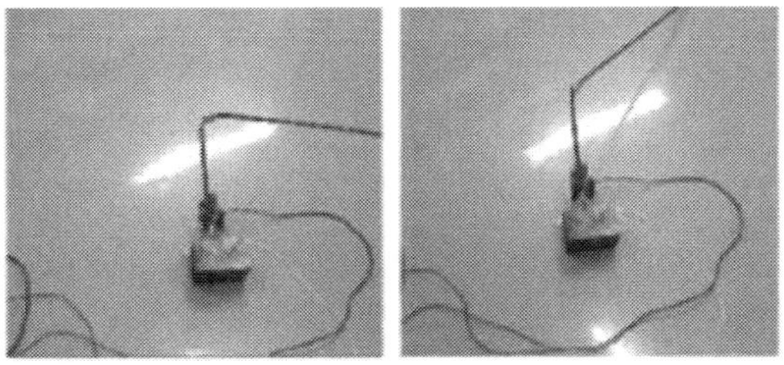

(c)

Figure 6. Deployment test of single antenna

Fig.7(a) shows drum before improvement and the two sides of drum are straight plates, so antenna tube can't be constrained in the deployment that swelled by itself elastic strain energy and bend under gravity. For solve this problem, the two sides of drum is improved into flanges which seem like concave plate. Thus they can constrain antenna tube for avoiding swell, as is shown in Fig.7(a). It was proved effective in deployment.

Three orthotropic antennas deployment test was performed successfully after single antenna tube deployed perfectly. But there is still a shortcoming in deployment time of three antennas. It can be deduced that heater mounting style and burn-wire winding style are the main reason that make the differences in deployment time. Fig.8 shows the three Orthotropic Tubular Antennas Deployment Mechanism in deployed status.

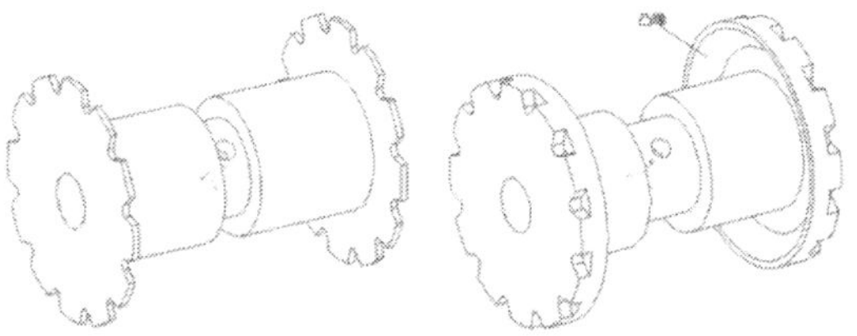

Figure 7. Structure of reel before and after improvement

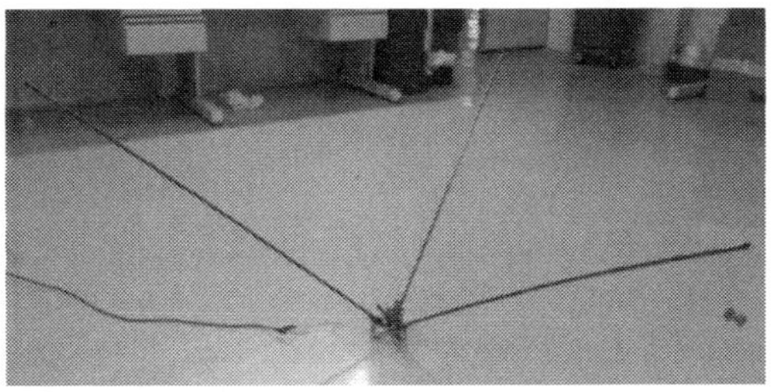

Figure 8. Three Orthotropic Tubular Antennas Deployment Mechanism in deployed status

4 Conclusion

Three orthotropic antennas deployment mechanism are developed which are comprised of upper plate, lower plate, drums and burn wire release element. There is a spring pre-loaded pull-bar for kick off for high reliability. The product is developed and deployment test is performed which proved deployment performance. Improvement on drum solve the problem that blossom phenomenon at the first of deployment, so the three antenna tubes can deploy smoothly and it also can decrease the speed by friction. It has smaller volume, lighter weight and lower power. The concept on antenna deployment mechanism can be applied in monopole and dipole antennas.

References

1. Gunnar Tibert. Deployable Tensegrity Structures for Space Applications[D].Royal Institute of Technology, Stockholm,Sweden,2002:10-16
2. Marco Straubel. Design and Sizing Method for Deployable Space Antennas[D]. Otto-von-Guericke University, Magdeburg, Germany,2012:14-21
3. Herzel, George G., Principal Contributor, "Tubular Spacecraft Booms (Extendible, Reel Stored)"[R], NASA SP 8065, Feb. 1971.
4. Chen liemin, et al. Space structure and mechanism[M]. Beijing: China Science and Technology Press, 2005:314-321(in Chinese)
5. J.M. Fernandez, M. Schenk, G. Prassinos, V.J. Lappas, S.O. Erb. Deployment Mechanisms of Gossamer Satellite Deorbiter[C]. 15th European Space Mechanisms & Tribology Symposium, ESMATS 2013 Noordwijk, The Netherlands, 25–27 September 2013
6. Fan Wenjie, Wu Lin. Design and mechanics Analysis on a X-type Cross-section Tubular Boom[J]. Chinese Journal of Space Science, Vol.(37)5:616-621(in Chinese)
7. MARCO Straubel, JOACHIM Block, MICHAEL Sinapius, et al. Deployable composite booms for various gossamer space structures[C]. The 52nd AIAA/ASME/ASCE/AHS/ASC Structures. Structural Dynamics and Materials Conference. Denver, Colorado, AIAA,2011
8. Hereck, L., et al. Development and test of deployable ultra-lightweight cfrp-booms for a solar sail[C]. European Conference on Spacecraft Structures, Material and Mechanical Testing 2002, CNES. Toulouse, France
9. F.P.J.RIMROTT, G.FRITZSCHE. Fundamentals of STEM Mechanism[C]. IUTAM-IASS Symposium on Deployable Structures: Theory and Applications, 2000:321-333

Dynamic positioning accuracy of dual-axis drive mechanism of satellite antenna with multi-clearance coupling

Chaoqun Qi[1], Huibo Zhang[1], Bindi You[2], Jizhuang Fan[3] and Shijie Dai[1]

[1]*Hebei University of Technology, Hebei Key Laboratory of Robot Sensing and Human-Robot Interaction, Tianjin, 300130, PR China*
[2]*School of Naval Architecture and Ocean Engineering, Harbin Institute of Technology at Weihai, Weihai 264209, Shandong, PR China*
[3]*State Key Laboratory of Robotics and System, Harbin Institute of Technology, Harbin 150001, Heilongjiang, PR China*

Abstract. The dual-axis drive mechanism of satellite antenna as a typical multi-joint aerospace mechanism consists of azimuth axis and pitch axis. The multi-clearance contained within those drive joint cause contact and impact in joint, and consequently affect the dynamic positioning accuracy. A nonlinear dynamic model of planetary gears with multiple clearances couplings is proposed to investigate the influence mechanism of clearances on the dynamic performances. The nonlinear factors such as radial clearance, backlash and time-varying meshing stiffness of the bearing are considered in the model. And the dynamic model of double-axis driving mechanism of satellite antenna with multi-clearance coupling is established. Then positional and rotational speed error curves of antenna reflector at different rotational speed are respectively analyzed. The numerical simulation results show that the influence of multi-clearance coupling on the dynamic positioning accuracy of double-axis drive mechanism of satellite antenna is remarkable.

1 Introduction

It is known that planetary gears are widely used in space mechanism. However, clearances in planetary gears are inevitable due to manufacturing, assembly errors, and wear. Those clearances cause contact and impact between gear sets, and consequently affect the control accuracy [1,2]. So the research on the vibration characteristics of planetary gear system are receiving more and more attention to obtain the high precision and high reliability of space mechanism.

From the modeling methodology point of view, several different methods have been developed. Kahraman [3-5] has presented a series of nonlinear time-varying dynamic models of planetary gears. Filipe [6] presented a new formulation to model spatial revolute joints with radial and axial clearances. Parker [7-9] analyzed the vibration mode of spur gear system using finite element method. Pan Bo [10] established a planetary gear dynamics model for the space manipulator joint considering the nonlinear factors. However, the current model ignores the effect of multi-clearance coupling on the vibration characteristics of planetary gear system. Multiple clearances are main factors that cause the performance and reliability decline of complicated mechanical systems. Those clearances cause contact and impact in joint, and consequently affect the control accuracy.

In this paper, a nonlinear dynamic model of planetary gears with multi-clearance couplings is proposed. Then the dynamic model of double-axis driving mechanism of satellite antenna is established. Finally, the effects of multi-clearance coupling on positional and rotational speed error curves of antenna reflector at different rotational speed are studied and discussed respectively by numerical simulation.

2 Dynamic model of system with multi-clearance coupling

2.1 Dynamic Model of Planetary Gear Drive Joint with multi-clearance coupling

2K-H planetary gear reducer as a typical joint system is the study object of current study. The system is composed of four parts: sun gear (s), ring gear (r), planet carrier (c), and a certain number of planet gears (p). It is known that a radial clearance exists in bearing, and a backlash occurs between the teeth. So that vibrations will occur in clearance during the course of movement, then the radial vibration at bearing lead to the change of gear's actual center distance, then the backlash will dynamically changed. The appearance of dynamic backlash will affect torsional vibration and even the whole vibration characteristics of gear system. Therefore, the dynamics model of planetary gear transmission with multi clearance coupling is established as shown in Figure 1.

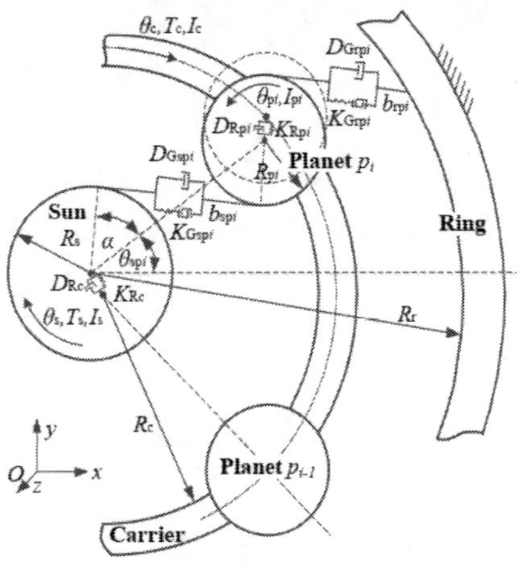

Figure 1. Dynamic model of planetary gear drive joint with multi-clearance coupling.

In Figure 1, there are n_p number of planetary gears. Each gear body j (j=s, p_i ,c) is modeled as a rigid gear radius R_j, angular displacement θ_j, and mass moment of inertia I_j. External torques T_s and T_c represent the input and output values. Between the planet gear p_i and gear n (s or r), K_{Gnpi}, D_{Gnpi} and b_{npi} are the periodically time-varying meshing stiffness, the viscous damper coefficient and backlash, respectively. α is theoretically pressure angle. K_{Rpi} and D_{Rpi} are the nonlinear stiffness and damping of the radial contact collision model at the planet gear bearing, respectively. θ_{spi} is the initial phase angle of planet gear relative to sun gear. And R_r is radius of ring gear. K_{Rc} and D_{Rc} are the nonlinear stiffness and damping of carrier output shaft and bearing radial contact collision model, respectively.

The equations of motion of planetary system are written as follows:

$$\begin{cases} I_s\ddot{\theta}_s(t)+R_s\sum_{i=1}^{n_p}\{D_{Gspi}\cdot\dot{g}_{tspi}(t)+K_{Gspi}\cdot f_{tspi}[g_{tspi}(t)]\}=T_s(t) \\ I_{cp}\ddot{\theta}_c(t)+I_p\sum_{i=1}^{n_p}\ddot{\theta}_{cpi}(t)-R_s\sum_{i=1}^{n_p}\{D_{Gspi}\cdot\dot{g}_{tspi}(t)+K_{Gspi}\cdot f_{tspi}[g_{tspi}(t)]\} \\ -R_r\sum_{i=1}^{n_p}\{D_{Grpi}\cdot\dot{g}_{trpi}(t)+K_{Grpi}\cdot f_{trpi}[g_{trpi}(t)]\}=T_c(t) \\ m_c\ddot{\delta}_{cx}(t)+\sum_{i=1}^{n_p}\{D_{Rpi}\cdot\dot{\delta}_{pix}(t)+K_{Rpi}\cdot f_{rpix}[\delta_{pix}(t)]\}-D_{Rc}\cdot\dot{\delta}_{cx}(t) \\ -K_{Rc}\cdot f_{rcx}[\delta_{cx}(t)]=0 \\ m_c\ddot{\delta}_{cy}(t)+\sum_{i=1}^{n_p}\{D_{Rpi}\cdot\dot{\delta}_{piy}(t)+K_{Rpi}\cdot f_{rpiy}[\delta_{piy}(t)]\}-D_{Rc}\cdot\dot{\delta}_{cy}(t) \\ -K_{Rc}\cdot f_{rcy}[\delta_{cy}(t)]=0 \\ I_p\ddot{\theta}_{cpi}(t)+I_{cp}\ddot{\theta}_c(t)-R_pD_{Gspi}\cdot\dot{g}_{tspi}(t)-R_pK_{Gspi}\cdot f_{tspi}[g_{tspi}(t)] \\ +R_pD_{Grpi}\cdot\dot{g}_{trpi}(t)+R_pK_{Grpi}\cdot f_{trpi}[g_{trpi}(t)]=0 \\ m_{pi}\ddot{\delta}_{pix}(t)-D_{Rpi}\cdot\dot{\delta}_{pix}(t)-K_{Rpi}\cdot f_{rpix}[\delta_{pix}(t)]=0 \\ m_{pi}\ddot{\delta}_{piy}(t)-D_{Rpi}\cdot\dot{\delta}_{piy}(t)-K_{Rpi}\cdot f_{rpiy}[\delta_{piy}(t)]=0 \end{cases} \tag{1}$$

In these equations, i=1, 2,..., n_p, and in Figure 1, $\theta_{cpi}(t)=\theta_{pi}(t)-\theta_c(t)$ represents rotation angle of planet gear with respect to carrier. Here, $I_{cp}=I_c+n_p(I_p+R_c^2m_p)$ is equivalent mass moment of inertia of carrier.

And in these equations, the radial collision stiffness K_{Rk} (k=c, p_i) could be obtained by collision experiment of two spheres, it is defined as:

$$K_{Rk}=\frac{4}{3\pi(\frac{1-\upsilon_{bk}^2}{\pi E_{bk}}+\frac{1-\upsilon_{jk}^2}{\pi E_{jk}})}\left[\frac{R_{bk}R_{jk}}{R_{bk}-R_{jk}}\right]^{1/2} \tag{2}$$

Where, υ_{ik} and E_{ik} (i=b, j) are Poisson's ratio and elastic modulus of center element i, respectively. The variable b represents bearing, and j represents shaft.

The damper coefficients of radial collision, namely, D_{Rc} and D_{Rpi} are defined as

$$D_{Rk}=\frac{3K_{Rk}(1-c_{ek}^2)\delta_k^m}{4\dot{\delta}_k^{(-)}} \tag{3}$$

Where c_{ek} is coefficient of restitution, $\dot{\delta}_k^{(-)}$ is initial relative velocity at the collision location.

The time-varying meshing stiffness K_{Gh} (h= spi, rpi)between gear teeth is given by

$$K_{Gh}(t)=k_{Gmk}+k_{Gah}\cos(\omega_{Gh}t+\varphi_{Gh}) \tag{4}$$

Where, k_{Gmh} is average meshing stiffness, k_{Gah} is time-varying meshing stiffness, ω_{Gh} is gear meshing frequency. φ_{Gh} is initial phase of variable stiffness (φ_{Gh}=0 in general).

D_{Gspi} and D_{Grpi} are the non-linear damper of gear meshing. In order to prevent the discontinuity of damping force during the meshing process and avoid the phenomenon in which the impact force of linear damping model is not equals to zero in critical contact state, the nonlinear damping force can be expressed as

$$D_{Gh}=\begin{cases} D_{Gmh} & f_{gh}(g_{th})\geq d_h \\ D_{Gmh}-\beta_h^2(3-2\beta_h) & 0\leq f_{gh}(g_{th})<d_h \\ 0 & f_{gh}(g_{th})<0 \end{cases} \tag{5}$$

Where D_{Gmh} is the maximum damping coefficient; d_h is the maximum embedding depth and values as 0.1mm, and β_h is defined as $\beta_h=f_{gh}(g_{th})/d_h$.

2.2 Dynamic Model of Dual-Axis Drive Mechanism of Satellite Antenna with Multi-Clearance Coupling

The dual-axis drive mechanism of satellite antenna, which be used in the inter-satellite link, is a typical multi-joint aerospace mechanism. It consists of azimuth axis and pitch axis, and the two axes are perpendicular to each other as shown in Figure 2.

In the current study, the driving mechanism of azimuth and pitch axis joints were converted into two plane rotation hinge, respectively. The model only considers the influence of radial clearance on the mechanism. And the kinetic equation is shown in equation (6):

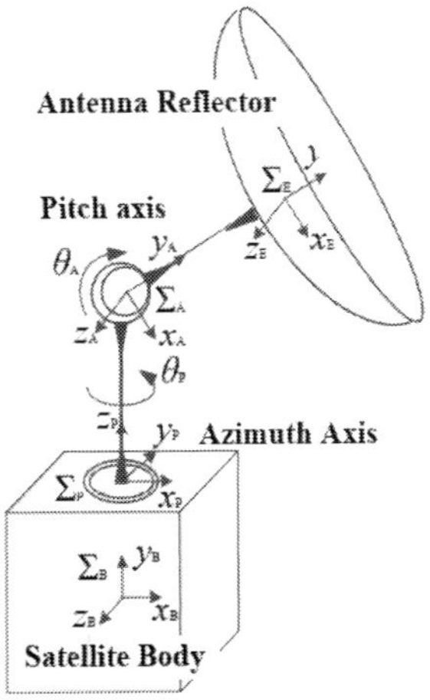

Figure 2. Dynamic model of dual-axis drive mechanism of satellite antenna with multi-clearance coupling.

$$\begin{cases} M\ddot{q} + C\dot{q} + Kq + \Phi_q^T \lambda = F \\ \Phi(q,t) = 0 \end{cases} \tag{6}$$

Where q is a generalized coordinate array, M is a generalized mass matrix of the gear rotor system, C and Φ_q^T are the damping matrix and jacobian of the constaints, respectively. K is a generalized stiffness matrix, λ refer to Lagrangian Multiplier vector, F is a generalized external force matrix.

Because a coupling relationship between radial clearance and backlash is existed. So after considering the multi-clearance coupling in the dynamic model of double-axis drive mechanism, the kinetic equation becomes as shown in equation (7):

$$M_p \ddot{q}_p = Q_F + Q_V + Q_C \tag{7}$$

Where M_p is a generalized mass matrix of all bodies in the double-axis drive mechanism (including the internal structure of a planetary gear drive joint),q_p is a generalized coordinate array. Q_F and Q_V are generalized external force matrix and the quadratic term related to velocity, respectively. And Q_c is the internal force matrix of the system at the clearance. Take into account the forces and moments inside the planetary gear joint, including the dynamic meshing torque T_{spi} between the sun gear and planetary gear i, the dynamic meshing torque T_{rpi} between planetary gear i and ring gear, the radial contact force between the planetary gear and the planet carrier is F_{Rpi}, and the radial contact force between the output axis of the planet carrier and the axle sleeve isF_{Rc}, the equations are expressed as:

$$\begin{bmatrix} m_{RR} & m_{R\theta} \\ m_{R\theta} & m_{\theta\theta} \end{bmatrix} \cdot \begin{bmatrix} \ddot{R} \\ \ddot{P} \end{bmatrix} = \begin{bmatrix} Q_{FR} \\ Q_{F\theta} \end{bmatrix} + \begin{bmatrix} Q_{VR} \\ Q_{V\theta} \end{bmatrix} + \begin{bmatrix} \sum_{j=1}^{2} \left(\begin{aligned} &\sum_{i=1}^{n_p} Q_{CR}(T_{spij}) + \sum_{i=1}^{n_p} Q_{CR}(T_{rpij}) \\ &+ \sum_{i=1}^{n_p} Q_{CR}(F_{Rpij}) + Q_{CR}(F_{Rcj}) \end{aligned} \right) \\ \sum_{j=1}^{2} \left(\begin{aligned} &\sum_{i=1}^{n_p} Q_{C\theta}(T_{spij}) + \sum_{i=1}^{n_p} Q_{C\theta}(T_{rpij}) \\ &+ \sum_{i=1}^{n_p} Q_{C\theta}(F_{Rpij}) + Q_{C\theta}(F_{Rcj}) \end{aligned} \right) \end{bmatrix} \tag{8}$$

Where, m_{RR} is the mass matrix of translational part of all bodies in the system generalized mass matrix. R and P are the position vector and the quaternion of all the bodies in the inertial coordinate system, respectively. And $m_{\theta\theta}$ is the inertial tensor of all the body rotating parts. Q_{FR} and $Q_{F\theta}$ are the generalized forces of the external forces corresponding to translation and rotation respectively. $m_{R\theta}$ means the inertial coupling of all body moving and rotating, and Q_{VR} and $Q_{V\theta}$ are the velocity dependent quadratic terms corresponding to translation and rotation, respectively.

3 Numerical simulation

In Figure 2, the azimuth axis is connected to the satellite body, which drives the entire drive mechanism and antenna reflector rotation during the movement, and the pitch axis is located between the azimuth axis and antenna reflector. Above all, the movement of azimuth axis and pitch axis not only affect each other, but also affect the dynamics positioning accuracy of antenna reflector. In this section, the dynamic response of antenna reflector of dual-axis drive mechanism with multi-clearance is studied for the simultaneous rotation of dual-axis at different rotation speeds.

In actual work, the dual-axis drive mechanism exists in which two axis rotate at the same time. But from the control point of view, the azimuth and pitch axis does not have a larger rotational speed difference , namely, the rotational speed of the azimuth and pitch axis is equal. Here, three typical joint output speeds are selected, that is, 0.49°/s, 0.98°/s, and 1.96°/s. And the radial clearance and backlash in the dual-axis transmission joints are taken as 100μm.

Figure 3 shows the positional error curves of satellite antenna reflector on the three directions X, Y and Z axis at three typical rotational speeds. It can be see from in Figure 2 that the positional error curves of satellite antenna reflector in the direction X axis is large, while in the direction Y and Z axis is nearly equal to each other. The main reason is that azimuth and pitch axis are perpendicular to each other, which leads to the position error superposition in the direction X axis. By comparing curve (a), (b) and (c) in Figure 3, it can be see that the amplitude of position error of X axis's direction is increases with an increase in rotational speed. This phenomenon is mainly due to the rotation speed increases, the output shaft of dual-axis and the sleeve is in state of continuous contact, as the collision contact deformation increases, the positional error of antenna reflector also fluctuates greatly during the movement.

Figure 4 shows rotational speed error curves of satellite antenna reflector on the three directions X, Y and Z axis at three typical rotational speeds. It is noteworthy that the antenna reflector has a larger speed error on the three directions axis due to the coupling of multi clearances, when dual-axis rotate simultaneously. Comparing to chart (a), (b) and (c), it can be conclude that the rotational speed error in three axial directions are

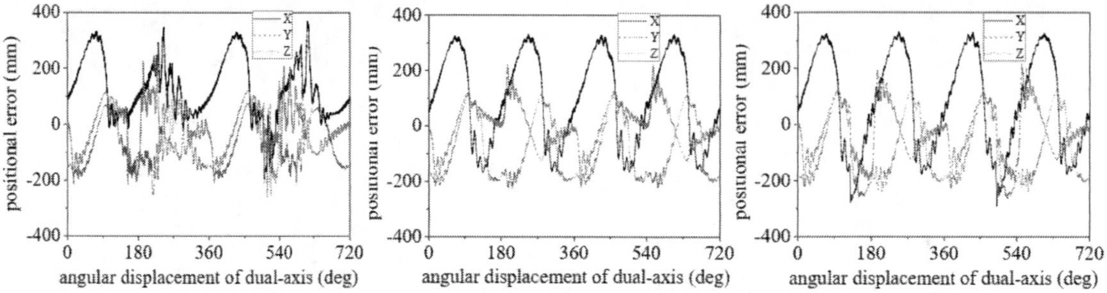

(a) Rotational speed 0.49°/s (b) Rotational speed 0.98°/s (c) Rotational speed 1.96°/s

Figure 3. Positional error curves of antenna reflector at different rotational speed.

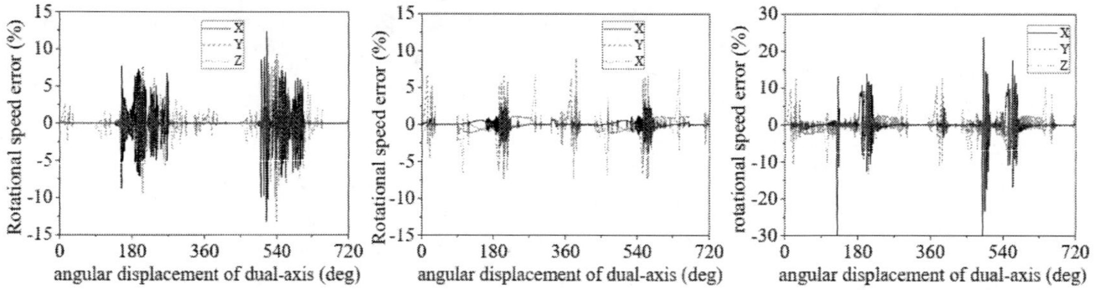

(a) Rotational speed 0.49°/s (b) Rotational speed 0.98°/s (c) Rotational speed 1.96°/s

Figure 4. Rotational speed error curves of antenna reflector at different rotational speed.

relatively small at the rotational speed of 0.98°/s, and the stability of system is better. However, when the rotational speed is 0.49°/s, the rotational speed error of satellite antenna reflector is not reduced with the lower rotational speed. And the amplitude of rotational speed error also fluctuates considerably at the rotational speed of 1.96°/s.

4 Conclusions

This analytical study on dynamic positioning accuracy of dual-axis drive mechanism of satellite antenna with multi-clearance coupling. First, this paper sets up a new nonlinear dynamic model of planetary gear joint, in which the radial clearance of bearing, dynamic backlash and time-varying meshing stiffness are considered. Second, the dynamic model of double-axis driving mechanism of satellite antenna with multi-clearance coupling is established. Finally, the effects of multi-clearance coupling on positional and rotational speed error curves of antenna reflector at different rotational speed are studied and discussed respectively by numerical simulation. It can be found that the amplitude of position error of X axis's direction is increases with an increase in rotational speed, while in the direction Y and Z axis is nearly equal to each other. For rotational speed errors curves of antenna reflector, the amplitude in the three axial directions are relatively small at the rotational speed of 0.98 °/s, and the stability of system is better.

5 Acknowledgment

This manuscript is funded by Science and Technology on Space Intelligent Control Laboratory, No.ZDSYS-2017 -08, and State Key Laboratory of Robotics and System (HIT), No.SKLRS-2017-KF-15, the National Natural Science Foundation of China (NSFC), No.51575126, and the Graduate Innovative Ability Training Project of Hebei Province, NO.CXZZSS2018025. The authors would like to thank the research grant from Spacecraft Dynamics Design and Simulation Lab (SDDSL) of HIT.

References

1. E. Salahshoor, S. Ebrahimi, M. Maasoomi. *Mech. Mach. Theory*. **105** 495-509 (2016)
2. H. Zhang, J. Tian, J. Zhou, Y. Zhao, B. You. *Chin. J. Mech. Eng.* **53** 29-37 (2017)
3. A. Alshyyab, A. Kahraman. *Proc. Inst. Mech. Eng. K. J. Multi-Body Dyn.* **221** 567-576 (2007)
4. P. Sondkar, A. Kahraman. *Mech. Mach. Theory*. **70** 157-174 (2013)
5. M. R. Kang, A. Kahraman. *J. Sound. Vib.* **350** 11-29 (2015)

6. M. Filipe, I. Fernando, D. Nuno, F. Paulo. *Mech. Mach. Theory.* **116** 123-144 (2017)
7. C. G. Cooley, R. G. Parker. *Mech. Syst. Signal. Process.* **40** 91-104 (2013)
8. Y.Guo, T. Eritenel, T. M. Ericson, R. G. Parker. *J. Sound. Vib.* **333** 5762-57852014
9. X.Dai, C. G. Cooley, R. G. Parker. *Mech. Mach. Theory.* **101** 60-74 (2016)
10. B. Pan, D. Yu, J. Sun. *J. Asrton.* **31** 2448-55 (2010)

Research on Shape Remain of Near-Earth Radial Collinear 4-Craft Coulomb Formation

SUN Yunlong , *YUAN Changqing* and LI Zhengguang

2222 Nanhu Road, Changchun City, Jilin Province, China

Abstract: This paper investigates the non-linear relative kinetic model of the collinear 4-craft Coulomb formation in geostationary orbit. Considering the fact that the formation remains statically fixed in the radial direction, using the equilibrium conditions to linearize the dynamic model. To keep the radial static stability of the collinear 4-craft Coulomb formation, the LQR controller is designed based on the linear dynamic model, taking the Coulomb force as the only control force. Considering the unmodeled disturbance force and the model error, the improved LQR control law is designed to improve the robustness of the traditional LQR controller. Simulation is executed by Matlab/Simulink, and numerical simulation results demonstrate the effectiveness of the dynamic model and the proposed control strategy.

1 Introduction

Satellites formation flying has obvious advantages in synthetic aperture radar and Spatial Interferometer. For recent years, utilizing Coulomb force to control satellite formation is a hot topic. The basic idea of Coulomb propulsion of free-flying vehicles is to control the spacecraft formation shape and size using the inter-spacecraft forces. This propellant-less thrusting is an attractive solution over conventional electric propulsion or chemical thrusting. It has very little electrical power requirements (one Watt or less) with a renewable energy source. And it causes no thruster plume contamination of the neighboring spacecraft, which increases the lifetime of the equipment and the probability of mission success.

Parker and King introduced the Coulomb propulsion concept in References 1 and 2 in 2002. It was proposed to control a cluster of free-flying spacecraft. Natarajan[3-5] designed feedback controller for 2-craft Coulomb formation at GEO based on relative distance and attitude dynamics equations. Inampudi and Schaub [6-7] designed charge feedback controller for 2-craft Coulomb formation at libration point based on dynamics equations.

2 Dynamics model

A 4-craft Coulomb formation is considered as shown in the Figure 1. It is assumed that the center of mass spin around the earth in GEO and the four charged satellites are nominally aligned along the radial direction with initial angle and separate distance error. In order to describe the relative motion of the satellite with respect to the formation center of mass, a rotating Hill orbit frame $\boldsymbol{H}:\{\boldsymbol{o}_r,\boldsymbol{o}_\theta,\boldsymbol{o}_h\}$ is chosen as shown in Figure 1. The relative position vector of the satellite is defined as $\boldsymbol{\rho}_i=[x_i \quad y_i \quad z_i]^T$.

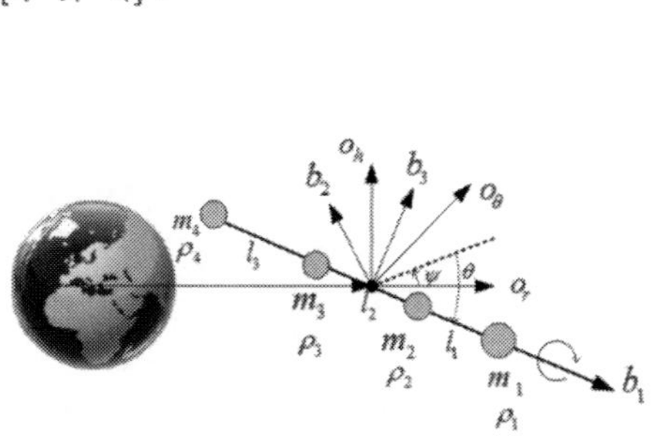

Figure 1. 4-craft Coulomb formation

The center of mass condition is defined as:

$$m_1\boldsymbol{\rho}_1+m_2\boldsymbol{\rho}_2+m_3\boldsymbol{\rho}_3+m_4\boldsymbol{\rho}_4=0 \tag{1}$$

Where m_i is the ith satellite mass.

Consider a coordinate frame $\boldsymbol{B}:\{\boldsymbol{b}_1,\boldsymbol{b}_2,\boldsymbol{b}_3\}$ as shown in Figure 1. l_i is the relative distance between satellites. The relationship between ρ_i and l_i can be expressed by:

$$\begin{cases}\rho_1-\rho_2=l_1\\ \rho_2-\rho_3=l_2\\ \rho_3-\rho_4=l_3\end{cases} \tag{2}$$

National Natural Science Foundation of China (11372353)

The direction cosine matrix that relates the $\boldsymbol{B}$ frame to $\boldsymbol{H}$ frame is given as $[\boldsymbol{BH}(\theta,\psi)]$. Using Eq. (1) (2) and $[\boldsymbol{BH}(\theta,\psi)]$, the position vector ${}^{H}\boldsymbol{\rho}_i$ can be solved. Take a derivative with respect to ${}^{H}\boldsymbol{\rho}_i$, we can get speed vector ${}^{H}\dot{\boldsymbol{\rho}}_i = [\dot{x}_i \;\; \dot{y}_i \;\; \dot{z}_i]^T$. The inertial velocity expressed in the $\boldsymbol{H}$ frame becomes

$$\boldsymbol{v}_i = \begin{pmatrix} \dot{x}_i - \Omega y_i \\ \dot{y}_i + \Omega(x_i + r_c) \\ \dot{z}_i \end{pmatrix} \quad (3)$$

The kinetic energy of the formation is: The kinetic energy of the formation is: $\boldsymbol{T} = \frac{1}{2}\sum_{i=1}^{4} m_i \|\boldsymbol{v}_i\|^2$. Using ${}^{H}\boldsymbol{\rho}_i$、${}^{H}\dot{\boldsymbol{\rho}}_i$ and Eq. (3), the $\boldsymbol{T}$ is rewritten as

$$T = \frac{1}{2\bar{m}}\left[\dot{\mathrm{X}} + \mathrm{X}\left(\dot{\theta}^2 + (\dot{\psi} + \Omega)^2 \cos^2\theta\right)\right] + \frac{\bar{m}}{2}\Omega^2 r_c^{\,2} \quad (4)$$

Where $\mathrm{X} = a_1 l_1^2 + a_2 l_2^2 + a_3 l_3^2 + 2a_4 l_1 l_2 + 2a_5 l_1 l_3 + 2a_6 l_2 l_3$,
$a_1 = m_1(m_2 + m_3 + m_4)$, $a_2 = (m_1 + m_2)(m_3 + m_4)$,
$a_3 = (m_1 + m_2 + m_3)m_4$, $a_4 = m_1(m_3 + m_4)$,
$a_5 = m_1 m_4$, $a_6 = (m_1 + m_2)m_4$, $\bar{m} = \sum_{i=1}^{4} m_i$.

The gravitational potential energy of the formation is:

$$V_g = -\frac{\mu\bar{m}}{r_c} - \frac{\mu\mathrm{X}}{2\bar{m}r_c^{\,3}}\left(3(\cos\theta\cos\psi)^2 - 1\right) \quad (5)$$

The Coulomb potential for the two-craft formation is:

$$V_e = k_c \sum_{i=1}^{3} q_i \sum_{j=i+1}^{4} \frac{q_j}{L} e^{-L/\lambda_d} \quad (6)$$

Where L is the distance between the two satellites. θ and ψ are the roll angle out of the orbital plane and the pitch angle in the orbital plane. q_i is the ith satellite charge.

The equations of motion are deduced from the Lagrangian function $\boldsymbol{L} = T - (V_g + V_e)$ in the following form:

$$\frac{d}{dt}\frac{\partial \boldsymbol{L}}{\partial \dot{q}} - \frac{\partial \boldsymbol{L}}{\partial q} = Q_i \qquad q \propto (\theta, \psi, l_1, l_2, l_3) \quad (7)$$

Where Q_i is the generalized force of the ith satellite excluding gravity. Using Eqs.(4), (5) and (6) in Eq. (7), we can get the nonlinear equations of motion governing the roll angle θ, the pitch angle ψ and the separation distance l_1、l_2、l_3.

The 4-craft formation remains statically fixed relative to the frame $\boldsymbol{H}$ in the radial direction, thus, the nonlinear equations satisfy the following equilibrium conditions:

$$\theta = \dot{\theta} = \ddot{\theta} = \psi = \dot{\psi} = \ddot{\psi} = \dot{l}_1 = \ddot{l}_1 = \dot{l}_2 = \ddot{l}_2 = \dot{l}_3 = \ddot{l}_3 = 0$$
$$l_1 = l_{1f}, l_2 = l_{2f}, l_3 = l_{3f} \quad (8)$$

The roll and pitch equations of motion are linearized about small roll and pitch angles respectively. In order to get the linearized equations of motion, assuming that the separation distance equations of motion are linearized about small variations δl_i and so as charge product δQ_{ij}. Shown as:

$$l_1 = l_{1f} + \delta l_1, l_2 = l_{2f} + \delta l_2, l_3 = l_{3f} + \delta l_3,$$
$$Q_{12} = Q_{12f} + \delta Q_{12}, Q_{13} = Q_{13f} + \delta Q_{13},$$
$$Q_{14} = Q_{14f} + \delta Q_{14}, Q_{23} = Q_{23f} + \delta Q_{23},$$
$$Q_{24} = Q_{24f} + \delta Q_{24}, Q_{34} = Q_{34f} + \delta Q_{34} \quad (9)$$

Where l_{if} and Q_{ijf} means the reference values. The charge product Q_{ij} should satisfy equations that utilize equilibrium conditions. In this paper, there are an infinite number of charge pairs can satisfy charge product requirements. Performing the necessary linearization and decouple equations about δl_i yields:

$$\ddot{\theta} + 4\Omega^2\theta = 0$$
$$\ddot{\psi} + 3\Omega^2\psi + 2\Omega\frac{X_1}{X_0}\delta\dot{l}_1 + 2\Omega\frac{X_2}{X_0}\delta\dot{l}_2 + 2\Omega\frac{X_3}{X_0}\delta\dot{l}_3 = 0$$
$$\delta\ddot{l}_1 - 2\Omega l_{1f}\dot{\psi} - 3\Omega^2\delta l_1 - \mu_1\frac{m_1 + m_2}{m_1 m_2}\delta Q_{12} + \frac{\mu_2}{m_2}\delta Q_{23}$$
$$-\frac{\mu_4}{m_1}\delta Q_{13} + \frac{\mu_5}{m_2}\delta Q_{24} - \frac{\mu_6}{m_1}\delta Q_{14} = 0$$
$$\delta\ddot{l}_2 - 2\Omega l_{2f}\dot{\psi} - 3\Omega^2\delta l_2 + \frac{\mu_1}{m_2}\delta Q_{12} - \mu_2\frac{m_2 + m_3}{m_2 m_3}\delta Q_{23} \quad (10)$$
$$+\frac{\mu_3}{m_3}\delta Q_{34} - \frac{\mu_4}{m_3}\delta Q_{13} - \frac{\mu_5}{m_2}\delta Q_{24} = 0$$
$$\delta\ddot{l}_3 - 2\Omega l_{3f}\dot{\psi} - 3\Omega^2\delta l_3 + \frac{\mu_2}{m_3}\delta Q_{23} - \mu_3\frac{m_3 + m_4}{m_3 m_4}\delta Q_{34}$$
$$+\frac{\mu_4}{m_3}\delta Q_{13} - \frac{\mu_5}{m_4}\delta Q_{24} - \frac{\mu_6}{m_4}\delta Q_{14} = 0$$

where $\mu_1 = \frac{k_c}{l_{1f}^2}$, $\mu_2 = \frac{k_c}{l_{2f}^2}$, $\mu_3 = \frac{k_c}{l_{3f}^2}$, $\mu_4 = \frac{k_c}{(l_{1f} + l_{2f})^2}$,
$\mu_5 = \frac{k_c}{(l_{2f} + l_{3f})^2}$, $\mu_6 = \frac{k_c}{(l_{1f} + l_{2f} + l_{3f})^2}$,
$\mathrm{X}_0 = a_1 l_{1f}^2 + a_2 l_{2f}^2 + a_3 l_{3f}^2 + 2a_4 l_{1f} l_{2f} + 2a_5 l_{1f} l_{3f} + 2a_6 l_{2f} l_{3f}$,
$\mathrm{X}_1 = a_1 l_{1f} + a_4 l_{2f} + a_5 l_{3f}$, $\mathrm{X}_2 = a_2 l_{2f} + a_4 l_{1f} + a_6 l_{3f}$,
$\mathrm{X}_3 = a_3 l_{3f} + a_5 l_{1f} + a_6 l_{2f}$, Ω is the orbit rate.

3 Controller Design

The ψ、δl_i are controlled with Coulomb force and θ is not controlled with Coulomb force. δQ_{ij} is chosen as control variable. Eq.(10) can be represented in the following form:

$$\dot{\boldsymbol{x}} = \boldsymbol{A}_1\boldsymbol{x} + \boldsymbol{B}_1\boldsymbol{u} \quad (11)$$

$$\boldsymbol{y} = \boldsymbol{C}_1 \boldsymbol{x} \tag{12}$$

Where the state $\boldsymbol{x} = \left[\psi, \dot{\psi}, \delta l_1, \delta \dot{l}_1, \delta l_2, \delta \dot{l}_2, \delta l_3, \delta \dot{l}_3\right]^T$.Using Coulomb force as an actuator mechanism, the $\boldsymbol{A}_1$、$\boldsymbol{B}_1$ and $\boldsymbol{C}_1$ matrices can be presented from Eq. (10) as:

$$\boldsymbol{A}_1 = \begin{bmatrix} 0 & 1 & 0 & 0 & 0 & 0 & 0 & 0 \\ -3\Omega^2 & 0 & 0 & -\frac{2\Omega\xi_1}{\xi_0} & 0 & -\frac{2\Omega\xi_2}{\xi_0} & 0 & -\frac{2\Omega\xi_3}{\xi_0} \\ 0 & 0 & 0 & 1 & 0 & 0 & 0 & 0 \\ 0 & 2\Omega l_{1f} & 3 & 0 & 0 & 0 & 0 & 0 \\ 0 & 0 & 0 & 0 & 0 & 1 & 0 & 0 \\ 0 & 2\Omega l_{2f} & 0 & 0 & 3 & 0 & 0 & 0 \\ 0 & 0 & 0 & 0 & 0 & 0 & 0 & 1 \\ 0 & 2\Omega l_{3f} & 0 & 0 & 0 & 0 & 3 & 0 \end{bmatrix}$$

$$\boldsymbol{B}_1 = \begin{bmatrix} 0 & 0 & 0 & 0 & 0 & 0 \\ 0 & 0 & 0 & 0 & 0 & 0 \\ 0 & 0 & 0 & 0 & 0 & 0 \\ \mu_1 \frac{m_1+m_2}{m_1 m_2} & -\frac{\mu_2}{m_2} & 0 & \frac{\mu_4}{m_1} & -\frac{\mu_5}{m_2} & \frac{\mu_6}{m_1} \\ 0 & 0 & 0 & 0 & 0 & 0 \\ -\frac{\mu_1}{m_2} & \mu_2 \frac{m_2+m_3}{m_2 m_3} & -\frac{\mu_3}{m_3} & \frac{\mu_4}{m_3} & \frac{\mu_5}{m_2} & 0 \\ 0 & 0 & 0 & 0 & 0 & 0 \\ 0 & -\frac{\mu_2}{m_3} & \frac{m_3+m_4}{m_3 m_4}\mu_3 & -\frac{\mu_4}{m_3} & \frac{\mu_5}{m_4} & \frac{\mu_6}{m_4} \end{bmatrix}$$

$$\boldsymbol{C}_1 = \begin{bmatrix} 1 & 0 & 0 & 0 & 0 & 0 & 0 & 0 \\ 0 & 0 & 1 & 0 & 0 & 0 & 0 & 0 \\ 0 & 0 & 0 & 0 & 1 & 0 & 0 & 0 \\ 0 & 0 & 0 & 0 & 0 & 0 & 1 & 0 \end{bmatrix}$$

The following feedback control is used to control the system with the feedback gain matrix, $\boldsymbol{K}$, computed using either the pole placement method or the linear quadratic regulator (LQR) method.

$$\boldsymbol{u} = -\boldsymbol{K}\boldsymbol{x} \tag{13}$$

Here the LQR methodology is applied to determine the optimal control, $\boldsymbol{u}$, such that the gain vector $\boldsymbol{K}$ minimizes the performance index.

$$J = \int_0^{\infty} \left(\boldsymbol{x}^T \boldsymbol{Q} \boldsymbol{x} + \boldsymbol{u}^T \boldsymbol{R} \boldsymbol{u}\right) \mathrm{dT} \tag{14}$$

where $\boldsymbol{u} = \left[\delta Q_{12} \quad \delta Q_{23} \quad \delta Q_{34} \quad \delta Q_{13} \quad \delta Q_{24} \quad \delta Q_{14}\right]^T$, $\boldsymbol{Q}$ and $\boldsymbol{R}$ are weighting matrices used to establish a faster response for control as design parameters.Eq.(10) can also be represented as:

$$\dot{\boldsymbol{x}} = \boldsymbol{A}_2 \boldsymbol{x} + \boldsymbol{B}_2 \boldsymbol{u} + \boldsymbol{\omega}(\boldsymbol{x}) \tag{15}$$

$$\boldsymbol{y} = \boldsymbol{C}_2 \boldsymbol{x} \tag{16}$$

where $\boldsymbol{x} = \left[\int \psi, \psi, \dot{\psi}, \int \delta l_1, \delta l_1, \delta \dot{l}_1, \int \delta l_2, \delta l_2, \delta \dot{l}_2, \int \delta l_3, \delta l_3, \delta \dot{l}_3\right]^T$.

$\boldsymbol{A}_2$、$\boldsymbol{B}_2$ and $\boldsymbol{C}_2$ matrices can be expressed as:

$$\boldsymbol{A}_2 = \begin{bmatrix} 0 & 1 & 0 & 0 & 0 & 0 & 0 & 0 & 0 & 0 & 0 & 0 \\ 0 & 0 & 1 & 0 & 0 & 0 & 0 & 0 & 0 & 0 & 0 & 0 \\ 0 & -3\Omega^2 & 0 & 0 & 0 & -\frac{2\Omega X_1}{X_0} & 0 & 0 & -\frac{2\Omega X_2}{X_0} & 0 & 0 & -\frac{2\Omega X_3}{X_0} \\ 0 & 0 & 0 & 0 & 1 & 0 & 0 & 0 & 0 & 0 & 0 & 0 \\ 0 & 0 & 0 & 0 & 0 & 1 & 0 & 0 & 0 & 0 & 0 & 0 \\ 0 & 0 & 2\Omega l_{1f} & 0 & 3 & 0 & 0 & 0 & 0 & 0 & 0 & 0 \\ 0 & 0 & 0 & 0 & 0 & 0 & 0 & 0 & 1 & 0 & 0 & 0 \\ 0 & 0 & 0 & 0 & 0 & 0 & 0 & 0 & 0 & 1 & 0 & 0 \\ 0 & 0 & 2\Omega l_{2f} & 0 & 0 & 0 & 0 & 3 & 0 & 0 & 0 & 0 \\ 0 & 0 & 0 & 0 & 0 & 0 & 0 & 0 & 0 & 0 & 1 & 0 \\ 0 & 0 & 0 & 0 & 0 & 0 & 0 & 0 & 0 & 0 & 0 & 1 \\ 0 & 0 & 2\Omega l_{3f} & 0 & 0 & 0 & 0 & 0 & 0 & 0 & 3 & 0 \end{bmatrix}$$

$$\boldsymbol{B}_2 = \begin{bmatrix} 0 & 0 & 0 & 0 & 0 & 0 \\ 0 & 0 & 0 & 0 & 0 & 0 \\ 0 & 0 & 0 & 0 & 0 & 0 \\ 0 & 0 & 0 & 0 & 0 & 0 \\ 0 & 0 & 0 & 0 & 0 & 0 \\ \mu_1 \frac{m_1+m_2}{m_1 m_2} & -\frac{\mu_2}{m_2} & 0 & \frac{\mu_4}{m_1} & -\frac{\mu_5}{m_2} & \frac{\mu_6}{m_1} \\ 0 & 0 & 0 & 0 & 0 & 0 \\ 0 & 0 & 0 & 0 & 0 & 0 \\ -\frac{\mu_1}{m_2} & \mu_2 \frac{m_2+m_3}{m_2 m_3} & -\frac{\mu_3}{m_3} & \frac{\mu_4}{m_3} & \frac{\mu_5}{m_2} & 0 \\ 0 & 0 & 0 & 0 & 0 & 0 \\ 0 & 0 & 0 & 0 & 0 & 0 \\ 0 & -\frac{\mu_2}{m_3} & \frac{m_3+m_4}{m_3 m_4} & -\frac{\mu_4}{m_3} & \frac{\mu_5}{m_4} & \frac{\mu_6}{m_4} \end{bmatrix}$$

$$\boldsymbol{C}_2 = \begin{bmatrix} 0 & 1 & 0 & 0 & 0 & 0 & 0 & 0 & 0 & 0 & 0 & 0 \\ 0 & 0 & 0 & 0 & 1 & 0 & 0 & 0 & 0 & 0 & 0 & 0 \\ 0 & 0 & 0 & 0 & 0 & 0 & 0 & 1 & 0 & 0 & 0 & 0 \\ 0 & 0 & 0 & 0 & 0 & 0 & 0 & 0 & 0 & 0 & 1 & 0 \end{bmatrix}$$

4 Numerical simulation

In this section, we present numerical simulation to illustrate the effectiveness of the proposed LQR controller. In the simulations,

$m_1 = m_2 = m_3 = m_4 = 50kg$, $l_{1f} = l_{2f} = l_{3f} = 20m$,

$k_c = 8.99 \times 10^9\ \mathrm{Nm^2/C^2}$, $\Omega = 2.66 \times 10^{-6}\ \mathrm{rad/sec}$,

$\delta\psi(0) = 0.1\mathrm{rad}$, $\delta l_1(0) = \delta l_2(0) = \delta l_3(0) = 0.6\ \mathrm{m}$,

$q_1 = 1.8697mC$, $q_2 = 42.445mC$, $q_3 = -46.569mC$, $q_4 = -195.103mC$.

And in the LQR control, control parameters matrices are:

$\boldsymbol{Q}_1 = diag\left[0.001 \quad 0.1 \quad 0.001 \quad 0.01 \quad 0.01 \quad 0.001 \quad 0.1 \quad 0.001\right]$

$\boldsymbol{R}_1 = diag\left[10^8 \quad 10^8 \quad 10^8 \quad 10^7 \quad 10^7 \quad 10^6\right]$

$\boldsymbol{Q}_2 = diag\left[1 \quad 10 \quad 30 \quad 0.5 \quad 0.01 \quad 0.01 \quad 0.1 \quad 0.01 \quad 0.01 \quad 0.1 \quad 0.001 \quad 0.1\right]$

$\boldsymbol{R}_2 = diag\left[10^7 \quad 10^8 \quad 10^7 \quad 10^7 \quad 10^7 \quad 10^7\right]$

The feedback gain K is solved by LQR method of MATLAB.

The dynamics of Coulomb formation system is simulated using the numerical differential equation solver in Matlab/Simulink for several cases. The relative tolerance is 10^{-6}.

The simulation figures shown as:

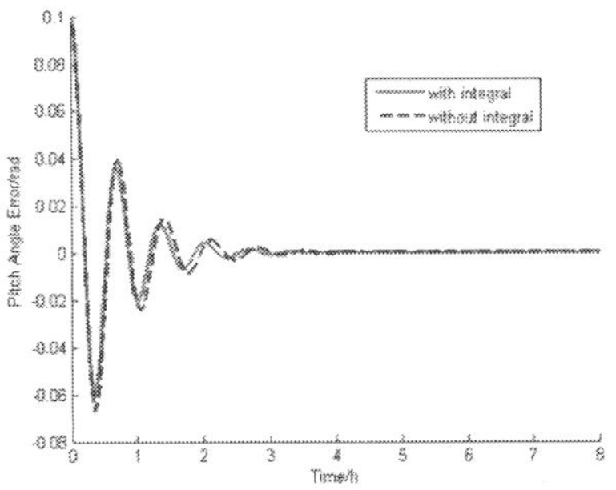

Figure 2. Time history of pitch angle $\delta\psi$

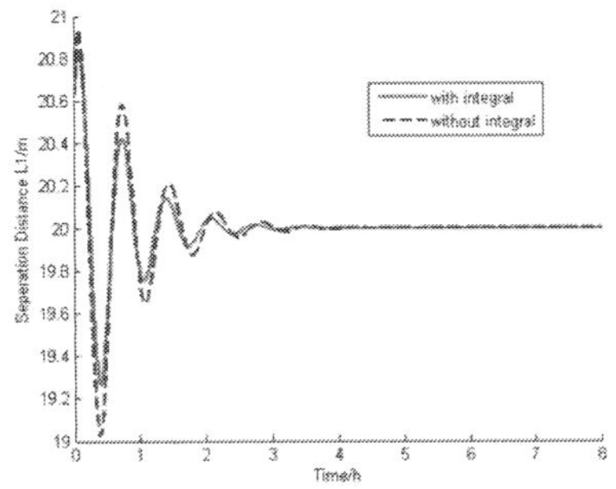

Figure 3. Time history of length variations l_1

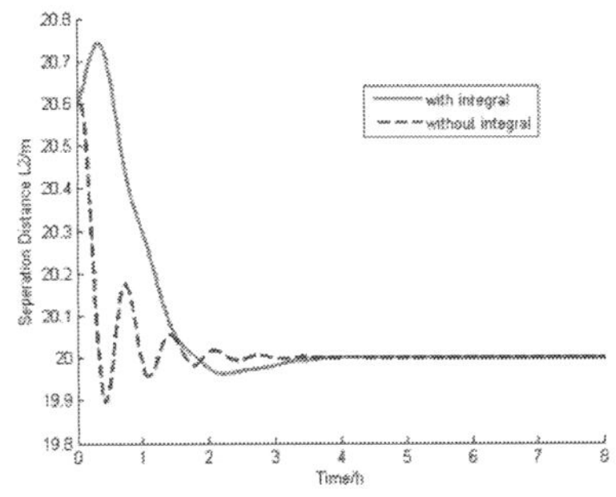

Figure 4. Time history of length variations l_2

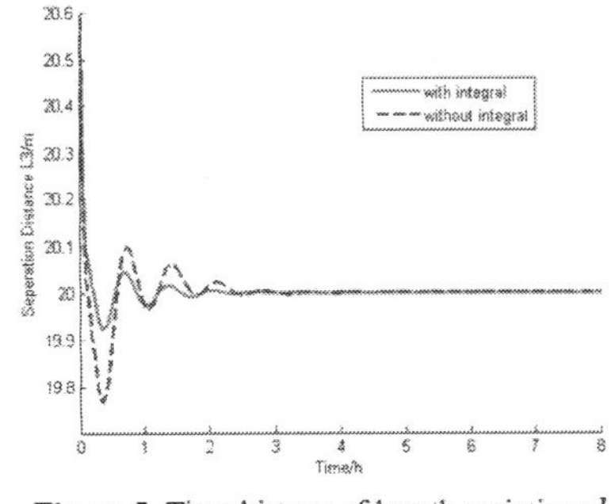

Figure 5. Time history of length variations l_3

From the simulation results we can see that the pitch angle error (Figure 2) of formation shows a amplitude attenuation and converge to zero at around 4h. The separation distances (Figure 3~ Figure 5) also converge to zero around 4h. All state variations converge to zero finally with given charge.

In each picture of Figure 2、Figure 3 and Figure 5, curves with improved LQR controller have less amplitude oscillation and converge faster than LQR controller without integral item. In Figure 4, the curve with integral part clearly shows a small amplitude variation and converge to desired distance with less oscillation frequency. It illustrates that the improved LQR controller gets better control results than the traditional LQR controller.

5 Conclusion

In this paper, a LQR control law for a 4-craft Coulomb formation at GEO is given. On the basis of dynamics model, considering the model error, the improved LQR control law is designed. Simulation results shows that the improved LQR controller gets better control results in controlling the attitude and separation distance to desired position in finite time. Numerical simulation results demonstrate the validity of the dynamic model and the effectiveness of the proposed control strategy.

References

1. King, L. B., Parker, G. G., Deshmukh, S., and Chong, J.-H., "Spacecraft Formation-Flying using Inter-Vehicle Coulomb Forces," Tech. rep., NASA/NIAC, January 2002, http://www.niac.usra.edu.
2. King, L. B., Parker, G. G., Deshmukh, S., and Chong, J.-H., "Study of Interspacecraft Coulomb Forces and Implications for Formation Flying," AIAA Journal of Propulsion and Power, Vol. 19, No. 3, May June 2003, pp. 497505.
3. Natarajan A, Schaub H. Linear Dynamics and Stability Analysis of a Coulomb Tether Formation. Journal of Guidance, Control, and Dynamics, 2006, 29(4): 831~839.
4. Natarajan A, Schaub H. Hybrid Control of Orbit Normal and Along-Track Two-Craft Coulomb Tethers. Aerospace Science and Technology, 2009,13:183~191
5. Natarajan A, Schaub H. Orbit-Nadir Aligned Coulomb Tether Reconfiguration Analysis. AAS/AIAA Spaceflight Mechanics Meeting, Galveston, TX, Jan. 27-31, 2008.
6. Inampudi R, Schaub H. Orbit Radial Dynamic Analysis of Two-Craft Coulomb Formation at Libration Points. AAS/AIAA Astrodynamics Specialist Conference, Toronto, Canada, Aug. 25, 2010.
7. Inampudi R, Schaub H. Orbit-Radial control of Two-Craft Coulomb Formation about circular orbits and Libration Points. 4th International Conference on Spacecraft Formation Flying Missions and Technologies. May 18-20, 2011.

Design and Experimental Verification of Double Star Sensor Bracket with Complex Space Angle for Microsatellites

Long Chen[1], Yu Xu[1], Xiaoli li[1], Tong Guo[1], Cheng Zhu[1], Rui Liu[1]

[1]*Innovation Academy for Microsatellites, Chinese Academy of Sciences, Shanghai 201203, China*

Abstract. In this paper, a double star sensor bracket design with complex space angle is proposed in order to improve the precision of microsatellites attitude measurement. This bracket is lightweight and is easy to manufacture. Finite element simulation and mechanical experiments are carried out. It is shown that the bracket satisfies the strength and stiffness requirements of microsatellites.

1 Introduction

Star sensor is a high-precision measurement device for satellite attitude, and it plays an important role in satellite attitude measurement and control system[1-4]. Star sensor has the advantages of high measurement accuracy, driftless system, etc. Therefore, it is widely used in satellite attitude measurement systems[5-6].

Generally star sensor is installed to the structural panel of the satellite via a particular bracket in order to keep its orientation fixed[7]. At present, the main used star sensor bracket of satellites is generally designed as a single independent bracket or a simple space angle double-star sensor bracket. Since the star sensor is generally the most precise measuring device when the satellite is in orbit, it is impossible to correct the measured value directly. To ensure the practicality of the compact layout remote sensing microsatellite's star sensor and improve the measurement precision of microsatellite attitude, in this paper, a novel two-frame design is proposed to achieve better efficiency and accuracy. This design can ensure that the two star sensors are in similar thermal and mechanical conditions in space, through the results measured by the two star sensors, the measurement precision of satellite attitude might be improved by mutual calibration and mutual correction.

In this paper, a double-star sensor embedding system with complex space angles was designed for the satellite attitude control subsystem. Mechanical analysis and production process optimization were performed. Experiments are carried out in order to validate the effectiveness of our design.

2 Model design

2.1 Design parameters

The angles between Optical axis of the star sensor and the satellite coordinate system are given in Table 1. Each star sensor has weight of 0.22 kg and envelope of 69mm×55mm×110mm. After installation, the sinusoidal response magnification at 0-100 Hz should be no more than 1.5 times acceleration input, and the RMS random response magnification at 20-2000 Hz should be no more than 2.5 times RMS input. The envelopes of the two sensors should occupy minimum space and the maximum total weight is limited to 0.25 kg. The deviation between star sensors and the main datum is less than 30″ before and after the vibration experiment.

Table 1. The angel between the star sensors' optical axis and the satellite's coordinate system

Items / Direction	X	Y	Z
Optical axis of star sensor A	67.48°	130°	131.56°
Optical axis of star sensor B	43.96°	70°	127.16°

2.2 Structural design

With the above-mentioned parameters and constraints, a bracket model is designed and is shown in Figure 1. The material used is 2A12T4 aluminum alloy, and the total weight is 0.245 kg.

The detailed design procedures are: a) draw the optical axes of star sensor A and star sensor B from the origin of the datum coordinate; b) draw star-sensors' mounting surfaces; c) establish normal planes of the two mounting surfaces; d) stretch the mounting surface entity based on the normal palne; e) cut the entity based on the mounting surface of the star sensor; f) draw bracket mounting surface; g) lightweight design and process optimization.

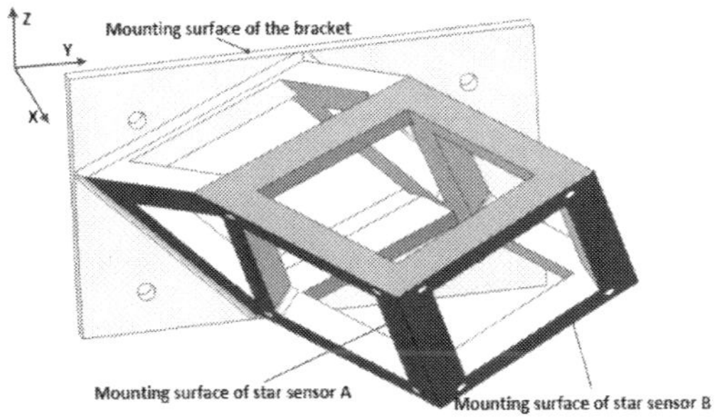

Figure 1. Model of the double star sensor bracket with complex space angle

3 Mechanical simulation

The finite element model of the double star-sensor bracket is shown in Figure 2.

The simulation of normal modes and frequency response adopts the same finite element model. The finite element model is modeling based on the following assumptions [8-9]:

(1) Assuming the model is a linear elastic structure system, geometric nonlinearity, contact nonlinearity caused by the clearance in structure assembly and complex mode of damping are not considered.

(2) The connection between the star sensor and the bracket is rigid.

(3) The bracket model is discretized using tetrahedral elements.

(4) Each star sensor is simplified as a lumped point to simulate its mass.

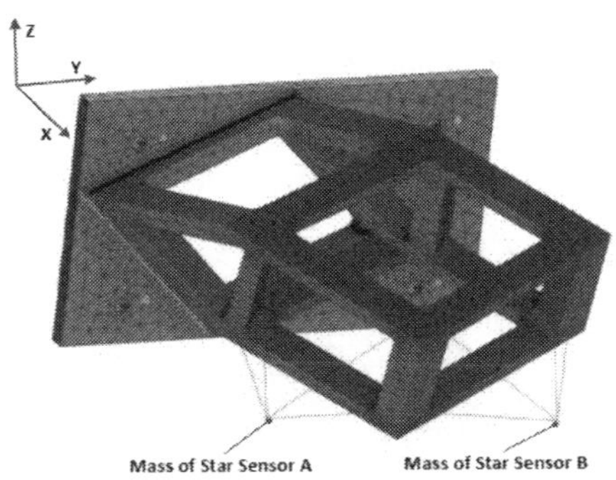

Figure 2. Finite element model of the bracket

3.1. Normal mode analysis

The mechanical parameters of the bracket model material are as follows: elastic modulus is 71 GPa, Poisson's ratio is 0.32, density is 2780 kg/m^3. The screw holes of the mounting surface of the bracket are fixed as boundary condition. The first three modes of the model are shown in Figure 3. The first-order eigenfrequency is 317.1Hz, which satisfies the design requirements of the bracket stiffness.

Figure 3. Mode 1: Freq. = 317.1Hz

Figure 4. Mode 2: Freq. = 488.8Hz

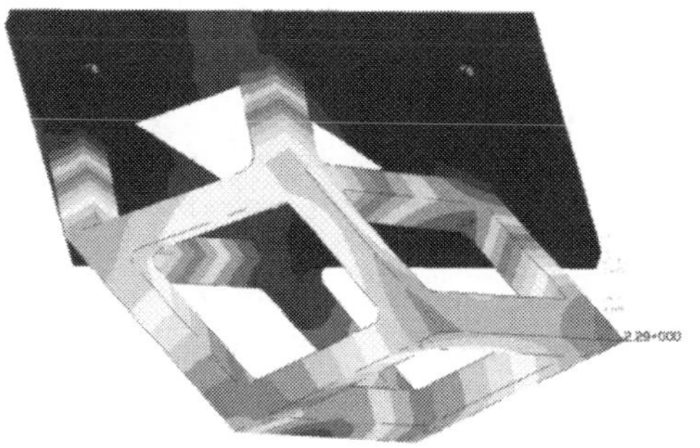

Figure 5. Mode 3: Freq. = 801.5Hz

3.2 Static analysis

According to mechanical environmental requirements of the rocket, the static loads applied along X, Y, and Z directions are 49 m/s^2, 49 m/s^2, and 157 m/s^2 respectively. By static simulation, the maximum stress in the bracket is 77.4MPa, which occurs around the installation screw hole. Since the allowable stress of 2A12T4 aluminium alloy is 255 MPa and the safety margin is 3.3, requirements in strength are satisfied.

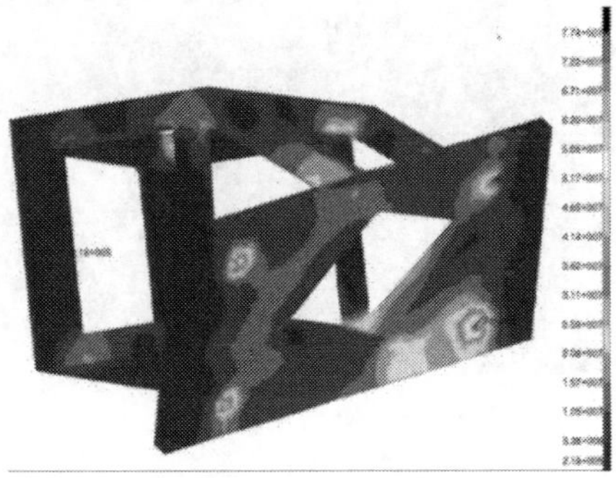

Figure 6. Von Mises Stress distribution of the bracket

3.3 Frequency response analysis

According to the rocket's mechanical environmental conditions, the sinusoidal response of the bracket at 0Hz-100Hz is mainly investigated. Since the first-order frequency of the bracket is 317.1Hz, which is greater than 100Hz, the sinusoidal response magnification in this frequency range is less than 1.2.

According to the rocket's mechanical environmental conditions, the random vibration input for the simulation is shown in Table 2. According to simulation results, in the range of 20Hz-2000Hz, the random response's RMS value magnifications of star sensor A in X, Y, and Z directions are respectively 1.27, 1.66 and 1.48, and those of star sensor B in X, Y, and Z directions are respectively 1.47, 1.86 and 1.96. As an example, the random vibration response curve of two star sensors in the Z direction is shown in Figure 7.

Table 2. Experimental conditions for random vibration

Random vibration experiment	Power spectral density	
Frequency（Hz）	Acceptance level	Identification level
20~150	+3 dB/oct.	+3 dB/oct.
150~1000	0.1 g^2/Hz	0.2g^2/Hz
1000~2000	-6 dB/oct.	-6 dB/oct.
Root-mean-square	11.9（grms）	16.9（grms）
Loading direction	X、Y、Z	X、Y、Z
Loading time	1 min in each direction	2 mins in each direction

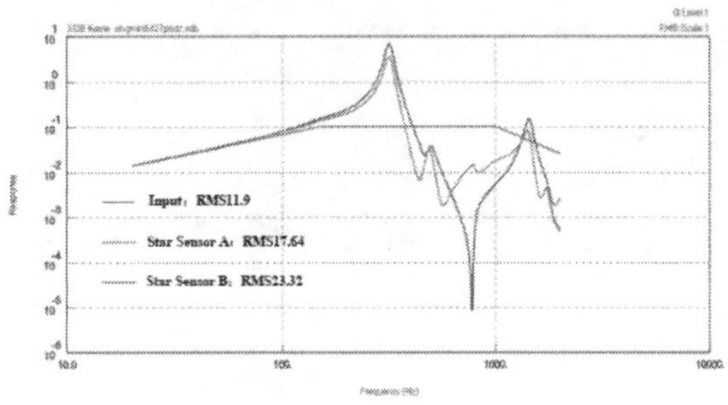

Figure 7. The simulation curve of random response of Z direction

4 Experimental verification

In order to verify rationality of the analysis results and the structure design, meanwhile expose defects in the manufacturing process, vibration experiments were carried out[10-11].

The montage of the bracket and star sensors is shown in Figure 8. Sinusoidal and random vibration experiments were performed along X, Y, and Z directions and the results are discussed as below:

1) The first-order eigenfrequency is 308.0 Hz. The sinusoidal response magnifications of the star sensors in X, Y, and Z directions are less than 1.2 for the range between 0 Hz to 100 Hz. The experiments curves in the three directions of X, Y, and Z are respectively shown in Fig. 9-11.

2) For the range of 20Hz-2000Hz, the random response's RMS value magnifications of the star sensor A in X, Y, and Z directions are respectively 1.69, 1.79 and 1.51, and those of the star sensor B are respectively 1.61, 2.22 and 1.52. The measured response curves for X, Y, and Z directions are shown respectively in Fig. 12-14.

3) Results of precision tests after the vibration experiments are shown in Table 3. The test results show that the deviation between the star sensors and the reference is less than 30″, which satisfies the design requirements.

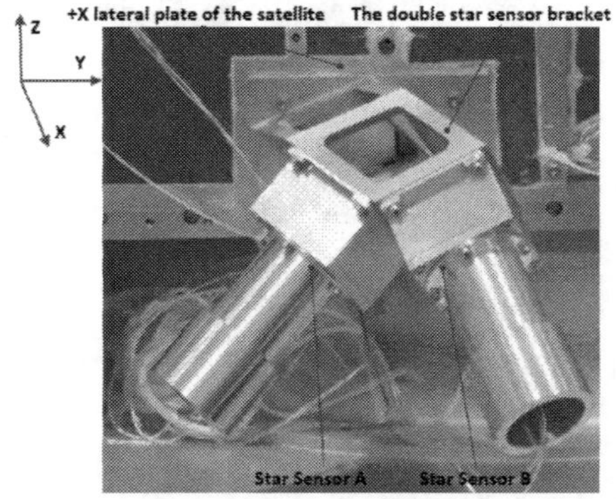

Figure 8. A montage of the combination for the bracket and star sensors

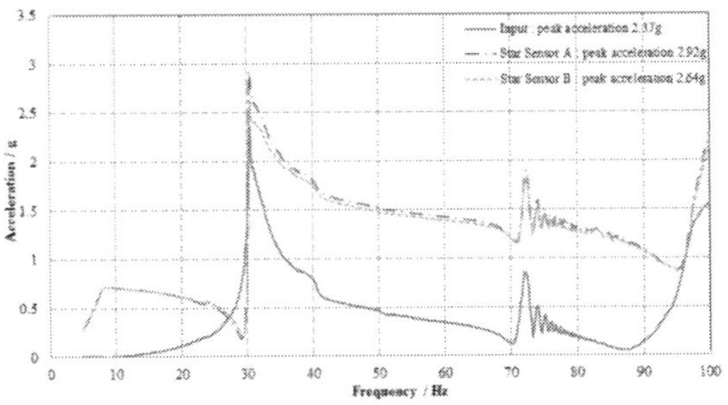

Figure 9. The sinusoidal vibration experiment curve of the bracket in X direction

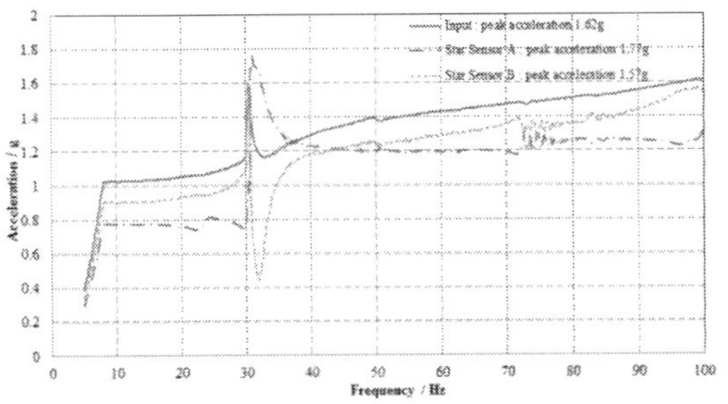

Figure 10. The sinusoidal vibration experiment curve of the bracket in Y direction

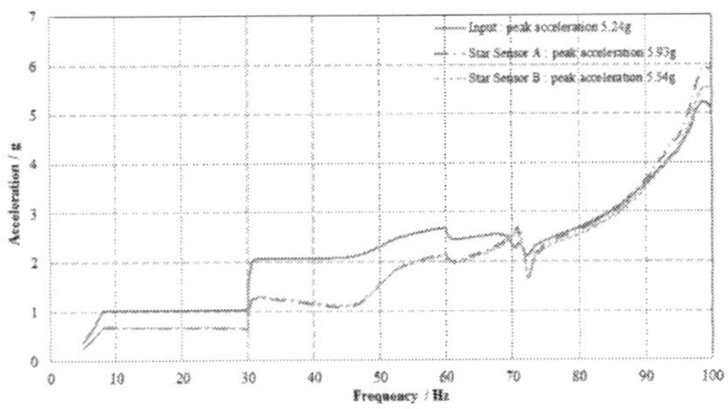

Figure 11. The sinusoidal vibration experiment curve of the bracket in Z direction

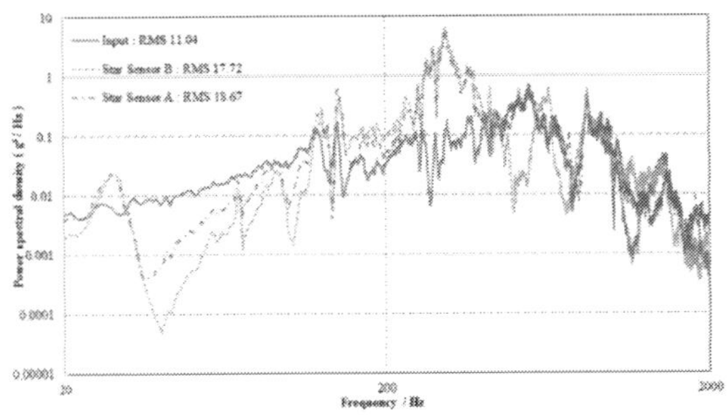

Figure 12. The random vibration experiment curve of the bracket in X direction

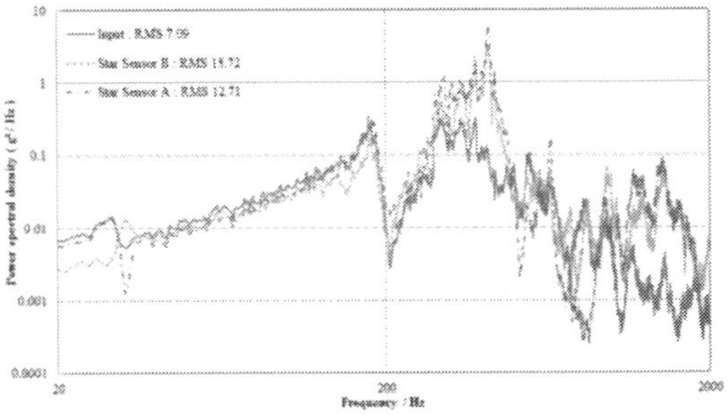

Figure 13. The random vibration experiment curve of the bracket in Y direction

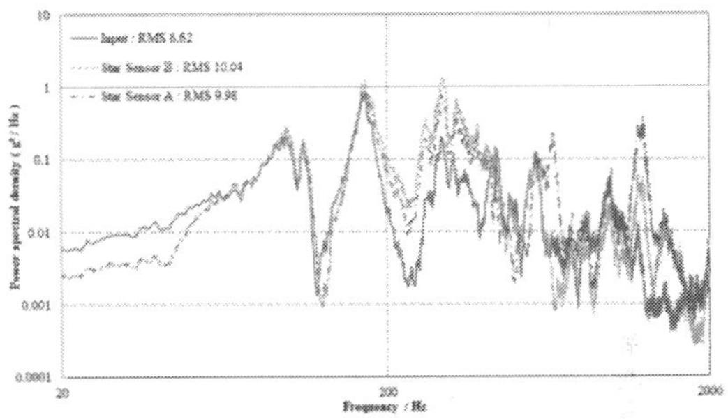

Figure 14. The random vibration experiment curve of the bracket in Z direction

Table 3. Results of precision tests after the vibration experiments

		Reference prism		
		X	Y	Z
Prism of star sensor A (-Y)	X′	-11.88″	3.24″	15.84″
	Y′	-9″	0	-5.4″
	Z′	-12.96″	7.56″	-13.32″
Prism of star sensor B (+Y)	X′	-20.16″	1.44″	24.84″
	Y′	11.16″	-12.24″	12.6″
	Z′	19.8″	-13.32″	16.56″

5 Conclusion

In this paper a double star sensor bracket with complex space angle is designed with aim of improving the precision of microsatellite attitude measurement and realizing mutual correction of the star sensors. Finite element simulation as well as mechanical experiments are carried out. The bracket has first-order eigenfrequency of 308 Hz. No obvious amplification was observed for sinusoidal vibrations. The random responses as well as the precision after vibration satisfy the design requirements.

References

1. Jorgensen J, Liebe C. The advanced stellar compass, development and operations [J]. Acta Astronautica, 1996 (11): 775-783.
2. Liao Zhibo, Fu Ruimin, Zong Xiaoyin. Optimal designing of baffle of star sensor [J]. Chinese Journal of Lasers, 2010, 37(4), 987-990.
3. Liu Haibo, Tan Jiyong, Shen Benjian. Thermal/ structure/ optical analysis of optical system of star sensor [J]. Journal of Astronautics, 2010, 31(3): 875-879.
4. Tu Shancheng. Attitue dynamics and control of satellite [M]. Beijing: China Astronautics Press, 2001.
5. Huang Xin. New APS star sensor [C]. Proceedings of Annual Congress of China association for Science and Technology, 2006: 49-55.
6. Liu Lei, Zhang Lu, Zheng Xin, et al. Current situation and development trends of star sensor technology [J]. Infrared and Laser Engineering, 2007, 36(9): 2529-2533.
7. Wang Haiming, Zhao Hua, Yang Wentao. Mechanical analysis of a star sensor's support of ZY-1 [J]. Spacecraft Environment Engineering, 2007, 24(3): 168-173.
8. Yuan Jiajun, et al. Design and analysis of satellite structures [M]. China Astronautics Publishing House, 2004.
9. Xu Benwen, Jiao Qunying. Mechanical vibration and modal analysis fundamentals [M]. China Mechanic Industry Press, 1998.
10. Chen Zhiping, Chen Zhiyuan, Yang Shimo. Modal analysis and experiment on the main truss of space solar telescope [J]. Opto-Electronic Engineering, 2004, 31(12).
11. Xia Lijuan, Yu Yin, Jin Xianding. Experimental research and optimal design for mode characteristic of the frame structure on satellite [J]. Journal of Shanghai Jiaotong University, 2004, 38(11).

A fast cold-start method of GPS receiver based on satellite orbit prediction

Ruihong Yang [1], Zhongguo Song [1], Jinsheng Zhang [2] and Xiaoli Xi [1]

[1] *Xi'an University of Technology, Xi'an, China*
[2] *Xi'an research inst of hi-tech, Xi'an, China*

Abstract. Time to first fix (TTFF) is one of the crucial indicators to evaluate the performance of a GPS receiver. In this paper, an orbit prediction algorithm to reduce the TTFF of GPS receivers without a network connection is presented. Satellite orbit is predicted by numerically integrating the satellite's equation of motion. Satellite's initial position and velocity value, as well as the Earth's polar motion parameters, used in prediction correspond to the locally collected previous broadcast ephemeris. As the solar radiation pressure (SRP) is one of the most critical factors causing orbit prediction error, an empirical SRP model with two parameters that change with the movement of satellite and Earth is also described. The presented algorithm is verified by GPS satellites using the initial conditions divided from three groups of broadcast ephemeris data. Simulation results show that, with the SRP model described in this paper the satellite's position error limited to 30 meters within 7 days prediction.

1 Introduction

Start navigation immediately is expected for people once their GPS receivers are turned on. However, it typically needs about 30 seconds for the receiver to provide the first position information even in an ideal receiving environment, during to the three first sub-frames which contain ephemeris data takes 18 seconds to transmit and the satellite sends it once every 30 seconds. If the signal is blocked by trees or buildings during the ephemeris collection and the TTFF can even take up to several minutes, which is unacceptable especially in an emergency or a special situation [1, 2].

Due to the time needed for the reception of the ephemeris data is the main reason for long TTFF, one method to reduce it is to provide satellite position information by orbit prediction. One technique of satellite orbit prediction that runs entirely on the receiver without any external assistance has been implemented in commercial products [3]. Reference [4] and [5] present a fully detailed description of an autonomous satellite orbit prediction method which only needs a few locally collected previous broadcast ephemeris data as a starting condition. This paper aims to develop a related method and make its efficiency and precision are competitive when compared to others.

In this paper, a practical orbit prediction algorithm to reduce the TTFF of a GPS receiver is presented. As the SRP is one of the most critical factors causing orbit prediction error, an enhanced sphere model that used for calculating SRP is adopted to improve the accuracy of orbit prediction. Also, to minimize the computational burden of GPS receiver, a gravity acceleration approximation function (GAAF) to calculate the Earth's gravitational acceleration is utilized to substitute the traditional spherical harmonic representation of the gravity field [6]. Simulation results show that the orbit prediction algorithm proposed in this paper can provide a trusted satellite position and significantly reduce the TTFF of a GPS receiver.

2 Force models

A Large number of different forces in the equation of motion will lead to a computationally heavier algorithm, which is not desired in a conventional GPS receiver. Therefore, only the four forces (including the Earth's gravitation, solar gravitation, lunar gravitation, and solar radiation pressure) that primarily affect the GPS satellite are taken into account.

The satellite's equation of motion used in this work is expressed as

$$\mathbf{a}_{Sat} = \mathbf{a}_{Earth} + \mathbf{a}_{Sun} + \mathbf{a}_{Moon} + \mathbf{a}_{SRP} \quad (1)$$

where $\mathbf{a}_{Sat}$ denotes the acceleration of the satellite, $\mathbf{a}_{Earth}$ is the acceleration due to the Earth's gravitation, $\mathbf{a}_{Sun}$ and $\mathbf{a}_{Moon}$ is the acceleration due to solar gravitation and lunar gravitation respectively. The term $\mathbf{a}_{SRP}$ is the acceleration caused by solar radiation pressure.

Once the satellite position and velocity at epoch t_0 is known for a satellite, one can predict the satellite orbit at an epoch t_k by integrating (1) with respect to time.

2.1 Earth's gravitation

For the Earth gravity, we utilized the GAAF method to compute the gravitational acceleration replacing the traditional spherical harmonic representation of the gravity field. The GAFF introduced an equation to represent gravity accelerations in terms of instant two-body acceleration for an instant pseudo-center for the earth. Pseudo-center location is represented by polynomials which as a function of height at fixed latitude and longitude. Pseudo-center at any point of interest can be calculated by classical six-point bivariate interpolation in latitude and longitude at a fixed height. Thus the calculation of gravitational acceleration is simplified to calculate the two-body acceleration. GAAF can significantly improve the computational efficiency in orbit prediction, more details about it can be found in [7]. In this work, we used the Earth gravitational model 2008 (EGM2008) geopotential up to degree and order 70 to generate the polynomial coefficient of pseudo-center calculation [8].

2.2 Solar and lunar's gravitation

Consider the Moon and the Sun as point masses. The acceleration due to a celestial body's gravitation is expressed as

$$\mathbf{a}_{\mathrm{Cb}} = GM_{Cb}\left(\frac{\mathbf{r}_{\mathrm{Cb}}-\mathbf{r}_{\mathrm{Sat}}}{\left|\mathbf{r}_{\mathrm{Cb}}-\mathbf{r}_{\mathrm{Sat}}\right|^3}-\frac{\mathbf{r}_{\mathrm{Cb}}}{\left|\mathbf{r}_{\mathrm{Cb}}\right|^3}\right) \quad (2)$$

where G is the gravitational constant, M_{Cb} and $\mathbf{r}_{\mathrm{Cb}}$ is the mass and position of the celestial body, $\mathbf{r}_{\mathrm{Sat}}$ is the position of the satellite.

2.3 Solar radiation pressure

The traditional sphere model of the SRP can be expressed as

$$\mathbf{a}_{\mathrm{SRP}} = -\nu\frac{C}{\left|\mathbf{r}_{\mathrm{s}}\right|^2}(1+\eta)\mathbf{e}_{\mathrm{D}} \quad (3)$$

where the factor ν is the shadow function based on the conical shadow model described in [9]. Vector $\mathbf{r}_{\mathrm{s}}$ points from the satellite to the Sun and $\mathbf{e}_{\mathrm{D}}$ is a unit vector from the satellite to the Sun. η is the reflectivity coefficient of the satellite. The constant C is

$$C = P_{Sun}\cdot\frac{A}{m}\cdot AU^2 \quad (4)$$

where AU is the astronomical unit, P_{Sun} is the solar radiation pressure at the distance of 1 AU from the Sun, A is the surface area of the satellite and m is the satellite's mass.

The simple sphere model described in (3) is widely used for calculating the SRP. However, it's accuracy need to be improved to meet the application with high precision requirements.

Taking the y-bias acceleration into consideration and then the SRP model we used in this paper can be written as

$$\mathbf{a}_{\mathrm{SRP}} = \nu\left(-\frac{C}{\left|\mathbf{r}_{\mathrm{s}}\right|^2}(1+\alpha_1)\mathbf{e}_{\mathrm{D}}+\alpha_2\mathbf{e}_{\mathrm{Y}}\right) \quad (5)$$

where $\mathbf{e}_{\mathrm{Y}}$ is the direction vector along the satellite's solar panel axis, and it can be obtained by

$$\mathbf{e}_{\mathrm{Y}} = \frac{\mathbf{r}_{\mathrm{Sat}}\times\mathbf{r}_{\mathrm{s}}}{\left|\mathbf{r}_{\mathrm{Sat}}\times\mathbf{r}_{\mathrm{s}}\right|} \quad (6)$$

α_1 and α_2 are SRP model parameters which need to be estimated. Although these two parameters change very little and are usually used as constants in a few days orbit prediction, in fact they are varying with the movement of satellite and Earth. Neglecting this variation will cause some errors.

The variation of SRP model parameters with the angle of Sun-satellite-Earth can be expressed by the following equation [10]:

$$\begin{cases}\alpha_1 = a+b\varepsilon^2+c\varepsilon^4+d\varepsilon^6 \\ \alpha_2 = e+f\varepsilon^2+g\varepsilon^4\end{cases} \quad (7)$$

where ε is the angle of the Sun-satellite-Earth, a □ g is the polynomial coefficients.

3 Procedures for orbit prediction

3.1. Estimation of SRP model parameters

Large amounts of data and computation power are needed in the estimation of the SRP parameters. Thus we have to estimate the SRP parameters offline on a server.

The parameters α_1 and α_2 in (5) are satellite-specific that need to be estimated separately for each satellite. Precise ephemeris from GPS weeks 1895-1925 is used as measurement data for this estimation [11]. For one satellite, a time series of parameter estimates are obtained by taking separate consecutive 1-hour arcs and estimating the parameters for each arc using an iterative least-squares minimization algorithm or the Extended Kalman Filter. For each arc the output of the estimator is taken as the estimated value. Then use (7) to fit the polynomial coefficients to all estimated values. By doing this processing for each satellite, the polynomial coefficients for calculating the SRP for all satellites are obtained.

In the estimation of SRP model parameters, the initial position and velocity of the satellite are taken from precise ephemeris, and the Earth Orientation Parameters (EOP) needed in coordinate system transformation are obtained from the homepage of International Earth Rotation and Reference Systems Service（IERS）[12].

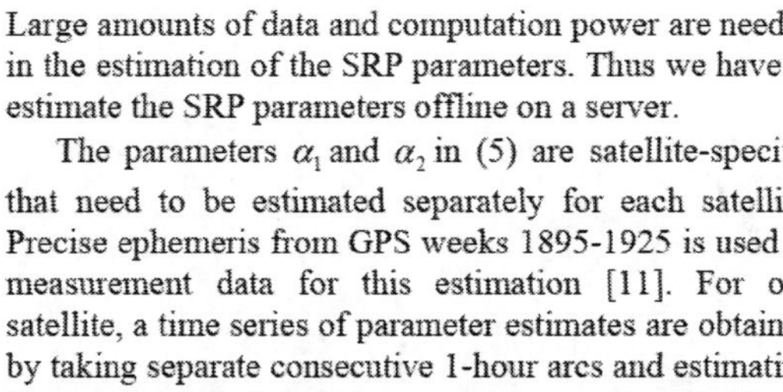

Figure 1. SRP parameters α_1 of GPS PRN2

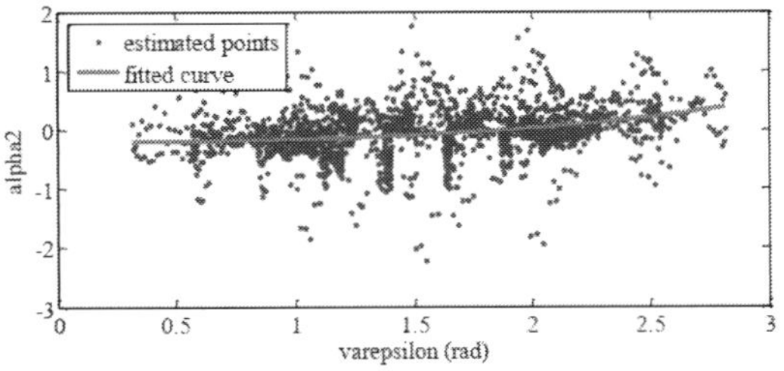

Figure 2. SRP parameters α_2 of GPS PRN2

Fig. 1 and 2 show the variation of α_1 and α_2 with ε angle for GPS PRN2.

3.2. Initialization

For orbit prediction, the forces model and the initial position and velocity significantly affect the prediction accuracy. Models and algorithms for calculating the forces have been discussed in the previous sections.

The initial position and velocity of satellite play a key role in orbit prediction. However, the position and velocity of satellite calculated by broadcast ephemeris are more inaccurate, and it cannot be used as an initial state directly. For this reason, the least squares algorithm is used to get an estimation of the initial state. In addition, the EOP used to calculate the matrix of coordinate system transformation in orbit prediction is unknown. So it has to be estimated together with the initial state from the broadcast ephemeris.

To minimize the number of parameters to be estimated, and improve the accuracy of the least squares solution, we only make a simple correction by antenna offsets for the initial position calculated by broadcast ephemeris. Furthermore, the time difference between Universal Time UT1 and Coordinated Universal Time is always set to zero. Thus only the initial velocity of satellite and polar motion parameters need to be estimated.

In this paper, we solve the unknown polar motion parameters and the initial velocity by fitting our model to multiple positions calculated by broadcast ephemeris. The variation of polar motion parameters in a few days orbit prediction is ignored.

For one satellite, the vector of unknowns is

$$\mathbf{x} = \begin{bmatrix} \mathbf{v}_0 \\ x_p(t_0) \\ y_p(t_0) \end{bmatrix} \qquad (8)$$

where $\mathbf{v}_0$ is the initial velocity of the satellite, $x_p(t_0)$ and $y_p(t_0)$ are the polar motion parameters at time instant t_0. As measurements we have the satellite's positions $\mathbf{y}_k$ at time instants $t_1, t_2, \ldots, t_m$ which can be described by equation

$$\mathbf{y}_k = \mathbf{h}_k(\mathbf{x}) + \boldsymbol{\varepsilon}_k, \qquad k = 1,2,\ldots,m. \qquad (9)$$

Vectors $\boldsymbol{\varepsilon}_k$ are the differences between the measured position at instant t_k and the position that was predicted using the function defined by equation

$$\mathbf{r}(t_k) = \mathbf{h}_k(\mathbf{x}) = \mathbf{f}(t_0, t_k, \mathbf{r}_0, \mathbf{v}_0, x_p(t_0), y_p(t_0)) \qquad (10)$$

where $\mathbf{f}$ is the function that predicts satellite states by integrating satellite's equation of motion and carrying out the required transformations between the Earth-fixed, Earth-centered reference frame and the inertial reference frame. Then we need to solve the nonlinear least squares fitting problem

$$\hat{\mathbf{x}} = \arg\min_x \left(\sum_k \boldsymbol{\varepsilon}_k^T(\mathbf{x}) \cdot \boldsymbol{\varepsilon}_k(\mathbf{x}) \right) \qquad (11)$$

3.3 Orbit prediction

After getting the estimated initial state and the earth orientation parameters, satellite position and velocity at any time can be calculated by integrating numerically. Various numerical integration methods have been used to solve the satellite's equation of motion. In this paper, the 8th order Runge-Kutta and 12th order Adams-Cowell methods are adopted to solve the satellite's equation of motion [13], because they are generally regarded as one of the most effective methods in satellite orbit prediction. These predicted orbit values can be used directly or fitted to broadcast ephemeris provided to the receiver.

4 Simulation result

The position errors for 7 days orbit prediction for GPS PRN 1 are shown in Fig. 3. The red line is the result with the SRP model which parameters varying with ε angle as described in (7). And the black line is the

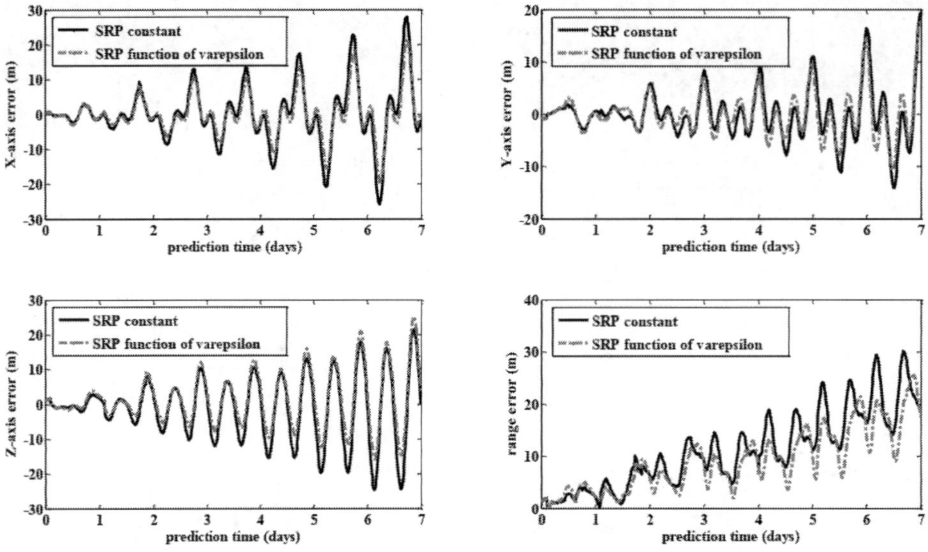

Figure 3. 7-days orbit prediction for GPS PRN1.

result of same orbit prediction, but using the constant parameter SRP model introduced in [5]. The initial points of prediction are calculated by each broadcast from 08:00:00 to 12:00:00 (GPS time) on 13 Jan 2017. The orbit information of initial points are improved by an least squares algorithm, then predict for 7 days with our propagator, and finally compare the predicted orbit with the precise orbit product to obtain the orbit prediction error.

Fig. 3 shows that with the SRP parameters varying as ε angle change, orbit prediction result became better than using constant SRP parameters, and the peak values of error are smaller.

The orbit prediction results show in Fig. 3 are summarized in Table 1.

Table 1. 7-days orbit prediction error for GPS PRN1.

SRP Model	Average error(m)	Variance of error(m²)	Maximum error(m)
SRP constant	11.74	64.44	37.00
SRP function of ε	9.52	40.44	26.40

5 Conclusion

The TTFF of receivers can be significantly reduced if the satellite's position and velocity information can be obtained by some alternative ways instead of reading the navigation message send by the satellite. This paper presents an orbit prediction algorithm which can run on receivers whose network connections are nonexistent.

Satellite orbit can be predicted by numerically integrating the satellite's equation of motion. In this work, only the four forces (including the Earth's gravitation, solar gravitation, lunar gravitation, and solar radiation pressure) that primarily affect the GPS satellite are taken into account. Satellite's initial position and velocity value, as well as the Earth's polar motion parameters, used in prediction correspond to the locally collected previous broadcast ephemeris. In this paper, an empirical SRP model with two parameters that change with the movement of satellite and Earth is presented to improve the accuracy of the orbit prediction. The procedure of estimating the SRP parameters, and its curve fitting were described in detail.

The orbit prediction algorithm is verified by GPS satellites using the initial position and velocity as well as polar motion parameters divided from three groups of the broadcast ephemeris data. As an example, the results of orbit prediction for GPS PRN1 with the SRP model described in this paper were show. From it we can see that the satellite's position error limited to 30 meters within 7 days prediction. What's more, with the SRP parameters varying with ε angle, orbit prediction result became better than using constant SRP parameters, and the peak values of error are smaller.

Acknowledgment

This work was supported in part by the National Natural Science Foundation of China (51707152), in part by the China Postdoctoral Science Foundation funded project (2014M562522XB), in part by the Shaanxi Province Natural Science Foundation of China (2016JQ4023), in part by the Scientific Research Program Funded by Shaanxi Provincial Education Department (15JK1523), and in part by the Scientific Research Project Foundation of Xi'an University of Technology (2016TS024).

References

1. Y. Li, Z. Nie, S. Chen, Y. Gao, *Proceedings of the ION 2015 Pacific PNT Meeting*, 117-130(2015)
2. M. Lytvyn, A. Kemetinger and P. Berglez, *Satellite Navigation Technologies & European Workshop on GNSS Signals and Signal Processing*, 1-6(2012)
3. P.G. Mattos, *ION GNSS 21st. International Technical Meeting of the Satellite Division*, 204–211(2008)
4. H. Lu, J. Zhou, C. Yan and A. Peng, *2017 11th IEEE International Conference on Anti-counterfeiting, Security, and Identification (ASID)*, 150-153(2017)
5. J. Ala-Luhtala, M. Seppänen, R. Piché, *Position Location and Navigation Symposium*, 568-575(2012)
6. Wu Z G, Yang J Z, Peng H J, Sci Sin Tech, 1232–1241(2016)
7. R. S. Hujsak, Adv. Astronaut. Sci,**93**,335-349(1996)
8. NGA EGM2008 model coefficients, Available online: http://earthinfo.nga.mil/GandG/wgs84/gravitymodel/egm2008/first_release.html
9. O. Montenbruck and E. Gill, *Satellite Orbits: Models, Methods, and Applications*(2012)
10. X. Zhang and R. Piché, *2016 International Conference on Localization and GNSS (ICL-GNSS)*, 1-6(2016)
11. NGA GPS Ephemeris/Station/Antenna Offset Documentation, Available online: http://earthinfo.nga.mil/GandG/sathtml/gpsdoc201009a.html
12. P. Gérard and B. Luzum, IERS Conventions, Available online: https://www.iers.org/IERS/EN/Publications/TechnicalNotes/tn36.html
13. Q. Zhang and L. Liu. *Publications of the Purple Mountain Observatory*, 19-27(1998)

AUTHOR INDEX

AUTHOR INDEX

EDP Sciences – Web of Conferences
17, Avenue du Hoggar
Parc d'Activité de Courtabouf
BP 112
F-91944 Les Ulis Cedex A
France

ISBN 978-1-5108-7085-7

9 781510 870857

CURRAN ASSOCIATES INC.
proceedings.com

Thirteenth Advanced International Conference on Telecommunications (AICT 2017)

Held at DataSys 2017

Venice, Italy
25 - 29 June 2017